12·95

7 day loan

McGraw-Hill Series in Electrical Engineering

MODERN COMMUNICATIONS AND
SPREAD SPECTRUM

Communications and Signal Processing

Consulting Editor
Stephen W. Director, Carnegie-Mellon University

MODERN COMMUNICATIONS AND SPREAD SPECTRUM

George R. Cooper

Clare D. McGillem

School of Electrical Engineering
Purdue University

McGraw-Hill Book Company

New York   St. Louis   San Francisco   Auckland   Bogotá   Hamburg
Johannesburg   London   Madrid   Mexico   Montreal   New Delhi
Panama   Paris   São Paulo   Singapore   Sydney   Tokyo   Toronto

MODERN COMMUNICATIONS AND SPREAD SPECTRUM
INTERNATIONAL EDITION

Copyright © 1986
Exclusive rights by McGraw-Hill Book Co., Singapore
for manufacture and export. This book cannot be re-exported
from the country to which it is consigned by McGraw-Hill.

1st Printing 1986

This book was set in Times Roman by Doyle Photosetting.
The editor was Sanjeev Rao;
the production supervisor was Leroy A. Young.
The cover was designed by Nadja Furlan Lorbek.
Project supervision was done by Cobb/Dunlop Publisher Services
Incorporated.

Library of Congress Cataloging in Publication Data

Cooper, George R.
  Modern communications and spread spectrum.

  (McGraw-Hill series in electrical engineering.
Communications and information theory)
  Includes bibliographies and index.
  1. Telecommunication.   2. Signal processing
3. Spread spectrum communications.   I. McGillem, Clare D.
II. Title.   III. Series.
TK5102.5.C66   1986   621.38   84-20092
ISBN 0-07-012951-7

When ordering this title use ISBN 0-07-100159-X

Printed and bound in Singapore by
Chong Moh Offset Printing Pte Ltd

CONTENTS

Chapter 9   Analysis of Direct-Sequence Spread-Spectrum Systems   289

Chapter 10   Analysis of Avoidance-Type Spread-Spectrum Systems   319

PREFACE

Although continuous evolution marks the technology and practice of communication engineering, a large segment of basic theory remains unchanged, and this provides the basis for evaluating new techniques and new systems. One of the main objectives of this text is to furnish an introduction to the basic principles of communication systems that is appropriate for a first-year graduate-level course. This proves to be more difficult than it might seem upon first consideration. The main problem is the diverse backgrounds of the students who are likely to take such a course. If the students have just completed their undergraduate program, they undoubtedly have had an introductory course in communication engineering that covers various modulation techniques. However, the degree of mathematical and statistical sophistication employed varies greatly with the undergraduate text used in the course, and with whether or not the students have had any previous exposure to the theory of probability and random processes. Compounding the problem is the likelihood that a number of students may be transferring from another specialty or another discipline, and although quite mature academically, may not be familiar with communication system principles.

The first seven chapters of this text are intended to address this problem. They were designed to cover topics relating to analog and digital communications in such a manner that they can be followed by students who have never taken a communications course and still provide sufficient breadth and depth of coverage to complement and extend the material normally presented in an undergraduate course. The text is written on the assumption that the reader understands something of modulation theory and is familiar with such techniques of linear system analysis as Fourier transforms and convolutions, and also is acquainted with probability and random processes. These techniques and concepts are reviewed as part of the text but it will be necessary for a student unfamiliar with them to do additional outside reading. It has been found that graduate students who are interested in this material generally will do the extra work necessary to overcome the lack of a previous communications course. The goal of this type of course is to bring the participating students to a more or less equal level of sophistication relative to the terminology, implementation, and analysis of communication systems so that subsequent courses can be based on a presumption that such an equality exists.

Because of the enormous variety of specialized modulation systems that have been developed over the years, it is not possible to provide coverage that is in any sense comprehensive. Rather the most widely used techniques and some of their variations are considered, and various systems are compared in terms of error probability, performance in the presence of noise, effects of system instabilities, and system complexity. The methods of analysis are intended to be correct and understandable, but no attempt has been made to demonstrate that they are mathematically rigorous.

A second purpose of this text is to provide a comprehensive introduction to the new and rapidly growing area of spread-spectrum communications. This material is presented at the same level as the first portion of the text, and is appropriate for a graduate-level course. The presentation is self-contained and offers a coherent development of the theory not previously available. References to a number of specialized topics are included to permit more in-depth study of spread-spectrum communications if desired.

A third purpose of this text is to furnish a reference source for practicing engineers from which they can gain ready access to many of the analytical results available for the analysis of communication systems. This should prove particularly valuable with regard to the material on spread-spectrum communications.

Relevant problems are included at the end of each chapter, and answers to a a number of them can be found at the end of the book.

George R. Cooper
Clare D. McGillem

MODERN COMMUNICATIONS AND SPREAD SPECTRUM

BASIC COMMUNICATION PRINCIPLES

1-1 INTRODUCTION

Communication systems of all types have many functional elements in common. Because of this it is often possible to adapt methods of analysis or implementation from one system to another. A very general type of communication system is illustrated in block diagram form in Fig. 1-1. Not all of the elements shown here would be present in every system, and there may be special elements in some systems (e.g., a diversity receiver) not represented here. However, most systems will contain most of the elements shown in Fig.1-1.

To permit the easy identification of where various specialized subsystems (discussed in detail later in the text) fit into an overall system, it is helpful to consider briefly the nature and function of each of the elements in the generalized system. The signal source and transducer together form the message source. The message

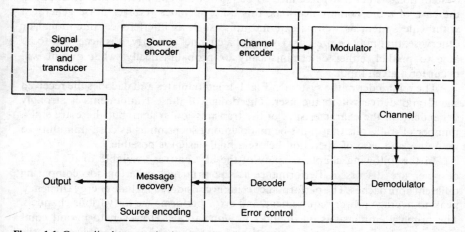

Figure 1-1 Generalized communication system.

might be an acoustic voice signal converted to an electrical waveform by a microphone, an optical image transduced by a television camera, or a measurement of some physical quantity converted to an electric signal by an appropriate sensor. Once the message is available, the source encoder modifies its structure for utilization in the particular communication system being used. This may include such operations as sampling and quantizing, companding, encrypting, or any of a variety of processes for dimensionality reduction.

The output of the source encoder goes to the channel encoder, which modifies the signal in such a way as to make its transmission through the channel more efficient. An example of a channel encoder is a processor that accepts a stream of binary signals and puts them into a series of groups with additional control symbols added so that the occurrence of an error during transmission and reception can be detected, or even corrected if desired. The channel encoder adds controlled redundancy to the transmitted message in order to accomplish error-free transmission through the channel. This added redundancy requires that extra information be transmitted along with the message and thus leads to a trade-off between speed of transmission (or bandwidth for a fixed speed) and error performance.

The modulator converts the encoded message into a signal suitable for transmission through the channel. For communications applications this channel can take many forms, ranging from free space to water to glass fibers. Generally the encoded message will be attached to a high-frequency carrier for transmission through the channel. The particular method of modulating the carrier with the message as well as the physical characteristics of the carrier itself are strongly dependent on the channel characteristics. In this book it is assumed that the channel is the atmosphere surrounding the earth and that the carrier is a radiated electromagnetic wave in the radio-frequency (RF) spectrum. However, many of the principles governing this type of carrier and channel carry over into other carriers such as acoustic or light waves and other channels such as water or waveguides. System noise is generally assumed to be part of the channel characterization. In the case of radio communications, this noise includes receiver noise generated within the system as well as external noise such as interference, intentional or otherwise, that may be present. To model a channel properly, it is necessary to be able to predict, either deterministically or probabilistically, what output will occur for a given input.

The remainder of the system in Fig. 1-1 demodulates and decodes the received signal and delivers it to the user. The design of these components is strongly dependent on the characteristics of the transmission system, but there are still a number of trade-offs that must be made among such things as cost, immunity to jamming, and ease of operation before a final design is possible.

In the following chapters, many of the alternatives available to the system designer are discussed. Performance is the principal criterion for comparing different approaches. The performance criteria considered include energy efficiency, peak-to-average power requirement, and error performance for various channels. The primary emphasis is on digital communications. However, to permit comparison with alternate and competitive analog communication systems, the general characteristics of these systems are also considered. Classical analog and digital systems are then discussed, followed by a detailed presentation of spread-spectrum systems as they are representative of the current state-of-the-art in modern digital communications.

1-2 BASEBAND SIGNAL CHARACTERISTICS

Baseband Signals

The term *baseband* designates the band of frequencies representing the original signal as delivered by the source. Most communications are concerned with transmission of acoustical, video, or data-type signals. In the case of voice signals, the baseband is a portion of the audio-frequency band extending to about 15 kHz. The precise portion included depends on the fidelity of transmission required. For a television image, the baseband is the video spectrum extending up to several megahertz or more, depending, again, on the fidelity required in the reproduction at the receiver. Data communication is concerned with the transmission of numbers, letters, and symbols; its baseband frequency is determined by how rapidly such data can be generated and it can extend to many megahertz. Some communication systems, such as local telephone service, operate at baseband; however, most services that operate over extended distances or are not between fixed points utilize *carriers* to convey the baseband signal information. The characteristics of the baseband signal vary significantly from source to source, and this in turn means that the characteristics of the modulated carrier vary from source to source. To see just how the baseband signal varies among different message sources, the characteristics of several typical sources are considered in detail.

Voice Signals

The sound waves produced by a speaker or heard by a listener are readily converted to and from electric signals by transducers such as microphones and loudspeakers. A typical speech wave is a sequence of discontinuous bursts of acoustic energy generated by various time-varying components of the human voice system. The spectral characteristics of speech vary over a wide range because of the different component sounds involved and because of wide differences among different speakers. Figure 1-2 shows the average energy-versus-frequency distribution for representative talkers.[1] It is seen from this figure that the energy is zero at frequencies near zero, rises to a peak at about 200 Hz, falls rapidly until about 400 Hz, and then decays more slowly to several thousand hertz. The importance of the various portions of the spectrum to speech understanding is not determined by this relative power distribution curve. In fact the large low-frequency components can be eliminated with little loss of speech intelligibility.

The frequency components required to obtain satisfactory speech intelligibility are determined experimentally by testing the subject's performance using

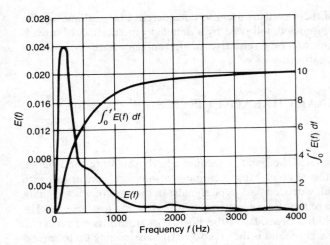

Figure 1-2 Average energy versus frequency for representative speakers. (Reproduced with permission from W. R. Bennett, *Introduction to Signal Transmission*, McGraw-Hill Book Co., Inc., N.Y., 1970.)

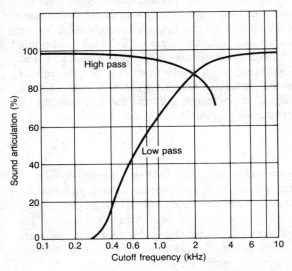

Figure 1-3 Experimentally determined articulation of filtered speech. (Reproduced with permission from W. R. Bennett, *Introduction to Signal Transmission*, McGraw-Hill Book Co., Inc., N.Y., 1970.)

sentences, words, and syllables. A typical test result is shown in Fig. 1-3.[1] In this test a group of listeners recorded the sound they thought they heard for speech having either the high frequencies or the low frequencies attenuated. The percentage of correct observations is called the "articulation." The curve marked "high pass" shows the result of removing the low-frequency components of the speech. It is seen that almost no effect is noticeable for removal of frequency components up to about 300 Hz. For the low-pass curve, the high-frequency components are attenuated and it is seen that components above about 3500 Hz contribute only slightly to improved performance. This type of test and other related tests have led to adoption of a voice baseband signal for communication purposes lying in the range of 300 to 3500 Hz.

In addition to the frequency band associated with voice communication, it is necessary to maintain a satisfactory signal-to-noise ratio (SNR) to ensure intelligibility. Fig. 1-4 shows the effect on word articulation in the presence of white noise.[2] It is evident from this figure that an SNR of the order of 10 dB or more is required for good performance. Another parameter of interest is the dynamic range of the speech signal. Fig. 1.5a shows the variation of the spectral density of speech with frequency.[2] The ordinate is in decibels per hertz and the abscissa is broken up into bands of equal contribution to the articulation index, which measures understandability of speech as shown in Fig. 1-5b. The 20 bands in Fig. 1-5a contribute equally to the articulation index when a full 30-dB range of maximum to minimum power is present in each band. If speech minimums are masked by noise, the contribution of any given band to articulation is reduced to a value given by (decibel level of peaks minus decibel level of noise)/6. For example, if the noise were only 12 dB below the peaks, the contribution to the articulation index for that particular band would be $12 \div 6 = 2$ percent. It is evident from this figure that the dynamic range of the baseband signal is of the order of 60 dB. By using a pre-emphasis network with a transfer function whose magnitude increases at about 8 dB per octave before transmission, followed by a deemphasis network after

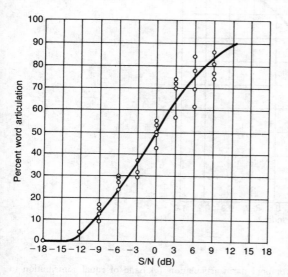

Figure 1-4 Articulation of words in the presence of white noise. (Reproduced with permission from *Reference Data for Radio Engineers*, Howard Sams & Co., Inc., Indianapolis, 1975.)

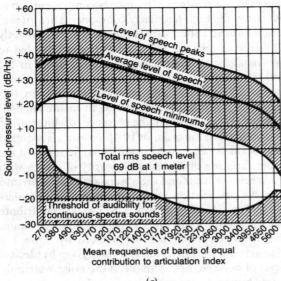

(a)

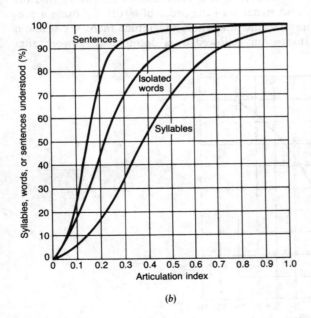

(b)

Figure 1-5 Importance of spectral components in articulation. (a) Band of equal contribution to articulation index; (b) relation of articulation index to understanding. (Reproduced with permission from *Reference Data for Radio Engineers*, Howard Sams & Co., Inc., Indianapolis, 1975.)

reception, the dynamic range can be reduced to 30 dB, thus greatly improving performance in a white noise environment.

The speech waveform can be modeled as a sample function from a random process. Unfortunately the speech probability density function (PDF) is significantly different from the widely used Gaussian probability density function. In speech peak excursions relative to the rms value occur more frequently than they would for a Gaussian process with the same power. In spite of the significant departure of the PDF from a Gaussian shape, it has been found that filters and processors designed on the basis of Gaussian statistics usually give good performance for speech waveforms.

Television

Another common communication signal is the video waveform associated with a television image. This signal is generated by scanning a scene in raster fashion using a light-sensitive sensor. The information consists of the brightness of the image as a function of the variables x, y, and t, where x and y are the image coordinates and t is the time. The transmitted signal is a single quantity such as voltage that varies with time. This transformation of three-dimensional information into a one-dimensional variable is an example of mapping between different signal spaces. In this case the mapping is made possible by setting up an independent reference, the raster, such that at any instant the voltage being transmitted corresponds to a specific position in the raster and this information is available at the receiver as a result of synchronization.

The principal characteristics of the baseband signal can be obtained by carrying out a Fourier analysis of the repetitive scanning of a fixed image. The process can be modeled by assuming an array of image fields recurring in both the x and y directions as shown in Fig. 1-6. The scanning aperture moves across the image with a slope equal to the ratio of the vertical and horizontal scan velocities. The continuous path of the straight line across the repeating array of image fields gives the same mathematical result as the discontinuous path in the single raster.

If the original image brightness function is $B(x, y)$ covering all values of x from 0 to a and all values of y from 0 to b, then the brightness function of the infinite array is just the doubly periodic extension of $B(x, y)$ such that

$$B(x, y) = B(x - a, y - b) \qquad \text{all } x, y \tag{1-1}$$

This periodic function can be expressed as a double Fourier series in the form

$$B(x, y) = \sum_{m=-\infty}^{\infty} \sum_{n=-\infty}^{\infty} \lambda_{mn} \exp\left[2\pi j \left(\frac{mx}{a} + \frac{ny}{b} \right) \right] \tag{1-2}$$

where the coefficients λ_{mn} are given by

$$\lambda_{mn} = \frac{1}{ab} \int_0^a \int_0^b B(x, y) \exp\left[-j2\pi \left(\frac{mx}{a} + \frac{ny}{b} \right) \right] dx \, dy \tag{1-3}$$

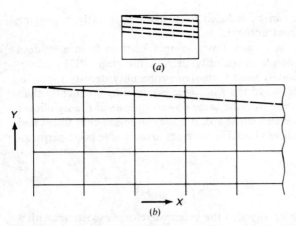

Figure 1-6 Raster scanning. (*a*) Single frame with instantaneous retrace; (*b*) replicated frames with continuous scan.

The position of the scanning aperture at any instant is given by

$$x = ut \tag{1-4}$$

$$y = vt \tag{1-5}$$

where u and v are the horizontal and vertical components of velocity respectively. The amplitude of the signal picked up by the sensor is proportional to the brightness and is therefore given by

$$s(t) = kB[x(t), y(t)] \tag{1-6}$$

$$= \sum_{k=-\infty}^{\infty} \sum_{n=-\infty}^{\infty} \lambda_{mn} \exp\left[j2\pi\left(\frac{mu}{a} + \frac{nv}{b}\right)t\right] \tag{1-7}$$

From Eq. (1-7) it is evident that $s(t)$ has a line spectrum in which the frequencies are $mu/a \pm nv/b$, where m and n take on integer values. The quantity u/a is the horizontal velocity divided by the image width and is, therefore, the line-scan frequency. Similarly, v/b is the frame rate. For a commercial black-and-white television image in the United States, there are 525 lines per image and 30 images per second. Therefore, the line-scanning frequency is $525 \times 30 = 15.75$ kHz. Accordingly, the line spectrum of $s(t)$ consists of a series of harmonics of 15.75 kHz around each of which is a cluster of components spaced 30 Hz apart. The variations in the amplitudes of the line-scanning harmonics are determined by the image parallel to the scanning direction; the higher the spatial frequency content of the image, the higher is the harmonic content of the signal. An attempt is made generally to have comparable spatial resolution in both the horizontal and vertical directions of the image. There are (theoretically) 525 horizontal lines in the image. Assuming that the highest spatial frequency components are half the sampling frequency, then the highest spatial frequency in the vertical direction would be $525/2b$. To obtain the same spatial frequency in the horizontal direction, the sampling frequency would have to be $(525a/2b) \times 15{,}750$ or $4.1 \times 10^6 a/b$ Hz. With an aspect ratio of $4:3$, this would require about a 5.6-MHz bandwidth. In practice bandwidths of the order of 3–4 MHz are employed because only a portion of the 525 lines is actually available as a result of blanking, which reduces the vertical resolution, and because of a need to conserve bandwidth. The rate at which the 30-Hz spaced components around the 15.75-kHz harmonics fall off is determined by the vertical spatial frequencies in the image. Because the vertical scanning velocity is 1/525 that of the horizontal velocity, the vertical spatial frequency components appear as correspondingly lower signal-frequency components. As a result the 30-Hz spaced clusters occupy only a narrow range around the line-scanning harmonics, and there are gaps in the spectrum that have negligible signal power. Removal of noise from these gaps by comb filtering can improve SNR with virtually no loss in image quality.

To obtain good image quality, it is necessary to preserve both the amplitudes and phases of the baseband signal components in the transmission and reception process. These frequencies must extend all the way to dc since it is this component that determines the average brightness of the image.

Data and Digital Computers

The baseband signals corresponding to data can take on many forms, depending on the actual data source and the preprocessing operations carried out prior to data transmission. The trend is very strong toward transmission of analog data as binary digits corresponding to the amplitude or some other parameter of the signal at a particular time. With the availability of digital output transducers and inexpensive, small microprocessors, it is likely that this trend will continue. The baseband signals for this situation correspond to groups of symbols consisting of anywhere from 2 to 20 or more binary digits. The rate of occurrence of these groups or words can range from a fraction of a word per second to millions of words per second. The words are made up of sequences of binary symbols such as two different levels of voltage occurring in accordance with the information stream of the source. In general, the shape of the pulses making up the bit stream does not need to be preserved in order to preserve the information content. Rather it is only necessary to identify which of the two possible symbols occurred. Thus the bandwidth requirement is basically that necessary to distinguish between two different possible signals taking place at a particular repetition frequency. This bandwidth is generally something of the order of the bit rate of the data stream.

1-3 NOISE IN COMMUNICATION SYSTEMS[3,4]

Sources of Noise

The degree to which it is possible to communicate reliably between different locations is determined by the capability of the receiver to distinguish among the various possible messages that might have been sent by the transmitter. One thing that limits this capability to distinguish between possible messages is noise. The noise can come from many sources and take many forms. In the broadest sense, noise encompasses such things as interfering signals from other sources, fluctuations in the channel transmission characteristics, variation in signal voltage levels due to the presence of discrete charge carriers in the conductors, and various other effects that cause deviation of the received waveform from the transmitted waveform. For purposes of analysis, it is convenient to group noises into certain broad categories that can be described by appropriate sets of parameters. The most common characteristic of noise is its random nature; that is, the fact that its magnitude cannot be predicted in advance. Because of this randomness, it is customary to describe noise in terms of a probabilistic model. Such a model describes the noise amplitude (or other parameter) by means of a probability density function $p(v)$. Figure 1-7 shows a typical PDF for a noise voltage. The magnitude of $p(v)$ represents the probability that v lies in the incremental interval dv around v and the probability that the voltage lies in the interval from V_1 to V_2 is given by the area under $p(v)$ between these values; that is,

$$P[V_1 < v \leq V_2] = \int_{V_1}^{V_2} p(v)dv \qquad (1\text{-}8)$$

For many important types of noise, the PDF can be determined theoretically, and for others it has been estimated empirically.

Certain properties of the noise can be determined directly from the PDF. In particular the average or dc value is given by the mean of the density function

$$\bar{v} = \int_{-\infty}^{\infty} vp(v)dv \qquad (1\text{-}9)$$

The mean square value is given by

$$\overline{v^2} = \int_{-\infty}^{\infty} v^2 p(v)dv \qquad (1\text{-}10)$$

and the rms value is

$$v_{\text{rms}} = \sqrt{\overline{v^2}} \qquad (1\text{-}11)$$

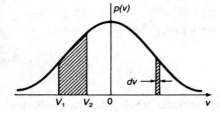

Figure 1-7 Probability density function of a noise voltage.

The PDF does not describe a noise waveform sufficiently to permit determination of its effects on performance of a communication system. To do this it is necessary to know how the noise changes with time. This information is generally obtained in terms of the frequency spectrum of the noise waveform. Since the waveform is not time limited, it does not have a Fourier transform in the ordinary sense. However, it is possible to compute a mean square voltage spectrum, called the power spectrum or spectral density, which represents the distribution of signal power as a function of frequency. The spectral density $S_n(f)$ of a noise waveform $n(t)$ is usually defined as the Fourier transform of the autocorrelation function of the time waveform; that is,

$$S_n(f) = \mathscr{F}\{R_n(\tau)\} \tag{1-12}$$

$$= \int_{-\infty}^{\infty} R_n(\tau) \exp[-j2\pi f\tau]\, d\tau \tag{1-13}$$

where

$$R_n(\tau) = E\{n(t)n(t-\tau)\} \tag{1-14}$$

and $E\{\ \}$ is the expected value operator. Letting $n(t) = n_1$ and $n(t-\tau) = n_2$, this can be written as

$$R(\tau) = \int_{-\infty}^{\infty} \int_{-\infty}^{\infty} n_1 n_2 p(n_1, n_2)\, dn_1\, dn_2 \tag{1-15}$$

where $p(n_1, n_2)$ is the joint probability density of n_1 and n_2.

The actual forms of these quantities and the methods of using them in noise calculations are considered in the following paragraphs and in more detail in Chap. 2.

Single-Frequency Sinusoid

A single-frequency interfering signal present simultaneously with the desired waveform represents a particular type of noise. It might seem that such an interfering signal could be eliminated by adding in a locally generated sine wave 180 degrees out of phase. However, usually the interfering signal originates externally and is likely to have an amplitude, frequency, and phase that are subject to unpredictable variations. For this reason such a signal is often considered a random noise with a sinusoidal distribution of amplitude versus time.

Assuming that the amplitude of the sine wave is A and the phase is a random variable uniformly distributed over the interval $(-\pi, \pi)$, the PDF of the instantaneous voltage can be shown to be

$$p(v) = \frac{1}{\pi\sqrt{A^2 - v^2}} \qquad -A < v < A$$

$$= 0 \qquad\qquad \text{elsewhere} \tag{1-16}$$

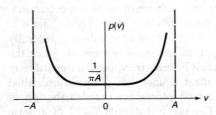

Figure 1-8 Probability density function of the amplitude of a random phase sinusoid.

This density function is sketched in Fig. 1-8. The mean and rms values for this random waveform are

$$\bar{v} = 0 \tag{1-17}$$

$$v_{rms} = \frac{1}{\sqrt{2}} A \tag{1-18}$$

The autocorrelation function for this waveform can be obtained from the PDF of the random phase angle θ as follows:

$$R_v(\tau) = E\{v(t)v(t+\tau)\}$$

The only random variable is the phase angle θ and so the expectation is computed with respect to it.

$$R_v(\tau) = \int_{-\pi}^{\pi} A \sin(2\pi f_0 t + \theta) A \sin[2\pi f_0(t+\tau) + \theta] \frac{d\theta}{2\pi}$$

$$= \frac{A^2}{2} \cos 2\pi f_0 \tau \tag{1-19}$$

The Fourier transform of $R_v(\tau)$ gives the spectral density as

$$S_v(f) = \frac{A^2}{4} [\delta(f + f_0) + \delta(f - f_0)] \tag{1-20}$$

Thermal Noise

The Brownian motion associated with the electrons in a conductor gives rise to fluctuations in the voltage across it as current flows through the conductor. Because of the high velocities and large number of charge carriers present, the noise extends to very high frequencies and takes on a very regular character in a statistical sense. Quantum theory predicts that the thermal noise power available from any conducting source has a spectral density given by

$$S_n(f) = \frac{0.5hf}{\exp(hf/kT) - 1} \tag{1-21}$$

where h is Planck's constant (6.625×10^{-34} J-s), f is the frequency (Hz), k is

Boltzmann's constant $(1.38 \times 10^{-23} \text{ J/K})$ and T is the absolute temperature (K). For cases where $hf \ll kT$; that is, $f \ll 2 \times 10^{10}T$ or about 1200 GHz at room temperature, this expression is closely approximated by replacing the exponential term by the first terms of the power series expansion giving

$$S_n(f) = \frac{kT}{2} \quad \text{W/Hz} \tag{1-22}$$

Thus the spectral density of thermal noise is essentially uniform and this type of noise is commonly referred to as white noise. The thermal noise power available in a bandwidth B is given by the product of $S_n(f)$ and $2B$ or

$$P_n = kTB \quad \text{watts} \tag{1-23}$$

Substituting $kT = 1.38 \times 10^{-23} \times 290 = 4 \times 10^{-21}$, this equation can be put into the form

$$P_n = 4 \times 10^{-21}B \quad \text{watts}$$
$$= -174 + 10 \log B \quad \text{dBm} \tag{1-24}$$

where dBm refers to decibels above 1 milliwatt. For a 1-MHz bandwidth, the noise power would be -114 dBm. Thermal noise represents an irreducible minimum of noise present in a system at room temperature. In virtually all communication systems, the noise is significantly greater than this.

Because of the many contributing elements in the generation of thermal noise (and other types of noise such as shot noise, flicker noise, and galactic noise), the PDF is very accurately represented by a Gaussian distribution. The density function for a Gaussian probability distribution is given by

$$p(v) = \frac{1}{\sigma\sqrt{2\pi}} \exp\left[-\frac{(v - \bar{v})^2}{2\sigma^2} \right] \tag{1-25}$$

where σ^2 is the variance and $\bar{v}$ is the mean. Two examples of this distribution are sketched in Fig. 1-9.

The mean $\bar{v}$ corresponds to the dc value of the voltage, which is zero for thermal noise. It is readily shown that for a zero mean noise, the mean square voltage is equal to the variance of the distribution σ^2, and, therefore, the rms voltage is equal

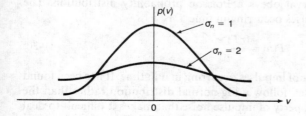

Figure 1-9 Gaussian PDF for two different standard deviations.

to the standard deviation σ. A useful measure of the distribution of amplitudes of a noise voltage is the peak factor, which is the ratio of peak to rms voltage. For voltages that theoretically have peaks extending to infinity, this definition is modified to be the ratio of the value exceeded by the noise 0.01 percent of the time to the rms value. For a Gaussian noise, this quantity is $3.89\sigma_n$; that is, $|v| > 3.89\sigma_n$ less than 0.01 percent of the time.

Shot Noise and Flicker Noise

Because of the discrete nature of the charge carriers in active devices, there are irregularities associated with current flow that appear as noise in the output. The characteristics of this noise are similar to those of thermal noise except for two important differences.

1. The magnitude of thermal noise is proportional to absolute temperature whereas shot noise does not change with temperature.
2. The magnitude of shot noise is proportional to the square root of current and thus is related to signal amplitude whereas thermal noise is not.

Flicker noise, or what is commonly called $1/f$ noise because its power spectrum is proportional to $1/f$, is associated with contact and surface irregularities in cathodes and semiconductors. It is believed to be due to fluctuation in the conductivity of the medium. Recent advances in device fabrication have greatly reduced this type of noise and devices can be built that have negligible $1/f$ noise above 1 kHz. This noise has been found to extend down to a fraction of a hertz. The PDF is Gaussian for such noise.

Impulse Noise

Impulse noise consists of short duration bursts of energy. Such signals have essentially flat spectra over wide frequency intervals. This type of noise arises from switching transients, corona discharges, and various types of arc discharge. Such noises are particularly damaging to certain types of digital systems that cannot distinguish between noise impulses and signal pulses. When such pulses occur randomly in time and independently of each other, the number arriving within a specified time interval obeys a Poisson probability distribution. The probability of exactly n impulses occurring in time T is

$$P(n) = \frac{(rT)^n e^{-rT}}{n!} \tag{1-26}$$

where r is the average number of impulses occurring in unit time. It has been found empirically that impulses often follow a log-normal distribution rather than the Poisson distribution. The property of impulse noise that makes it difficult to deal with is that the duration of the impulses is short relative to the time between them. A wide-band receiving system resolves such signals as individual events. If the impulses are applied to a narrow-bandwidth system, the system will cause the impulses to be smeared out and merged into a more uniform type of noise; however, this noise may have a significant amplitude. By preceding the band-limiting portion of the circuit by a wide-band peak clipping circuit, a substantial reduction in the noise level can be achieved.

Other Noise

Several other sources of noise are of importance in communication system analysis. Among them are background noise arising from natural phenomena such as solar and galactic radiation. The magnitude of such noise is generally expressed in terms of equivalent noise temperature. In the case of a thermal noise source, the noise temperature is the same as the physical temperature. For some sources of noise, there is often no direct relation between physical temperature and the noise power available from the source. This is true, for example, of galactic noise. It is convenient in this case, and in many others also, to describe an arbitrary noise source in terms of the temperature of a thermal noise source that would produce the same amount of available noise power. If a given source were to produce a noise having a spectral density $S_n(f)$, then the equivalent noise temperature of this source would be $T(f) = 2S_n(f)/k$. If $S_n(f)$ varies with frequency, then the noise temperature will also vary with frequency.

This concept is particularly useful for characterizing radiation noise sources such as galactic noise or atmospheric noise. When an antenna is pointed toward a radiation noise source with a noise temperature of T_n, the output noise at the antenna terminals can be computed by assuming that the radiation resistance of the antenna has a temperature equal to T_n and calculating the corresponding thermal noise. For example, if an antenna were directed toward the sky at a point where the background radiation has an equivalent noise temperature of 100K, then the spectral density of the noise power from the antenna terminals would be $50 \times 1.38 \times 10^{-21} = 6.9 \times 10^{-20}$ W/Hz. Fig. 1-10a is a composite showing the equivalent noise temperature for several different background noise sources,[2] and Fig. 1-10b shows data on typical atmospheric noise.[3]

Another type of noise that is particularly difficult to deal with is so-called multiplicative noise. In this type of noise, a received signal is of the form

$$r(t) = \alpha(t)s(t) \tag{1-27}$$

where $\alpha(t)$ is a sample function from a random process. This type of noise is produced by a variable attenuation in a channel and usually requires special techniques for processing the received signal to reduce its effects.

Noise Figure and Noise Calculations

The noise performance of a component or a system is frequently described by means of a quantity called the noise figure. This quantity expresses the system noise that is actually present relative to that which would be present if only the

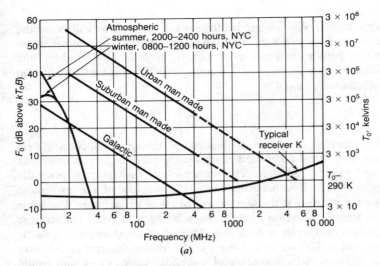

Figure 1-10a Equivalent noise temperature for several background sources. (Reproduced with permission from *Reference Data for Radio Engineers*, Howard Sams & Co., Inc., Indianapolis, 1975.)

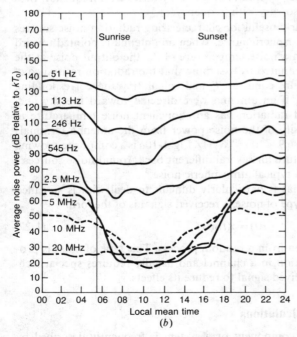

Figure 1-10b Average noise power of atmospheric noise as a function of local time and frequency for a north latitude site. (Reproduced with permission from E. N. Skomal, *Manmade Radio Noise*, Van Nostrand Reinhold Co., N.Y., 1975.)

(irreducible) thermal noise were present. The noise figure can be defined at a particular frequency (spot noise figure) or averaged over a band of frequencies (integrated noise figure). It is this latter noise figure that is considered here. If, over the frequency band of interest, the system gain is not uniform or the noise spectrum is not flat, then the determination of the noise figure will require carrying out an integration to evaluate the output power. However, for most communication applications, the system gain can be assumed constant over an effective noise bandwidth and the spectrum of the noise can be assumed constant. In the following analysis, it is assumed that the impedances of the sources and loads are matched. For this case the noise figure is defined as

$$F = \frac{N_{ao}}{GkT_0B_n} \tag{1-28}$$

where N_{ao} is the available output noise power of the system, G is the system power gain, k is Boltzmann's constant, T_0 is the standard temperature (290K), and B_n is the equivalent noise bandwidth. When the source connected to the amplifier has a noise temperature T_0, the same as that of the receiver, the noise figure can be expressed as

$$F = \frac{\text{SNR at input}}{\text{SNR at output}} = \frac{(\text{SNR})_i}{(\text{SNR})_o} \tag{1-29}$$

$$(\text{SNR})_i = \frac{P_s}{P_n} = \frac{P_s}{kT_0B_n} \tag{1-30}$$

$$(\text{SNR})_o = \frac{GP_s}{G(kT_0B_n + \text{internally generated noise power})}$$

$$F = \frac{P_s}{kT_0B_n} \cdot \frac{(kT_0B_n + \text{internally generated noise power})}{P_s}$$

$$= 1 + \frac{\text{internally generated noise power}}{kT_0B_n} \tag{1-31}$$

It follows that the internally generated noise power (referred to the input) is $(F-1)kT_0B_n$. The total noise power (referred to the input) for a system connected to a source having equivalent noise temperature T_s is

$$P_n = kT_sB_n + (F-1)kT_0B_n \tag{1-32}$$

The effective noise temperature T_e of a system expresses the internally generated noise power that is available in terms of the temperature of a thermal noise source that would produce the same power. If such a system is connected to an ideal measuring instrument, it will lead to a total available noise power of $kT_sB_n + kT_eB_n$. Using Eq. (1-32) the noise figure of such a system can be expressed as

$$F = \frac{P_n - kT_sB_n}{kT_0B_n} + 1 = 1 + \frac{T_e}{T_0} \tag{1-33}$$

or the effective noise temperature is given by

$$T_e = (F-1)T_0 \tag{1-34}$$

For example, a system with effective noise temperature $T_e = 1000K$ would have a noise figure of

$$F = 1 + \frac{1000}{290} = 4.45 \rightarrow 6.48 \text{ dB} \tag{1-35}$$

When systems are cascaded, the noise generated in the later stages contributes less to the overall output noise than that generated in the earlier stages. If a sequence of systems characterized by noise figures $F_1, F_2, \ldots$, gains $G_1, G_2, \ldots$, and effective noise temperatures $T_{e1}, T_{e2}, \ldots$, are connected in cascade, the overall system noise figure and overall effective noise temperature are given by

$$F = F_1 + \frac{F_2-1}{G_1} + \frac{F_3-1}{G_1 G_2} + \cdots \tag{1-36}$$

$$T_e = T_{e1} + \frac{T_{e2}}{G_1} + \frac{T_{e3}}{G_1 G_2} + \cdots \tag{1-37}$$

It is evident that by having high gain in early stages of a system, the contribution to system noise by later stages can be made negligible.

The presence of a resistive attenuator in a cascade of networks alters the noise figure by reducing the signal while still generating the same amount of thermal noise. If the attenuator loss factor is $L(L>1)$, then for components at room temperature, the attenuator noise figure and effective temperature are

$$F = L \tag{1-38}$$

$$T_e = (L-1)T_0 \tag{1-39}$$

When a source at temperature $T_s \neq T_0$ is connected to a component having noise figure F, the output SNR is given by

$$(\text{SNR})_0 = \frac{(\text{SNR})_i}{1 + (F-1)(T_0/T_s)} = \frac{(\text{SNR})_i}{1 + (T_e/T_s)} \tag{1-40}$$

In dealing with noise, it is often of interest to calculate the voltages or currents present in the circuit elements. This is readily done by considering the power relations that are present in an equivalent circuit. Consider the resistive circuit shown in Fig. 1-11. The noise generator has an rms voltage v_n. The available noise power from this resistor in a bandwidth B_n is kTB_n, where T is the resistor temperature and k is Boltzmann's constant. This power is obtained by connecting a resistance of equal value across the terminals as shown. The voltage v_n must provide the necessary power, thus

$$\left(\frac{v_n}{2}\right)^2 \frac{1}{R} = kTB_n$$

$$v_n = \sqrt{4RkTB_n} \tag{1-41}$$

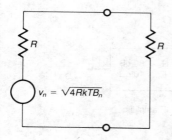

Figure 1-11 Resistor and equivalent noise source.

This is the open-circuit voltage across resistor R. The voltage appearing across a matched load would be one half of this value. It is a voltage with a Gaussian PDF with variance $4RkTB_n$ and a uniform spectral density with amplitude $2RkT$. The left side of the circuit in Fig. 1-11 is thus the Thévenin's equivalent circuit for a thermal noise source.

A similar calculation for current shows that Norton's equivalent circuit for a thermal noise source has a constant current generator with an rms value of

$$i_n = \frac{v_n}{R} = \sqrt{\frac{4kTB_n}{R}} \tag{1-42}$$

As an example of noise-figure calculation, consider the system shown in Fig. 1-12. The overall noise figure of this system is given by

$$F = 4 + \frac{2-1}{10} + \frac{15.85-1}{10 \times 0.5} = 7.1 \rightarrow 8.5 \text{ dB}$$

As a further example, consider the high-performance satellite receiving system shown in Fig. 1-13. The high-gain directional antenna viewing background radiation has an equivalent noise temperature $T_e = 15K$ and is connected to a low noise maser amplifier by a low-loss waveguide. The maser amplifier is followed by a traveling-wave-tube (TWT) amplifier, a mixer and an intermediate-frequency (IF) amplifier. It is desired to compute the effective noise temperature of the system. The effective noise temperature of the receiver is given by

$$T_e = T_{e1} + \frac{T_{e2}}{G_1} + \frac{T_{e3}}{G_1 G_2} + \frac{T_{e4}}{G_1 G_2 G_3}$$

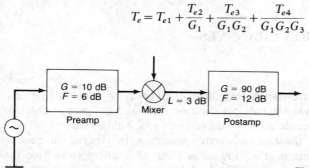

Figure 1-12 Noise figure example.

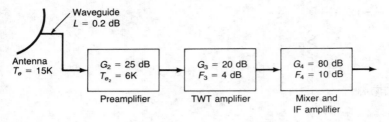

Figure 1-13 High-performance satellite receiver.

where T_{e1}, T_{e2}, T_{e3}, and T_{e4} are the effective noise temperature of the waveguide, the maser, the TWT, and the mixer-IF amplifier, and G_1, G_2, and G_3 are the gains of the same elements. The noise figures are converted to effective input noise temperature by the expression

$$T_e = T_0(F - 1) \tag{1-43}$$

Therefore,

$$T_{e3} = 290(2.5 - 1) = 438$$

$$T_{e4} = 290(10 - 1) = 2610$$

The waveguide loss factor is 1.047 ($G_1 = 0.955$) giving an effective temperature of

$$T_{e1} = 290(1.047 - 1) = 13.7$$

The effective input temperature of the receiver is then

$$T_e = 13.7 + \frac{6}{0.955} + \frac{438}{0.955 \times 316} + \frac{2610}{0.955 \times 316 \times 100}$$

$$= 21.5$$

The overall equivalent system noise temperature is given by

$$T_{\text{sys}} = 15° + 21.5° = 36.5\text{K}$$

1-4 COMMUNICATION CHANNELS

Free Space Channels

Communication signals are transmitted through a variety of different channels. For present purposes, however, only propagation through free space or through the atmosphere will be considered. Also, the carrier signal will be assumed to be an electromagnetic wave in the radio-frequency spectrum. In general, an electromagnetic wave that originates at a point spreads out as it moves away from the originating point. This is illustrated in Fig. 1-14. If it is assumed that the power P is

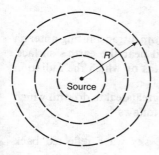

Figure 1-14 Electromagnetic wave emanating from a point.

radiated by an isotropic radiator, then the power density at a distance R is given by

$$\frac{P_t}{4\pi R^2} \quad \text{W/m}^2 \tag{1-44}$$

If the transmitting antenna has a power gain of G_1 in a particular direction, then the power density in that direction will be increased by the factor G_1, giving a value at a distance R of

$$\frac{P_t G_1}{4\pi R^2} \quad \text{W/m}^2 \tag{1-45}$$

If now a receiving antenna with effective receiving cross section A_e intercepts the radiated signal, it will extract a power equal to

$$P_r = \frac{P_t G_1 A_e}{4\pi R^2} \quad \text{watts} \tag{1-46}$$

It is a fundamental principle of field theory that the gain of an antenna and the effective area are related by wavelength λ according to the expression

$$A_e = \frac{G\lambda^2}{4\pi} \quad \text{m}^2 \tag{1-47}$$

Substituting this into the expression for the received power gives

$$P_r = \frac{P_t G_1 G_2 \lambda^2}{(4\pi R)^2} \quad \text{watts} \tag{1-48}$$

where G_2 is the gain of the receiving antenna. The factor $(4\pi R/\lambda)^2$ is often referred to as the free-space loss factor, L_{fs}, since it represents an attenuation present independent of any intervening medium. When atmospheric and other losses L_a are present, the received power will be

$$P_r = \frac{P_t G_1 G_2}{L_a L_{fs}} \quad \text{watts} \tag{1-49}$$

The loss factor between two antennas, therefore, is given by

$$L = \frac{P_t}{P_r} = \frac{L_a L_{fs}}{G_1 G_2} \tag{1-50}$$

Propagation Over the Earth's Surface

There are many paths by which radiation may travel from a transmitting antenna to a receiving antenna when the antennas are located near the earth's surface. Figure 1-15 illustrates this situation and shows the paths by which the radiation might reach the receiving antenna.

Actually the frequency of the radiated energy has a great deal to do with which paths are important and can be summarized briefly as follows:[4]

20–100 kHz   Low ground-wave attenuation, sky wave reflected back with little attenuation.
At large distance ground wave absorbed and only sky wave present.
No fading.
Uses: Navigation, time signals, long-range military communications.

100–500 kHz   Similar to very-low-frequency signals.
Ground wave attenuated faster.
Long-distance communication not possible in daytime.
Uses: Navigation and beacons.

500–2000 kHz   Ground wave strong at short distance from transmitter (primary service area) in daytime and nighttime.
Sky wave from *E*-layer very strong in nighttime gives large signal at long distances from transmitter.
Intermediate distances have severe fading at night because of sky-wave/ground-wave interference.
Very long distances have fading due to multiple sky waves.
Use: Broadcast.

2–30 MHz   No ground wave present, only direct wave or sky wave.
Skip may occur, giving no signal at intermediate distances.

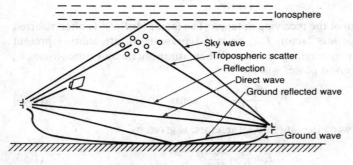

Figure 1-15 Propagation paths over the earth's surface.

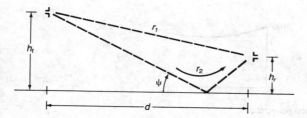

Figure 1-16 Effect of ground-reflected signal.

Uses: Short-wave broadcast, mobile, amateur, CB, marine, etc.

Above 30 MHz   Generally line of sight, or troposcatter for over-the-horizon communication.

Uses: TV, FM, satellite, mobile.

Of particular importance in many communication systems is the effect of the ground reflected wave arriving along with the direct wave. The geometry of the situation is shown in Fig. 1-16. The difficulty arises because the phases of the direct and reflected signals are different when they arrive at the receiving antenna and this produces interference effects. Using phasor notation, the received signal can be written as

$$E_r\underline{/\theta} = E_1\underline{/\theta_1} + \Gamma E_2\underline{/\theta_2} \qquad (1\text{-}51)$$

where $E_1\underline{/\theta_1}$ is the phasor corresponding to the direct path, $E_2\underline{/\theta_2}$ is the phasor resulting from propagation along the reflected signal path, and Γ is the reflection coefficient.

The reflection coefficient depends on the type of surface, the grazing angle, the frequency, and the polarization. However, for frequencies of general interest for communication purposes and for typical ground surfaces, the reflection coefficients behave in the manner shown in Fig. 1-17. For small grazing angles, it is seen that $|\Gamma| \simeq 1$ and $\angle\Gamma \simeq 180$ degrees for either polarization. This means that the phase is reversed by reflection.

The difference in path length for the direct and reflected waves is readily found to be

$$\Delta l = \sqrt{d^2 + (h_t + h_r)^2} - \sqrt{d^2 + (h_t - h_r)^2}$$

$$= d\left\{\left[1 + \frac{(h_t + h_r)^2}{d^2}\right]^{1/2} - \left[1 + \frac{(h_t - h_r)^2}{d^2}\right]^{1/2}\right\}$$

$$\simeq d\left\{1 + \frac{1}{2}\frac{(h_t + h_r)^2}{d^2} - 1 - \frac{1}{2}\frac{(h_t - h_r)^2}{d^2}\right\} \quad d \gg h_t, h_r$$

$$= \frac{2h_t h_r}{d} \qquad (1\text{-}52)$$

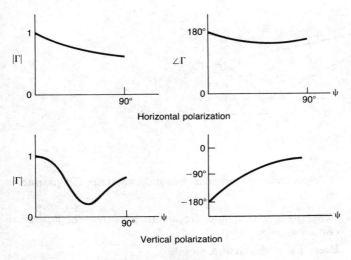

Figure 1-17 Reflection coefficient of the earth.[4]

This path length difference results in a phase difference of

$$\Delta\phi = \frac{2\pi\,\Delta l}{\lambda} = \frac{4\pi h_t h_r}{\lambda d} \tag{1-53}$$

The phase of the reflected signal relative to the direct signal is, therefore,

$$\Delta\theta = \pi + \frac{4\pi h_t h_r}{\lambda d} \tag{1-54}$$

Assuming that the signals radiated along the direct and reflected paths are equal, that is, $E_1 = E_2 = E$, then the received signal amplitude becomes

$$E_r = \sqrt{E^2 + E^2 + 2E^2 \cos(\Delta\phi + \pi)}$$

$$= 2E \sin(\Delta\phi/2) = 2E \sin\left(\frac{2\pi h_t h_r}{\lambda d}\right) \tag{1-55}$$

The value of E_r varies from 0 to $2E$ depending on $\Delta\phi$. When the distance d is very large compared with h_t and h_r, as in a microwave relay link, the sine can be replaced by its argument giving a further simplification. As an example consider a microwave system with $d = 50$ km, $\lambda = 50$ cm, $h_t = 25$ meters, and $h_r = 15$ meters. The ratio of the power received with the reflected signal compared with that with no reflection is given by

$$\frac{P_{\text{refl}}}{P_{\text{no refl}}} = \left[2 \sin \frac{2\pi h_t h_r}{\lambda d}\right]^2 = \left[2 \sin\left(\frac{2\pi \times 25 \times 15}{0.5 \times 50 \times 10^3}\right)\right]^2$$

$$= 0.0354$$

$$= -14 \text{ dB}$$

It is customary to write the field strength at the receiving antenna in terms of the field strength E_0 at unit distance from the transmitter. Using this convention the received field strength is

$$E_r = \frac{2E_0}{d} \sin \frac{2\pi h_t h_r}{\lambda d} \simeq \frac{4\pi h_t h_r}{\lambda d^2} E_0 \tag{1-56}$$

Antenna Gain and Beam Width

Two of the most important characteristics of antennas of interest to the communications engineer are power gain and the directivity. These quantities are generally functions of the physical dimensions of the antenna and the wavelength of the radiated energy. The gain of an antenna varies with angular orientation. However, the quantity of primary interest is the maximum gain, which can then be multiplied by the directivity factor to determine the gain in other directions. This gain is defined as

$$G_0 = \frac{\text{maximum radiation intensity produced by the antenna}}{\text{radiation intensity from an isotropic source}} \tag{1-57}$$

where the radiation intensity is measured in watts per square meter. As stated previously the antenna gain and the effective receiving area A_e are related by

$$G = \frac{4\pi A_e}{\lambda^2} \tag{1-58}$$

$$A_e = \frac{G\lambda^2}{4\pi} \tag{1-59}$$

where λ is the wavelength. For two-dimensional antennas, the effective area is related to the physical area by

$$A_e = \eta A \tag{1-60}$$

where A is the physical area and η is the illumination efficiency.

Table 1-1 lists the (maximum) gain and (maximum) effective area for a number of antennas. Any antenna that is short relative to the wavelength is considered a short dipole and, therefore, has a constant gain of 1.5, which corresponds to 1.76 dB. From the expression for the effective area it is evident why short antennas work so well at low frequencies.

As an example of a high-gain antenna, consider a parabolic reflector antenna 3 meters in diameter operating at a frequency of 6000 MHz. Let the efficiency factor be $\eta = 0.5$. The gain of this antenna is

$$G_0 = \frac{4\pi \times \pi \times (1.5)^2 \times 0.5}{(0.05)^2} = 17,765$$

$$\doteq 42.5 \text{ dB}$$

For most practical antennas, the gain varies with the direction of propagation of the radiated energy relative to the axes of the antenna. This is described mathe-

Table 1-1 Gain and effective area of typical antennas

Antenna	Maximum gain, G_0	Maximum effective area, A_e
Short dipole	1.5	$0.119\lambda^2$
$\dfrac{\lambda}{2}$ dipole	1.64	$0.13\lambda^2$
Optimum horn (area $= A$)	$\dfrac{10A}{\lambda^2}$	$0.81A$
Parabolic reflector (area $= A$)	$\dfrac{4\pi\eta A}{\lambda^2}$	$\eta A (\eta = 0.5 - 0.6)$
Broad-side array (area $= A$)	$\dfrac{4\pi A}{\lambda^2}$	A
Turnstile	1.15	$0.915\lambda^2$

matically by making the gain a function of the azimuth and elevation relative to the direction of maximum gain; thus

$$G(\theta, \phi) = \frac{4\pi P(\theta, \phi)}{P_t} \qquad (1\text{-}61)$$

where P_t is the total radiated power and $P(\theta, \phi)$ is the power per unit solid angle radiated in the direction of θ (elevation) and ϕ (azimuth). This function is the same whether the antenna is used for transmitting or for receiving. The gain function can be written in an equivalent form as

$$G(\theta, \phi) = G_0 g^2(\theta, \phi)$$

where $g(\theta, \phi)$ is called the pattern function or directivity of the antenna with a maximum of unity in the direction of maximum gain and G_0 is the maximum power gain. The gain function depends on the size of the antenna relative to the wavelength, the shape of the antenna, the amplitude, and the phase distribution over the antenna, and can be significantly affected by obstructions near the antenna.

Antenna gain functions are usually described by a graph of amplitude versus angle. The data are presented in either rectangular or polar coordinates as shown in Fig. 1-18.

Two important quantities are obtained from the antenna pattern: the antenna half-power beam width and the maximum side-lobe level. The half-power beam width is defined as the angular width of the main lobe between points at which the gain has fallen to one half the maximum value. These points correspond to a reduction in electric field strength of $1/\sqrt{2}$ relative to the maximum value. The side-lobe level is the amplitude of the largest lobe of the antenna pattern other than the main lobe and is usually measured in decibels relative to the main-lobe amplitude. For symmetrical antennas such as a vertical dipole or a turnstile antenna, the beam pattern in the azimuth direction is constant. In general the beam

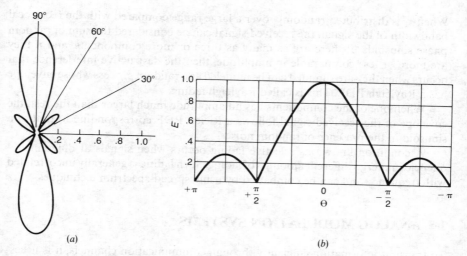

Figure 1-18 Antenna gain functions. (a) Polar plot; (b) rectangular plot.

width of an antenna is inversely proportional to the length of the antenna in the plane perpendicular to the direction of propagation. A rough approximation is that the beam width in radians is equal to the reciprocal of the length in wavelengths. In terms of degrees, this would be $57\lambda/d$ where λ is the wavelength and d is the linear extent of the aperture. For a typical circular parabola, the beam width is approximately $65\lambda/d$. For such an antenna, the first side lobe has an amplitude of -20 dB relative to the main lobe.

Multipath

Although only a single electromagnetic wave is radiated by a transmitting antenna, there are many instances in which that wave reaches the receiver by more than one path. The alternate paths may involve reflection from the ionosphere, reflection from the ground, or the presence of anomalous propagation paths in the atmosphere due to weather conditions. When more than one wave arrives at the receiving antenna, the net signal at the antenna terminals is the phasor sum of the separate waves. This signal will have the mathematical form

$$s(t) = \sum_k a_k s_0(t - t_0 - \tau_k)$$

where

$$a_k = \text{path transmission factor}$$

$$t_0 = \text{average propagation delay}$$

$$\tau_k = \text{differential delay along } k\text{th path}$$

When τ_k is distributed randomly over a large range compared with the reciprocal bandwidth of the signal, the received signal can be considered the sum of random phase sinusoids. If there are as many as three or more components, and if they are more or less comparable in amplitude, then the destructive interference that occurs when they are summed can be modeled as a random process whose envelope has a Rayleigh PDF. This is called Rayleigh fading.

When one of the components has an amplitude much larger than the rest, the envelope of the received signal follows a Rician PDF corresponding to that of a sinusoid in the presence of random noise.

A condition known as selective fading occurs when signals at different frequencies undergo different amounts of fading. Such fading is generally uncorrelated with frequency and can be combatted by using spread-spectrum techniques.

1-5 ANALOG MODULATION SYSTEMS

To transmit information efficiently through communication channels, it is necessary to employ a carrier signal. The information is attached to the carrier through the process of modulation. The carrier is then transmitted and received by the communication system and the information extracted by the process of demodulation. It is convenient and useful to classify communication systems by the methods of modulation employed.

In analog modulation systems, a continuous time waveform is used to vary one of the parameters of a carrier in synchronism with it. The major types of analog modulation are evident from the form of a general modulated sinusoidal carrier

$$s(t) = a(t)\cos[\omega_c t + \phi(t)]$$

where

$$a(t) = \text{amplitude of the carrier}$$

$$\omega_c = \text{angular frequency of carrier}$$

$$\phi(t) = \text{phase of carrier.}$$

The parameters that can be varied are the amplitude, the frequency, and the phase. Variation of the frequency or phase is referred to as angle modulation with specific designations of frequency modulation (FM) and phase modulation (PM). When such modulation is employed, the amplitude is held constant. Similarly, when the amplitude is varied, providing amplitude modulation (AM), the phase and frequency are held constant.

Linear Modulation

When the modulating signal directly multiplies the carrier amplitude, the procedure is called product modulation and the result is

$$s(t) = A m(t)\cos \omega_c t \qquad (1\text{-}62)$$

where A is the carrier amplitude, $m(t)$ is the message signal, and the phase angle of the carrier has been assumed to be zero for mathematical convenience. Some insight into the characteristics of this type of signal can be obtained by considering the case in which the modulating signal is a sinusoid. Let

$$m(t) = \cos \omega_m t \tag{1-63}$$

Then

$$s(t) = A \cos \omega_m t \cos \omega_c t$$

$$= \frac{A}{2} \cos(\omega_c - \omega_m)t + \frac{A}{2} \cos(\omega_c + \omega_m)t \tag{1-64}$$

The line spectrum of $x(t)$ is shown in Fig. 1-19.

A more general form is obtained by considering the Fourier spectrum obtained from the Fourier transform of $s(t)$. Thus

$$S(f) = \mathscr{F}\{x(t)\} = \mathscr{F}\{m(t)\cos 2\pi f_c t\}$$

$$= M(f) * \tfrac{1}{2}[\delta(f + f_c) + \delta(f - f_c)]$$

$$= \tfrac{1}{2}M(f + f_c) + \tfrac{1}{2}M(f - f_c) \tag{1-65}$$

where $M(f)$ is the Fourier transform of $m(t)$ and $*$ implies convolution. A representative continuous spectrum is shown in Fig. 1-20.

Note the absence of a discrete component at the carrier frequency in both Figs. 1-19 and 1-20. This type of signal is designated as *double-sideband suppressed carrier* (DSB/SC) modulation. It is evident from the figures that the band of frequencies occupied by the modulated carrier is twice as wide as that occupied by the modulation alone. In this type of signal, all of the power is contained in the

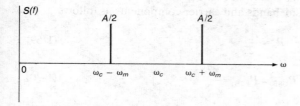

Figure 1-19 Line spectrum of $A \cos \omega_m t \cos \omega_c t$.

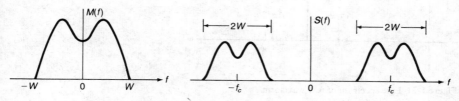

Figure 1-20 Continuous spectrum of $m(t)\cos \omega_c t$.

sidebands, designated as upper sideband and lower sideband by their position relative to the positive frequency carrier component.

As discussed in Chap. 3, the demodulation of a DSB/SC waveform requires the generation of a local carrier, which is then used in a product demodulator. By transmitting a phase-coherent carrier component with the DSB waveform, the demodulation process is greatly simplified and can be accomplished by a simple envelope detector consisting of a rectifier and R-C filter. Such a signal is called an *amplitude-modulation* (AM) waveform and has the following mathematical representation.

$$s(t) = A[1 + \alpha m(t)]\cos 2\pi f_c t \qquad (1\text{-}66)$$

where $m(t)$ is the modulating (information) signal normalized to have a maximum amplitude of unity, α is a constant called the modulation index, which has a magnitude ranging from 0 to 1, and A is a constant determining the power of the modulated carrier. The modulation index is often referred to in terms of percent modulation. Thus when $\alpha = 1$, the carrier is said to be 100 percent modulated, and when $\alpha = 0.8$, the carrier is said to be 80 percent modulated. If the modulation index exceeds unity, the carrier is said to be overmodulated. This is undesirable because overmodulated waveforms cannot be demodulated with an envelope detector without serious distortion. Amplitude-modulated signals are widely used where it is desired to keep the receivers as simple as possible; for example, in medium- and short-wave broadcasting.

The principal characteristics of an AM waveform are readily seen by considering sinusoidal modulation. If $m(t) = \cos \omega_m t$, then the AM waveform is

$$s(t) = A(1 + \alpha \cos \omega_m t)\cos \omega_c t \qquad (1\text{-}67)$$

$$= A \cos \omega_c t + \tfrac{1}{2}A\alpha \cos(\omega_c - \omega_m)t + \tfrac{1}{2}A\alpha \cos(\omega_c + \omega_m)t \qquad (1\text{-}68)$$

The line spectrum of this signal is shown in Fig. 1-21. The power in an AM waveform is distributed among the sidebands and carrier components as follows:

$$P_c = \frac{A^2}{2} \qquad (1\text{-}69)$$

$$P_{\text{USB}} = P_{\text{LSB}} = \frac{A^2\alpha^2}{8} \qquad (1\text{-}70)$$

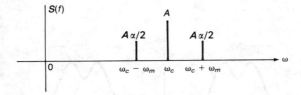

Figure 1-21 Line spectrum of AM waveform.

The total power is

$$P_{\text{TOT}} = \frac{A^2}{2} + \frac{2A^2\alpha^2}{8} = \frac{A^2}{2}\left(1 + \frac{\alpha^2}{2}\right) \tag{1-71}$$

For 100 percent modulation, $\alpha = 1.0$ and it is seen that the total power is $3A^2/4$ with two thirds of the power in the carrier component and one sixth of the power in each sideband.

More complex modulating waveforms can be analyzed using the Fourier transform representation. In this case the spectrum of an AM waveform is given by

$$s(t) = A[1 + \alpha m(t)]\cos 2\pi f_c t$$

$$S(f) = \frac{A}{2}[\delta(f + f_c) + \delta(f - f_c)] + A\frac{\alpha}{2}[M(f + f_c) + M(f - f_c)] \tag{1-72}$$

This is illustrated in Fig. 1-22.

In considering the DSB/SC and AM waveforms, the phase angle of the carrier has been assumed to be zero. However, there are some situations in which the phase angle is important (e.g., product demodulation) and it can be readily incorporated in the following manner. To simplify the notation, it will be assumed that the modulation index is unity.

$$s(t) = A[1 + m(t)]\cos (2\pi f_c t + \phi)$$

$$= A[1 + m(t)]\cos\left[2\pi f_c\left(t + \frac{\phi}{2\pi f_c}\right)\right] \tag{1-73}$$

$$S(f) = A[\delta(f) + M(f)] * \frac{1}{2}[\delta(f + f) + \delta(f - f_c)]e^{jf\phi/f_c}$$

$$= \frac{A}{2}[\delta(f) + M(f)] * [\delta(f + f_c)e^{-j\phi} + \delta(f - f_c)e^{j\phi}]$$

$$= \frac{A}{2}[\delta(f + f_c)e^{-j\phi} + \delta(f - f_c)e^{j\phi}] + \frac{A}{2}[M(f + f_c)e^{-j\phi} + M(f - f_c)e^{j\phi}] \tag{1-74}$$

The first two terms in Eq. (1-74) are the carrier components and the second two terms are the modulation components.

Modulation can also be carried out using a random process. Let $m(t)$ be a

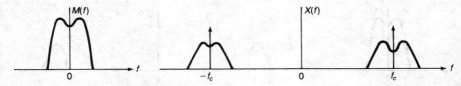

Figure 1-22 Fourier spectrum of AM waveform.

sample function of a zero mean random process, and let the modulated signal be given by

$$s(t) = A[1 + \alpha m(t)]\cos(2\pi f_c t + \phi) \tag{1-75}$$

where ϕ is a fixed but random phase angle uniformly distributed over the interval $0-2\pi$. The spectral density of the modulated signal can be obtained from the auto-correlation function of $x(t)$ as follows:

$$R_s(\tau) = E\{A[1 + \alpha m(t)]\cos(2\pi f_c t + \phi) \cdot A[1 + \alpha m(t + \tau)]\cos(2\pi f_c(t + \tau) + \phi)\}$$
$$= A^2[1 + \alpha \overline{m(t)} + \alpha \overline{m(t + \tau)} + \alpha^2 \overline{m(t)m(t + \tau)}][\overline{\cos(2\pi f_c t + \phi)\cos(2\pi f_c(t + \tau) + \phi)}]$$

$$\tag{1-76}$$

Assuming $m(t)$ and ϕ are statistically independent, the two factors can be considered separately. The first factor becomes $A^2[1 + \alpha^2 R_m(\tau)]$ and the second factor becomes

$$\int_0^{2\pi} \left\{ \frac{1}{2}\cos 2\pi f_c \tau + \frac{1}{2}\cos[2\pi f_c(2t + \tau) + 2\phi] \right\} \frac{d\phi}{2\pi} = \frac{1}{2}\cos 2\pi f_c \tau \tag{1-77}$$

The resulting correlation function is

$$R_s(\tau) = \frac{A^2}{2}\cos 2\pi f_c \tau + \frac{\alpha^2 A^2}{2} R_m(\tau)\cos 2\pi f_c \tau \tag{1-78}$$

The spectral density is obtained by taking the Fourier transform of Eq. (1-78).

$$S_s(f) = \frac{A^2}{4}[\delta(f + f_c) + \delta(f - f_c)] + \frac{\alpha^2 A^2}{4}[S_m(f + f_c) + S_m(f - f_c)] \tag{1-79}$$

where $S_m(f)$ is the spectral density of the random process.

When one of the sidebands of a DSB/SC waveform is eliminated, the result is called *single sideband* (SSB), or sometimes single-sideband suppressed carrier (SSB/SC). There is no loss of information with this type of modulation because each sideband contains all of the modulating waveform. Figure 1-23 illustrates the spectrum of such a waveform. The mathematical expression for an SSB modulated carrier is quite involved and will be considered in Chap. 3 in connection with the demodulation of SSB signals.

Single sideband has the advantage of requiring only half the bandwidth of DSB/SC or AM waveforms and, therefore, makes more efficient use of the spectrum.

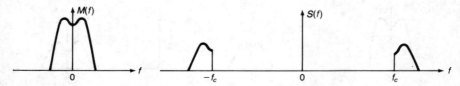

Figure 1-23 Continuous spectrum of SSB waveform.

It is particularly attractive for voice communication because frequency components below about 300 Hz need not be transmitted. This allows the SSB signal to be generated directly from the DSB signal by a simple filter as shown in Fig. 1-24.

The problem is more difficult when low-frequency components and direct current are present in the modulating waveform. One technique that is widely used employs phase-shifted signals to cancel one sideband. This procedure is shown in Fig. 1-25.

It is possible to avoid the problem of sharp cutoff filters by permitting part of the second sideband to be retained. This is referred to as *vestigial sideband* transmission and is the procedure employed for transmission of the video signal in commercial television. By using a filter that has a symmetrical rolloff around f_c, a modulated signal is generated that permits exact recovery of the modulation. The vestigial sideband filter is illustrated in Fig. 1-26. The basic requirement is for the filter to leave exactly the proper amount of the unwanted sideband to compensate for that portion of the desired sideband that is lost by the filter rolloff extending into the frequency range of the desired sideband. The lost information will then be recovered during synchronous demodulation. The transmittance function of the vestigial sideband filter must satisfy the following relationship.

$$H(\omega_c - \omega) + H(\omega_c + \omega) = 2H(\omega_c) \qquad \omega_c - \omega_m \le \omega < \omega_c + \omega_m \qquad (1\text{-}80)$$

The power relations in vestigial sideband transmission are essentially the same as SSB; the bandwidth being only slightly greater.

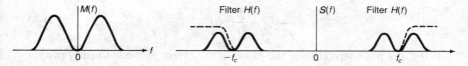

Figure 1-24 Generation of SSB using a filter.

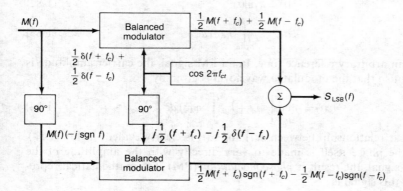

Figure 1-25 Generation of SSB by phase cancellation.

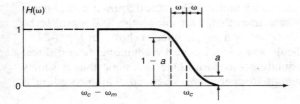

Figure 1-26 Vestigial sideband filter.

Angle Modulation

Modulating the phase or frequency of the carrier is referred to as angle modulation. This modulation process is not a linear operation and such things as superposition do not apply. As a consequence the detailed analysis of angle modulation is very difficult except in relatively simple cases where few modulating components are involved. For a sinusoidal carrier, the instantaneous frequency is defined as $1/2\pi$ times the derivative of the phase of the carrier. Thus if

$$s(t) = A \cos[\omega_c t + \phi(t)]$$

the instantaneous frequency is

$$f_i = \frac{1}{2\pi} \frac{d}{dt} [\omega_c t + \phi(t)]$$

$$= \frac{\omega_c}{2\pi} + \frac{1}{2\pi} \dot{\phi}(t) \tag{1-81}$$

When the time derivative of the phase of the carrier is made linearly proportional to the amplitude of the baseband (modulating) signal, the process is called *frequency modulation* (FM). Using the previous notation, this would be

$$\dot{\phi}(t) = k_f m(t) \tag{1-82}$$

$$\phi(t) = k_f \int_{t_0}^{t} m(\xi) \, d\xi \tag{1-83}$$

where t_0 is an arbitrary reference time. In an FM signal, the carrier amplitude is held constant so that the modulated waveform is given by

$$s(t) = A \cos\left[\omega_c t + k_f \int_{t_0}^{t} m(\xi) \, d\xi \right] \tag{1-84}$$

The nonlinear relationship between $m(t)$ and $x(t)$ is clearly evident in Eq. (1-84).

When the phase itself is made to vary directly with the amplitude of the modulating signal, the result is *phase modulation* (PM). The mathematical representation of this signal is

$$s(t) = A \cos[\omega_c t + k_p m(t)] \tag{1-85}$$

Angle-modulated waveforms contain frequencies other than those of the upper and lower sidebands present in linear modulation. To illustrate this consider a sinusoidally modulated waveform.

$$s(t) = A \cos(\omega_c t + \beta \sin \omega_1 t) \tag{1-86}$$

The parameter β is called the modulation index. For phase modulation $\beta = k_p$, and for frequency modulation, $\beta = k_f/\omega_1$. The expression for $x(t)$ cannot be expanded in terms of elementary functions but can be expanded using the following Bessel function identity.

$$\cos(\omega t + a \sin \theta) = \sum_{n=-\infty}^{\infty} J_n(a)\cos(\omega t + n\theta) \tag{1-87}$$

where $J_n(\cdot)$ is the nth-order Bessel function. Applying this to Eq. (1-86) gives

$$s(t) = A \sum_{n=-\infty}^{\infty} J_n(\beta)\cos(\omega_c + n\omega_1)t \tag{1-88}$$

It is seen that $s(t)$ contains an infinite number of sidebands. The amplitudes of the various sidebands can be determined from a table of Bessel functions or from a graphical representation of Bessel functions as shown in Fig. 1-27. As an illustration consider a simple case in which $\beta = 5$. From Fig. 1-27 the amplitudes of $J_n(5)$ are as shown in Table 1-2. The Bessel functions have the symmetry property

$$J_{-n}(x) = (-1)^n J_n(x) \tag{1-89}$$

so that values of $J_{-n}(5)$ can be readily found from Table 1-2 and correspond to phase reversals of the odd harmonics. The line spectrum of this waveform is shown in Fig. 1-28.

The sidebands of an angle-modulated waveform theoretically extend to $\pm \infty$; however, as a practical matter, the amplitudes decrease very rapidly for the high harmonics. An important question is what bandwidth is required to permit good fidelity to be retained in the demodulation process. From Fig. 1-27 it is seen that $J_n(\beta)$ decreases very rapidly for $n > \beta$. Therefore, for good fidelity the sidebands must extend over the interval $-\beta < n < \beta$. This gives a bandwidth of

$$\text{BW} = 2n_{\max}\omega_1 = 2\beta\omega_1 \qquad \text{rad/sec} \tag{1-90}$$

For small β the bandwidth must be at least as large as $2\omega_1$ since otherwise no modulation component would be present in the waveform. From these require-

Table 1-2 Amplitudes of $J_n(5)$

n	$J_n(5)$	n	$J_n(5)$
0	−0.18	4	0.39
1	−0.33	5	0.26
2	0.05	6	0.13
3	0.36	7	0.06

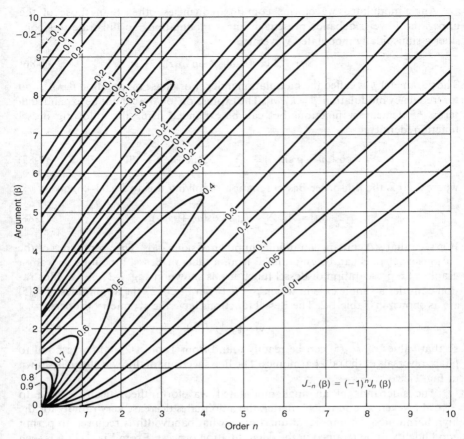

Figure 1-27 Bessel functions of the first kind. Curves of $J_n(\beta) = $ constant.

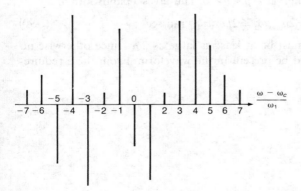

Figure 1-28 Line spectrum of sinusoidally modulated FM waveform.

ments it is evident that the RF bandwidth of a sinusoidally angle-modulated carrier must be

$$\Omega_r = 2[\beta\omega_1 + \omega_1] = 2\omega_1(\beta + 1) \qquad \text{rad/sec} \tag{1-91}$$

Calling $\beta\omega_1 = \omega_{pk}$ the maximum or peak frequency deviation and ω_b the maximum baseband frequency, it follows that

$$\Omega_r = 2(\omega_{pk} + \omega_b) \tag{1-92}$$

In words Eq. (1-92) states that for an FM waveform the RF or IF bandwidth must be twice the sum of the peak frequency deviation and the maximum baseband frequency.

It is often convenient to consider the modulating waveform $m(t)$ to be a sample function of a stationary random process. Assume that $p(m)$ is the PDF of $m(t)$. It can be shown that under the condition that the peak frequency deviation is large compared with the maximum baseband frequency, the spectral density of the FM modulated carrier is[5, 6]

$$S_s(\omega) = \frac{A^2}{4k_s}\left[p\left(\frac{\omega+\omega_c}{k_f}\right) + p\left(\frac{\omega-\omega_c}{k}\right) \right] \tag{1-93}$$

where k_f is the angular frequency deviation constant. In terms of frequency f, this becomes

$$S_s(f) = \frac{A^2}{4f_\Delta}\left[p\left(\frac{f+f_c}{f_\Delta}\right) + p\left(\frac{f-f_c}{f_\Delta}\right) \right] \tag{1-94}$$

where k_f is the angular frequency deviation constant. In terms of frequency f, this and (1-94) are sometimes referred to as Woodward's theorem.

As an example of a randomly modulated FM waveform, consider the case where $m(t)$ has a Gaussian PDF; that is,

$$p(m) = \frac{1}{\sigma_m\sqrt{2\pi}}\exp\left[\frac{-m^2}{2\sigma_m^2}\right] \tag{1-95}$$

Let it be further assumed that $m(t)$ is band-limited to $\pm W$ Hz. It cannot be guaranteed that $|m(t)| < 1$, but by choosing σ_m small enough, the probability of this occurring can be made very small. For this case the spectral density becomes

$$S_s(f) = \frac{A^2}{4f_\Delta}\frac{1}{\sigma_m\sqrt{2\pi}}\left\{ \exp\left[\frac{-(f+f_c)^2}{2f_\Delta^2\sigma_m^2}\right] + \exp\left[\frac{-(f-f_c)^2}{2f_\Delta^2\sigma_m^2}\right] \right\} \tag{1-96}$$

$$= \frac{A^2}{4\sqrt{2\pi}\sigma_f}\left\{ \exp\left[\frac{-(f+f_c)^2}{2\sigma_f^2}\right] + \exp\left[\frac{-(f-f_c)^2}{2\sigma_f^2}\right] \right\} \tag{1-97}$$

where $\sigma_f = f_\Delta\sigma_m$. Note that $S_s(f)$ is not dependent on the spectrum of $m(t)$. It consists of Gaussian-shaped spectra located at $\pm f_c$ with standard deviations of σ_f.

The bandwidth requirements of general types of modulating signals can be approximated in a manner similar to that for sinusoidal modulation. The modulation index β is not defined for nonsinusoidal signals and a new quantity, the deviation ratio D, is defined as

$$D = \frac{f_{pk}}{W_m} \qquad (1\text{-}98)$$

Using this quantity in place of β in Eq. (1-91) gives

$$B = 2(D+1)W_m \qquad \text{Hz} \qquad (1\text{-}99)$$

This is referred to as Carson's rule and is widely used to estimate FM bandwidth requirements.

The power in an angle-modulated signal is determined by the amplitude of the carrier and not by the amplitude of the modulating signal as in linear modulation. Increasing the amplitude of the modulating signal only increases the peak frequency deviation of the signal.

Linear Pulse Modulation

A pulse-amplitude-modulated (PAM) signal is obtained by varying the amplitude of a pulse train in proportion to the amplitude of a baseband signal. As an example let $g(t)$ be a periodic pulse train given by

$$g(t) = \frac{g_0}{2} + \sum_{n=1}^{\infty} g_n \cos(n\omega_c t + \theta_n)$$

For linear or product modulation, the modulated signal is

$$s(t) = m(t)g(t)$$

Assuming $m(t) = A_1 \cos \omega_1 t$, this becomes

$$s(t) = A_1 \cos \omega_1 t \left[\frac{g_0}{2} + \sum_{n=1}^{\infty} g_n \cos(n\omega_c t + \theta_n) \right]$$

$$= \frac{A_1}{2} \left\{ g_0 \cos \omega_1 t + \sum g_n \cos[(n\omega_c + \omega_1)t + \theta_n] \right.$$

$$\left. + \sum g_n \cos[(n\omega_c - \omega_1)t + \theta_n)] \right\} \qquad (1\text{-}100)$$

It is evident from Eq. (1-100) that there is a pair of sidebands around every harmonic of the pulse repetition frequency. Also, there is an unshifted signal component if $g_0 \neq 0$; that is, if the pulse train has a nonzero average value. To prevent sideband overlap, it is necessary that

$$n\omega_c + \omega_1 < (n+1)\omega_c - \omega_1$$

or

$$\omega_c > 2\omega_1$$

That is, the pulse repetition frequency must be greater than twice the frequency of the modulating signal.

As an illustration consider the case of a rectangular pulse train as shown in

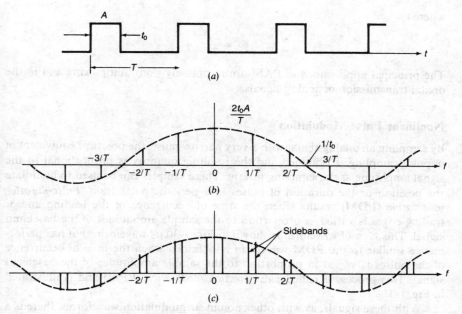

Figure 1-29 Pulse-amplitude modulation. (a) Pulse train; (b) line spectrum of pulse train; (c) line spectrum of PAM signal.

Fig. 1-29a. This signal can be written in terms of the rect(t) function which is defined as

$$\text{rect}(t) = 1 \qquad |t| < \frac{1}{2} \tag{1-101}$$

$$= 0 \qquad |t| > \frac{1}{2}$$

Thus,

$$g(t) = \sum_{n=-\infty}^{\infty} A \, \text{rect}\left(\frac{t - nT}{t_0}\right)$$

and

$$g_n = \frac{2}{T} \int_{-T/2}^{T/2} A \, \text{rect}\left(\frac{t}{t_0}\right) \cos\left(\frac{2\pi n t}{T}\right) dt$$

$$= \frac{2 t_0 A}{T} \, \text{sinc}\left(\frac{n t_0}{T}\right) \tag{1-102}$$

where $\text{sinc}\, t = \sin(\pi t)/\pi t$. The line spectrum of $g(t)$ is shown in Fig. 1-29b and the line spectrum of the sinusoidally modulated pulse train is shown in Fig. 1-29c. Around the first harmonic, the sinusoidally modulated signal would be

$$\frac{A_1 t_0 A}{T} \, \text{sinc}\left(\frac{t_0}{T}\right) \left[\cos(\omega_c + \omega_1)t + \cos(\omega_c - \omega_1)t\right] \tag{1-103}$$

where

$$\omega_c = \frac{2\pi}{T}$$

The principal applications of PAM are in time-division multiplexing and in the digital transmission of analog signals.

Nonlinear Pulse Modulation

By sampling an analog signal with a very narrow pulse, the practical equivalent of impulse sampling is obtained and the resulting quantity is proportional to the signal amplitude at the sampling instant. These samples can be used to modulate the position or the duration of pulses in a periodic pulse train. *Pulse-duration modulation* (PDM) results when the time of occurrence of the leading and/or trailing edges is varied in proportion to the sample amplitudes of the baseband signal. This type of waveform is shown in Fig. 1-30. A waveform that has performance similar to the PDM waveform is obtained when the time of occurrence of the pulses is varied in proportion to the sample amplitudes of the baseband signal. This process is called *pulse-position modulation* (PPM) and is illustrated in Fig. 1-31.

With these signals, as with other nonlinear modulation waveforms, there is a trade-off possible between bandwidth and the SNR of the demodulated signal. However, it is much less for these waveforms than for others (e.g., PCM) and the analysis is very involved.[7]

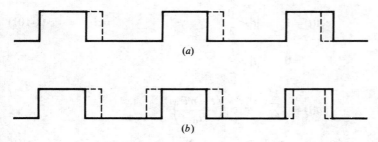

(a)

(b)

Figure 1-30 Pulse-duration modulation. (a) Leading edge fixed, trailing edge modulated; (b) both edges modulated.

Figure 1-31 Pulse-position modulation.

1-6 DIGITAL MODULATION SYSTEMS

Pulse-Code Modulation

In digital modulation systems, the baseband signal is converted to a set of numbers by quantizing the amplitudes obtained by periodically sampling the baseband signal. These numbers are then converted to a stream of binary digits for transmission by an appropriate binary signal set. The received signal is reconstructed from the decoded binary representations of its samples. The basic system of this type is *pulse-code modulation* (PCM). In PCM the baseband signal is sampled at a frequency at least twice that of the highest frequency component present. This is necessary to prevent so-called aliasing distortion that results from sidebands around one harmonic of the sampling frequency overlapping the frequency band occupied by sidebands around a different harmonic. These samples are then quantized in amplitude by assigning their actual value to one of L preselected possible values. The various quantum levels are encoded into numbers and transmitted as the signal level at that sampling instant. This process is illustrated in Fig. 1-32.

Errors that result from assigning continuous sample values to the nearest discrete quantization level are called quantizing noise. In the case of PCM, the signal is reconstructed by means of a digital-to-analog (D/A) converter. The direct regeneration leads to a stairstep representation as shown in Fig. 1-33. By passing this waveform through a low-pass filter to remove components above one half the sampling frequency, a smooth continuous representation of the analog signal is obtained.

The attractive feature of PCM is that all that is necessary for exact reproduction of the quantized samples at the receiver is for the binary transmitted signals to be properly detected; that is, system noise does not affect the demodulated waveform unless it is large enough to cause an error in detecting whether a received signal corresponds to a 0 or a 1. This means that the system can operate with a much lower SNR than would be required, for example, by an AM system where the demodulated waveform is directly affected by additive noise. In many systems the

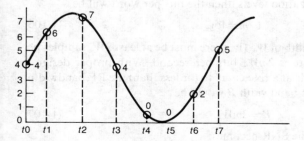

Figure 1-32 Sampling and quantization to eight levels.

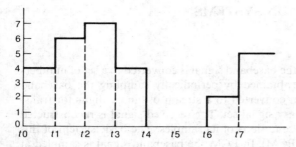

Figure 1-33 Digital-to-analog conversion of the samples in Fig. 1-32.

quantization noise is the largest component present and the channel noise can be neglected.

The magnitude of the quantization noise can be readily determined. Calling the quantizing error e and the quantizer step size Δ, then assuming that the errors are uniformly distributed over the step intervals, the mean squared error is

$$\overline{e^2} = E\{e^2\} = \int_{-\Delta/2}^{\Delta/2} e^2 \cdot \frac{1}{\Delta} \, de = \frac{\Delta^2}{12} \tag{1-104}$$

For a full-amplitude PCM signal, all steps will be used. Assuming there are L steps and a sinusoidal waveform is to be quantized, we have

$$\text{Sine-wave amplitude} = \frac{L\Delta}{2}$$

$$\text{Mean square value of sine wave} = \frac{L^2\Delta^2}{8}$$

$$\text{SNR} = \frac{L^2\Delta^2/8}{\Delta^2/12} = \frac{3}{2} L^2 \tag{1-105}$$

Thus the signal-to-quantization-noise ratio depends only on the number of steps, assuming a full-amplitude signal. If the sample values are transmitted by binary words and there are L quantization levels, then the bits per word will be

$$m = \log_2 L \tag{1-106}$$

For a baseband signal bandwidth of W_m Hz, there must be at least $2W_m$ samples per second. Therefore, the bit rate is $2mW_m$ bits per second. With proper design a digital system can transmit bits at a rate equal to (or less than) the RF bandwidth. For this case the transmission bandwidth B would be

$$B = 2mW_m \tag{1-107}$$

In terms of bits per word m, the SNR becomes

$$\text{SNR} \frac{3L^2}{2} = \frac{3}{2} 2^{2m} \tag{1-108}$$

Substituting for m from Eq. (1-107) gives

$$\text{SNR} = \frac{3}{2} 2^{B/W_m} \qquad (1\text{-}109)$$

From Eq. (1-109) it is seen that the SNR improves exponentially with the ratio of transmission bandwidth to baseband bandwidth. This is much greater than FM, which follows a square law. Sampling and quantization are discussed in more detail in Chap. 2.

Differential Pulse-Code Modulation

Speech and many other types of signals contain enough structure that when sampled at the Nyquist rate (i.e., twice the highest frequency present), it is found that there is significant correlation among adjacent samples. If the samples of a signal are designated as $x_1, x_2 \ldots x_r, x_{r+1}, \ldots$, then the first difference is defined as

$$D_r(1) = x_r - x_{r-1} \qquad r = 1, 2, \ldots \qquad (1\text{-}110)$$

The mean square value of the first difference is

$$\overline{D_r^2(1)} = \overline{(x_r - x_{r-1})^2} = \overline{x_r^2 + x_{r-1}^2 - 2x_r x_{r-1}} \qquad (1\text{-}111)$$

For a zero mean stationary process, $\overline{x_r^2} = \overline{x_{r-1}^2} = \sigma_x^2$ and $\overline{x_r x_{r-1}} = \rho \sigma_x^2$ where σ_x^2 is the variance and ρ is the correlation coefficient of adjacent samples. Substitution into Eq. (1-111) gives

$$D_r^2(1) = 2\sigma_x^2(1 - \rho) \qquad (1\text{-}112)$$

For $\rho > 1/2$, the variance of the first difference is seen to be less than that of the signal. Therefore, for a given number of quantization steps, better performance can be obtained by quantizing and transmitting the first differences of the samples rather than the samples themselves.

This technique is called *differential pulse-code modulation* (DPCM). The procedure is to encode the difference between the present value of the signal and the predicted value based on the correlation with the previous samples. The receiver also makes a prediction based on previous samples and then uses the received difference signal to correct its estimate. In the case of speech, using three previous samples in the prediction has been shown to permit reductions of the order of 1.5 to 2 bits per sample over PCM for the same quality of reproduction.

The weight used in the predictor can be continuously readjusted by measuring statistics of the incoming signal over a short interval (5–25 ms). This technique is called *adaptive differential pulse-code modulation* (ADPCM) and has shown to have the potential of providing reductions in bit rate of up to five over PCM.

Delta Modulation

By sampling very rapidly (i.e., greater than the Nyquist rate), only small changes in signal amplitude can occur between samples. In this case it is possible to track the

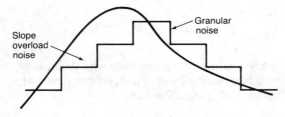

Figure 1-34 Delta-modulation noise.

signal by transmitting only a single bit that conveys the direction of change; that is,

0—change upward

1—change downward

This is called *delta modulation* (DM). When the step size is changed in accordance with the data (e.g., when a specified number of corrections in the same direction occur), the system is called *adaptive delta modulation* (ADM).

There are two kinds of intrinsic noise that arise in delta-modulation systems. These are slope overload noise, which occurs when the signal is changing faster than the tracking capability of the system, and granular noise, which occurs because of the finite step size present. Figure 1-34 illustrates the nature of these noises.

One advantage of DM is its ability to handle impulse noise. No matter how large an impulse is or where in a time sequence it occurs, it can only result in a small change in the signal amplitude.

1-7 REFERENCES

1. Bennett, W. R.: *Introduction to Signal Transmission*, McGraw-Hill Book Company, New York, 1970.
2. *Reference Data for Radio Engineers*, Howard Sams & Company, Inc., Indianapolis, 1975.
3. Skomal, E. N.: *Manmade Radio Noise*, Van Nostrand Reinhold Company, New York, 1975.
4. Terman, F. E.: *Radio Engineers Handbook*, McGraw-Hill Book Company, New York, 1943.
5. Blackman, N.: Calculation of the Spectrum of an FM Signal Using Woodward's Theorem, *IEEE Trans. Comm. Tech.*, August 1969.
6. Taub, H., and D. Schilling: *Principles of Communication Systems*, McGraw-Hill Book Company, New York, 1971.
7. Black, H. S.: *Modulation Theory*, Van Nostrand, Princeton, N.J., 1953.

1-8 PROBLEMS

1.2-1 Sound levels are generally measured in terms of rms sound pressure relative to that at the threshold of hearing for a 1000-Hz tone; viz., 0.0002 dyne/cm^2. Thus a sound level of 80 dB would correspond to a sound pressure of 2 dynes/cm^2. The threshold of hearing varies with frequency. Similarly, the sensation of loudness varies with frequency. One way of expressing this variability is to plot contours of equal loudness versus frequency. Table P1.2-1 gives values for the sound pressure at the threshold of hearing and for loudness equal to that of a 70-dB, 1-kHz tone versus frequency. The 70-dB level corresponds to the sound level of ordinary speech. Plot these values in decibels versus frequency. What additional gain would be required to produce the same threshold level at 400 Hz as at 4 kHz?

Table P1.2-1 Threshold of hearing and 70-dB equal-loudness contour

Frequency (Hz)	Threshold of hearing (dynes/cm^2)	70-dB loudness contour (dynes/cm^2)
50	84.3×10^{-3}	2.52
100	14.2×10^{-3}	1.50
200	3.17×10^{-3}	1.00
400	0.632×10^{-3}	0.670
900	0.200×10^{-3}	0.632
1000	0.200×10^{-3}	0.632
1100	0.200×10^{-3}	0.632
2000	0.126×10^{-3}	0.564
4000	0.080×10^{-3}	0.423
5000	0.100×10^{-3}	0.597
6000	0.200×10^{-3}	1.12
7000	0.317×10^{-3}	1.78
8000	0.399×10^{-3}	2.22
10000	0.564×10^{-3}	2.52

1.2-2 It is desired to reduce the transmission bandwidth of speech from 3200 Hz (300–3500 Hz) to 2000 Hz. What band of frequencies should be chosen to give the highest articulation index for average speech? How much loss of articulation would occur relative to the original band?

1.2-3 Independent noise sources are generally considered to be incoherent so that their noise powers add rather than their sound pressures. Assume that there are three white noise sources with sound pressure levels of 50 dB, 66 dB, and 63 dB measured relative to 0.0002 dyne/cm^2. How would the articulation index of normal speech (sound pressure level = 69 dB) be affected by the presence of this noise?

1.2-4 It is desired to design a slow-scan television system that would be able to operate over a channel capable of handling a bandwidth of 5 kHz. Assume that no interlacing is to be used, a square raster with 100×100 pixels is to be employed, and 10 percent of each frame is lost due to blanking for retrace and synchronization. What frame rates and scan line frequencies could be used?

1.3-1 Find the rms thermal noise voltage in a bandwidth of 100 kHz appearing across the 1-MΩ resistor in the accompanying figure.

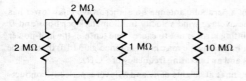

1.3-2 It is found that changing the source temperature of a system from $T_s = 290$ K to $T_s = 500$ K produces a 25 percent increase in the system output noise power. Find the system noise figure.

1.3-3 A resistive attenuator with loss factor $L(L > 1)$ is connected to matched input and output resistances. If the attenuator has a temperature equal to the source temperature T_s, what are the effective noise temperature and noise figure of the cascade of the source resistance and attenuator? Show that if the attenuator is at room temperature T_0, the noise figure is $F = L$.

1.3-4 Find the overall noise figure for the receiving system shown. Find the rms value of the noise voltage at the output. Assume all components are at room temperature (290K).

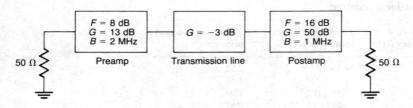

1.4-1 Determine the SNR at the output of the receiver IF amplifier for the system shown. The system parameters are as follows:

$f_c = 100$ MHz
$P_t = 20$ watts
Transmitter to antenna loss $= 0.2$ dB
Transmitting antenna $=$ short dipole
Distance to receiver $= 80$ km
Receiving antenna $=$ short dipole
Receiving antenna to mixer loss $= 0.3$ dB
Mixer noise figure $= 8$ dB
Mixer conversion loss $= 3$ dB
IF amplifier noise figure $= 6$ dB
Receiver noise bandwidth $= 4$ kHz

What IF amplifier gain would be required to give an output voltage of 1 volt rms for the specified transmitter power assuming an IF amplifier output impedance of 300 ohms?

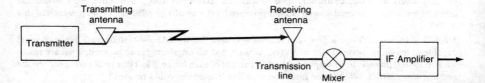

1.4-2 A microwave communication system operating at 5 GHz is to have a parabolic antenna with a power gain of 50 dB. Given that the aperture loss factor is 0.54, find the half-power beam width in degrees and the antenna diameter in meters.

1.4-3 Assume that the angular pointing error of a parabolic antenna on a microwave relay tower has an rms value of $VD/100$ degrees, where V is the average wind velocity in kilometers per hour and D is the antenna diameter in meters. If the rms pointing error is not to exceed one tenth of the half-power beam width for any wind velocity up to 100 km/h, what is the maximum antenna gain (dB) that can be obtained? Assume an antenna loss factor of 0.54 and an operating frequency of 5 GHz.

1.4-4 A microwave transmission system has antennas at heights of h_T and h_R over a perfectly conducting, flat earth and they are separated by D.

    (a) Derive an approximate general expression for the lowest frequency at which the direct and reflected waves add in phase, under the assumption that D is very much greater than either h_T or h_R.

(b) Evaluate this optimum frequency assuming $h_T = 100$ meters, $h_R = 10$ meters, and $D = 4$ km.

(c) What is the closest frequency above that found in (b) at which the resultant signal is zero?

1.4-5 A satellite in synchronous orbit is transmitting to a ground station over a slant range of 40,000 km. The satellite antenna has a gain of 6 dB and the operating frequency is 4 GHz. The ground station has a parabolic antenna 15 meters in diameter with an aperture efficiency of 0.54. What transmitted power at the satellite is required to achieve a received signal power of 10^{-12} watts?

1.4-6 A satellite-to-ground link operates at a frequency of 3000 MHz with a free-space propagation loss of 290 dB. The satellite antenna has a gain of 10 dB and it is desired to achieve a received signal power of $4\pi \times 10^{-14}$ watts. If satellite transmitter power costs $100,000 per watt, and the effective area of the ground station antenna costs $10,000 per square meter, what is the cheapest combination of transmitter power and antenna area that will achieve the desired results?

1.5-1 The AM signal

$$s(t) = [1 + a \cos 2\pi f_a t] \cos 2\pi f_c t$$

is detected by an ideal envelope detector. When $a > 1$, overmodulation occurs and distortion is produced. For the case of $a = 1.5$, find the total distortion and express it as a fraction of the signal component. (*Hint:* Use a Fourier series expansion.)

1.5-2 It is desired to make a superheterodyne receiver to cover the AM band from 550 to 1650 kHz. An intermediate frequency of 455 kHz is to be used. If the local oscillator is below the reception band, what frequency range must it cover? What bands of frequencies might cause spurious reception if they reached the mixer? What could explain the occurrence of a spurious high-frequency audio tone when a station at 920 kHz is tuned in?

1.5-3 Narrow-band FM has a carrier and sidebands much the same as an AM signal. Write out the corresponding expressions for sine-wave modulation in both cases and determine the differences. What would be the result of using an envelope detector to detect a narrow-band FM signal?

1.5-4 Nonlinear resistors are available in which the current–voltage relationship is

$$i = a_1 v + a_2 v^2$$

How can these be used to obtain a product modulator? An amplitude modulator?

1.5-5 A 10-MHz carrier is frequency modulated by a 400-Hz sinusoidal signal with a modulation index $\beta = 1$. The modulated signal is passed through a 1000-Hz bandpass filter centered at 10 MHz. Sketch the line spectrum of the output.

1.5-6 An FM and a PM modulator are each adjusted so that for a single tone input the output line spectra are identical. Assuming the frequency of the modulating tone is increased, describe how the two spectra will vary. How will an increase in the amplitude of the modulating tone affect the two spectra?

1.5-7 Commercial FM broadcasting has a restriction of 75 kHz on the maximum frequency deviation that can be employed. Assuming a signal frequency of 15 kHz, what is the bandwidth according to Carson's rule? What is the amplitude of the harmonics 15 kHz on either side of the band edge determined by Carson's rule and how do they compare with the amplitude of the largest harmonic?

1.5-8 Multitone FM can be analyzed by means of trigonometric expansion of the components involved. Let the modulating signal be

$$a(t) = a_1 \cos \omega_1 t + a_2 \cos \omega_2 t$$

Then the modulated signal will be

$$s(t) = A \cos (\omega_c t + \beta_1 \sin \omega_1 t + \beta_2 \sin \omega_2 t)$$

where

$$\beta_1 = \frac{a_1 k}{\omega_1}$$

$$\beta_2 = \frac{a_2 k}{\omega_2}$$

Show that the modulated signal can also be expressed as

$$m(t) = A \sum_{n=-\infty}^{\infty} \sum_{m=-\infty}^{\infty} J_n(\beta_1)J_m(\beta_2)\cos[\omega_c + n\omega_1 + m\omega_2]t$$

Tabulate the frequencies of the components of the modulated waveform assuming $f_1 = 1000$, $f_2 = 1200$, and $n, m = 0, \pm 1, \pm 2$.

1.5-9 A PAM signal employs an unmodulated pulse train as shown.

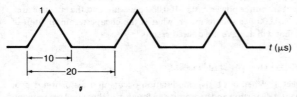

Sketch the line spectrum of the pulse train and the line spectrum of a PAM signal where the modulating signal is $m(t) = A \sin 2\pi(10^4)t$ and assuming that the modulated signal is of the form $x(t) = m(t)g(t)$.

1.5-10 Frequently a sample-and-hold circuit is employed to generate PAM signals. In this case the pulse amplitude corresponds to the signal at the sampling instant and the problem is most readily handled by assuming impulse sampling followed by pulse-amplitude modulation. The PAM signal is then

$$s(t) = \left[m(t) \sum_{n=\infty}^{\infty} \delta(t - nT) \right] * p(t)$$

where $m(t)$ is the modulating signal and $p(t)$ is the pulse shape. Assuming that $p(t)$ is a rectangular pulse of duration t_0 and unit amplitude, sketch the spectrum for modulation by $A \cos(2\pi t/4T)$. What scale factor is required to receover the modulation from a low-pass filtered version of $s(t)$? What processing is required to recover a modulating signal having a continuous spectrum extending from $0 \rightarrow 1/4T$? Under what conditions would it be possible to omit the equalization filter?

1.6-1 In a remote sensing system measuring only positive signals, the data are sampled and quantized to 8 bits. Under the assumption that the signal components are sinusoidal and cover the entire dynamic range, what is the SNR that can be obtained? If the signal components are square waves covering the quantization levels from 5 to 30 (out of the range 0–255), what is the SNR?

1.6-2 Assume that the bandwidth required to transmit a signal digitally equals the number of binary digits per second in the sampled and quantized waveform. What bandwidth would be required to transmit a speech signal (BW = 3.5 kHz) with a signal-to-quantization-noise ratio of 30 dB? A television video signal (BW = 4.0 MHz) with a signal-to-quantization-noise ratio of 40 dB?

1.6-3 A delta-modulation system employs a step size of 0.1 volt and a sampling interval of 10^{-4}. What is the maximum amplitude that a 1-kHz sinusoid can have and not produce slope overload? What is the maximum frequency that a 1-volt sinusoid can have and not produce slope overload?

FUNDAMENTALS OF SIGNAL PROCESSING

2-1 LINEAR SYSTEM ANALYSIS[1]

To predict the performance of communication systems and to synthesize processors to perform specified operations, it is necessary to be able to calculate how a particular system affects signals transmitted through it. When the system is linear—that is, when the variables are interrelated by a set of linear differential equations—it is usually possible to make a complete analysis. Such analyses are usually carried out in the time domain using the convolution integral or in the frequency domain using transform methods. Frequently a combination of methods is used. To illustrate the procedures, and in fact for most communication system analysis, it is sufficient to consider systems with a single input and a single output. Also, the most widely encountered systems may be satisfactorily modeled as having zero initial stored energy and as being time invariant. The output of such a system depends only on the excitation applied, and for a delayed excitation the response is delayed by the same amount. Linear system analysis is also useful in connection with nonlinear systems; for small variations around a particular operating point, such systems can often be approximated by a linear model.

Impulse Response

In a linear time-invariant system with no initial stored energy, the output time function is related to the input time function by the convolution integral. This relationship makes use of the impulse response $h(t)$ of the system, which is the time function occurring at the system output in response to an input of a unit impulse. If a general input $x(t)$ is applied to a system with impulse response $h(t)$, then the system output $y(t)$ is given by the convolution of $x(t)$ and $h(t)$. Mathematically this can be expressed as

$$y(t) = x(t) * h(t) \tag{2-1}$$

$$= \int_{-\infty}^{\infty} x(\tau)h(t - \tau)\, d\tau \tag{2-2}$$

49

$$= \int_{-\infty}^{\infty} x(t-\tau)h(\tau)\, d\tau \tag{2-3}$$

where the centered asterisk (∗) indicates convolution. As an example consider the low-pass RC filter shown in Fig. 2-1. The impulse response for this circuit can be expressed in terms of the product $1/RC = \alpha$ as

$$h(t) = \alpha e^{-\alpha t} u(t) \tag{2-4}$$

If a step input $x(t) = u(t)$ were applied to this filter, the output would be given by

$$y(t) = \int_{-\infty}^{\infty} u(\tau)\alpha e^{-\alpha(t-\tau)} u(t-\tau)\, d\tau \tag{2-5}$$

$$= \alpha e^{-\alpha t} \int_0^t e^{\alpha \tau}\, d\tau = 1 - e^{-\alpha t} \qquad t > 0$$

$$= 0 \qquad\qquad\qquad\qquad\quad t < 0$$

$$y(t) = (1 - e^{-\alpha t}) u(t) \tag{2-6}$$

Frequently convolution operations are carried out when one of the functions in the integrand is an impulse (or delta function). In such cases the integral is usually quite simple to evaluate because of the sifting property of impulses. The following examples illustrate such evaluations.

$$x(t) * \delta(t) = \int_{-\infty}^{\infty} x(\tau)\delta(t-\tau)\, d\tau = x(t) \tag{2-7}$$

$$x(t) * \delta(t - t_0) = x(t - t_0) \tag{2-8}$$

$$x(at) * \delta(t - t_0) = x[a(t - t_0)] \tag{2-9}$$

$$x(t) * \delta(at - t_0) = x(t) * \frac{1}{|a|} \delta\left(t - \frac{t_0}{a}\right)$$

$$= \frac{1}{|a|} x\left(t - \frac{t_0}{a}\right) \tag{2-10}$$

$$\delta(at - t_1) * \delta(bt - t_2) = \frac{1}{|a||b|} \delta\left(t - \frac{t_1}{a} - \frac{t_2}{b}\right) \tag{2-11}$$

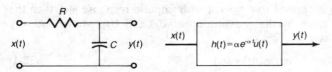

Figure 2-1 RC low-pass filter.

Transfer Function

The analysis of time-invariant linear systems can also be carried out by transform techniques. This is generally referred to as frequency-domain analysis. The most useful transformation is the Fourier transform since the variable of this transformation is the frequency of a sinusoidal waveform and corresponds to an easily measured and interpreted quantity. Other transformations, such as the Laplace and z-transforms, are also useful, but frequently must be converted to the equivalent Fourier representation for proper interpretation. Both the Fourier series, which is actually a time representation, and the Fourier transform are useful. The Fourier transform of a time function is defined as follows:

$$\mathscr{F}\{x(t)\} = X(f) = \int_{-\infty}^{\infty} x(t)e^{-j2\pi ft} \, dt \tag{2-12}$$

Similarly, the variable $\omega = 2\pi f$ may also be used, leading to the equivalent expression:

$$\mathscr{F}\{x(t)\} = X(\omega) = \int_{-\infty}^{\infty} x(t)e^{-j\omega t} \, dt \tag{2-13}$$

Conventionally capital letters are used to represent the transformed functions and the use of f or ω is evident from the variables in the expression.

$$\mathscr{F}^{-1}\{X(f)\} = x(t) = \int_{-\infty}^{\infty} X(f)e^{j2\pi ft} \, df \tag{2-14}$$

$$\mathscr{F}^{-1}\{X(\omega)\} = x(t) = \frac{1}{2\pi}\int_{-\infty}^{\infty} X(\omega)e^{j\omega t} \, d\omega \tag{2-15}$$

It is generally not necessary to carry out the actual integrations. Tables of simple transforms and rules for their manipulation are sufficient to handle most of the problems of interest in communications engineering. Such tables are provided in Appendix A and include basic signal and transform definitions as well as a summary of transforms of operations and elementary functions. The transforms are given in terms of both the variable f and the variable ω since often it is convenient to switch between the two. One major advantage of transform analysis is that many of the techniques of steady-state ac circuit analysis carry over directly by replacing circuit elements by their impedances and replacing the excitation and response signals by their Fourier transforms. This is particularly useful in computing system transfer functions; that is, the ratio of the transform of the output function to the transform of the input function. This is defined as

$$H(f) = \frac{Y(f)}{X(f)} \tag{2-16}$$

$$H(\omega) = \frac{Y(\omega)}{X(\omega)} \tag{2-17}$$

and is used to determine the output when the input is given. This can be illustrated by repeating the analysis of the RC circuit considered previously. In this case the circuit is represented in terms of the impedances of its elements as shown in Fig. 2-2. The transfer function is obtained by considering the circuit as a simple voltage divider as follows:

$$Y(\omega) = \frac{\dfrac{1}{j\omega C} X(\omega)}{R + \dfrac{1}{j\omega C}} = \frac{\dfrac{1}{RC}}{j\omega + \dfrac{1}{RC}} X(\omega) \tag{2-18}$$

Since $H(\omega) = Y(\omega)/X(\omega)$, it follows from Eq. (2-18) that the transfer function of the RC circuit is

$$H(\omega) = \frac{Y(\omega)}{X(\omega)} = \frac{\dfrac{1}{RC}}{j\omega + \dfrac{1}{RC}} = \frac{\alpha}{j\omega + \alpha} \tag{2-19}$$

where $\alpha = 1/RC$. For the step input previously considered, the output can be found as follows using the table of transforms.

$$Y(\omega) = H(\omega)X(\omega)$$

$$= \frac{\alpha}{j\omega + \alpha}\left[\pi\delta(\omega) + \frac{1}{j\omega}\right]$$

$$= \pi\delta(\omega) + \frac{\alpha}{j\omega(j\omega + \alpha)}$$

$$= \pi\delta(\omega) + \frac{1}{j\omega} - \frac{1}{j\omega + \alpha} \tag{2-20}$$

$$y(t) = \mathscr{F}^{-1}\{Y(\omega)\}$$

$$= u(t) - e^{-\alpha t}u(t)$$

$$= (1 - e^{-\alpha t})u(t)$$

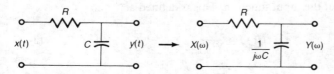

Figure 2-2 Transformation of RC circuit.

Relationship Between Impulse Response and Transfer Function

There is a simple relationship between the transfer function and the impulse response. Consider the case of a unit impulse input to a system having a transfer function $H(f)$.

$$Y(f) = X(f)H(f) \tag{2-21}$$

$$= 1 \cdot H(f)$$

$$y(t) = \mathcal{F}^{-1}\{H(f)\} = h(t) \tag{2-22}$$

$$H(f) = \mathcal{F}\{h(t)\} \tag{2-23}$$

Thus the impulse response is the inverse Fourier transform of the transfer function and the transfer function is the Fourier transform of the impulse response.

As an illustration of the utility of the Fourier transform to deal with what might otherwise be very difficult systems to analyze, consider the impulse response of an ideal low-pass filter as shown in Fig. 2-3. Such a filter is defined in terms of its frequency response as

$$H(f) = 1 \quad -W < f < W$$

$$= 0 \qquad \text{elsewhere}$$

$$H(f) = \text{rect}\left(\frac{f}{2W}\right) \tag{2-24}$$

$$h(t) = \mathcal{F}^{-1}\left\{\text{rect}\frac{f}{2W}\right\} \tag{2-25}$$

From the table of transforms

$$\text{rect } t \leftrightarrow \text{sinc } f = \frac{\sin \pi f}{\pi f} \tag{2-26}$$

and

$$\mathcal{F}(t) \leftrightarrow f(-f) \tag{2-27}$$

$$f(-t) \leftrightarrow \mathcal{F}(-f) \tag{2-28}$$

$$af(at) \leftrightarrow \mathcal{F}(f/a) \tag{2-29}$$

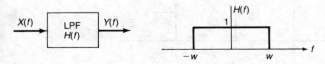

Figure 2-3 Ideal low-pass filter.

Therefore, it follows that

$$a \, \text{sinc}(at) \leftrightarrow \text{rect}(f/a) \tag{2-30}$$

and accordingly

$$h(t) = 2W \, \text{sinc}(2Wt) \tag{2-31}$$

This waveform is sketched in Fig. 2-4. It is evident from the figure that this is a noncausal time function; that is, there is an output prior to application of the impulse at $t = 0$. What this means is that to obtain this ideal type of filtering, it is necessary to have the entire signal—both past and future values. If only past values are available, something less than the ideal performance of Fig. 2-3 will be the best that can be obtained. This is, however, a useful transfer function as it often provides a bound on the best performance possible and a simple and reasonable estimate of actual performance. The noncausality of the impulse response is the result of the frequency function (Fourier transform) being nonzero over only a finite interval along the frequency axis.

These results for a low-pass filter can be extended readily to the bandpass case. The transfer function for an ideal bandpass filter is shown in Fig. 2-5. From the table of transforms it is seen that

$$F(f - f_0) \leftrightarrow f(t)e^{j2\pi f_0 t}$$

Therefore, it follows that the impulse response of the bandpass filter can be obtained in the following manner.

$$H_{\text{BP}}(f) = H_{\text{LP}}(f + f_0) + H_{\text{LP}}(f - f_0)$$

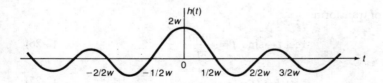

Figure 2-4 Impulse response of an ideal low-pass filter.

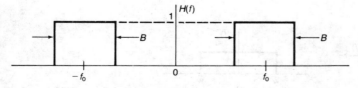

Figure 2-5 Transfer function of a bandpass filter.

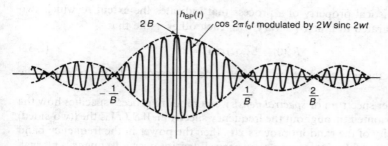

Figure 2-6 Impulse response of ideal bandpass filter.

$$h_{BP}(t) = h_{LP}(t)e^{-j2\pi f_0 t} + h_{LP}(t)e^{j2\pi f_0 t}$$

$$= 2h_{LP}(t)\cos 2\pi f_0 t$$

$$= 2B \text{ sinc } Bt \cos 2\pi f_0 t$$

This waveform is shown in Fig. 2-6.

2-2 TRANSMISSION OF RANDOM PROCESSES THROUGH LINEAR SYSTEMS[2]

A noise waveform is a sample function of a random process. Most such waveforms are modeled as being of infinite extent and consequently having infinite energy. Except in special cases where the power is concentrated in discrete components, as with sinusoids, such power signals do not have Fourier transforms and frequency-domain methods cannot be used in the same manner as with deterministic signals. For most purposes the waveshape of the output noise signal is not required but rather some of its statistical properties, such as its mean, variance, autocorrelation function, and power spectrum or spectral density. In some cases the probability density function is also of importance.

The processes of most interest in communications, because they are most often encountered, are wide-sense stationary processes. If a sample function from such a process is designated $x(t)$, then wide-sense stationarity requires that (1) the mean of the random variable $x(t_1)$ must be independent of the choice of t_1, and (2) the correlation of two random variables $x(t_1)$ and $x(t_2)$ depends only on the time difference $t_2 - t_1$—that is, $E\{x(t_1)x(t_2)\} = R_x(t_2 - t_1)$. These conditions guarantee that the mean, variance, mean square, and correlation coefficient of any two variables are constants independent of the time origin. Thermal and other intrinsic noise occurring in receiving systems may be considered to be wide-sense stationary. Unless otherwise stated it will be assumed in subsequent discussions that all random processes being considered are wide-sense stationary.

The *autocorrelation function* of a random process is defined as

$$R_x(\tau) = E\{x(t)x(t+\tau)\} \tag{2-32}$$

This is a statistical property of a process that indicates the extent to which two variables separated in time by τ vary with each other. Note that

$$R_x(0) = E\{x(t)x(t)\} = \overline{x^2} \tag{2-33}$$

Thus $R_x(0)$ is the mean square value of $x(t)$ and corresponds to the power of the process.

The power spectrum or *spectral density* of a random process specifies how the power is distributed throughout the frequency spectrum. If $S_x(f)$ is the (two-sided) spectral density of the random process $x(t)$, then the power in the frequency band Δf at f_1 is $S(f_1) \Delta f$. By integrating the spectral density over a frequency interval, the power contained in that interval can be determined. The spectral density is a real, positive, even function of frequency, and to obtain the power in a specified frequency interval, it is necessary to include both positive and negative frequency components. For the frequency band extending from f_1 to f_2, this would be

$$\int_{-f_2}^{-f_1} S_x(f)\, df + \int_{f_1}^{f_2} S_x(f)\, df = 2\int_{f_1}^{f_2} S_x(f)\, df \tag{2-34}$$

where the final result follows from the symmetry properties of $S_x(f)$.

The spectral density is defined most simply as the Fourier transform of the autocorrelation function of the process under consideration. It follows, therefore, that

$$S_x(f) = \mathscr{F}\{R_x(\tau)\} = \int_{-\infty}^{\infty} R_x(\tau)e^{-j2\pi f\tau}\, d\tau \tag{2-35}$$

$$R_x(\tau) = \mathscr{F}^{-1}\{S_x(f)\} = \int_{-\infty}^{\infty} S_x(f)e^{j2\pi f\tau}\, df \tag{2-36}$$

The units of the spectral density for signal-processing applications are generally taken as volts squared per hertz. However, the spectral density can also be expressed in terms of the variable ω. With the ω variable, the relationships are

$$\text{Power in frequency interval } (\omega_1, \omega_2) = \frac{1}{\pi}\int_{\omega_1}^{\omega_2} S_x(\omega)\, d\omega \tag{2-37}$$

$$S_x(\omega) = \int_{-\infty}^{\infty} R_x(\tau)e^{-j\omega\tau}\, d\tau \tag{2-38}$$

$$R_x(\tau) = \frac{1}{2\pi}\int_{-\infty}^{\infty} S_x(\omega)e^{j\omega\tau}\, d\omega \tag{2-39}$$

One random process that is frequently encountered in communication problems is *white noise*. This process is characterized by having a spectral density that is constant with frequency. It is this type of process that is used to represent thermal noise. By conventional usage such a noise spectrum is considered to have an amplitude

$$S_n(f) = \frac{N_0}{2} \tag{2-40}$$

The autocorrelation function of this process is

$$R_n(\tau) = \mathscr{F}^{-1}\left\{\frac{N_0}{2}\right\} = \frac{N_0}{2}\,\delta(\tau) \tag{2-41}$$

Thus the mean square value or power of this noise is infinite. The spectral density and autocorrelation function of this noise are sketched in Fig. 2-7. It is important to note that for white noise the spectral density in terms of the variable ω is numerically the same as $S_n(f)$; that is, $S_n(\omega) = N_0/2$. This can be seen by making the change of variable $\omega = 2\pi f$ in the integral (2-35) when the integrand is a constant.

Another important random process involves the concept of band-limited white noise; that is, noise that has a constant spectral density throughout a specified band of frequencies. Such a noise process has a spectral density and autocorrelation function as follows:

$$S_n(f) = \frac{N_0}{2} \qquad |f| < W \tag{2-42}$$

$$= 0 \qquad \text{elsewhere}$$

$$R_n(\tau) = WN_0 \; \text{sinc} \; 2W\tau \tag{2-43}$$

These quantities are shown in Fig. 2-8.

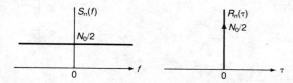

Figure 2-7 Spectral density and autocorrelation function of white noise.

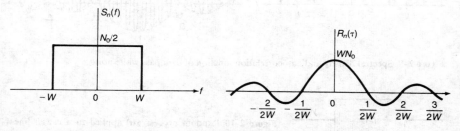

Figure 2-8 Spectral density and autocorrelation function of band-limited white noise.

This concept can be extended readily to include band-limited white noise centered at a carrier frequency f_c. For this case the spectral density and autocorrelation function are

$$S_n(f) = \frac{N_0}{2} \qquad |f - f_c| < \frac{B}{2} \qquad\qquad (2\text{-}44)$$

$$= 0 \qquad\quad \text{elsewhere}$$

$$R_n(\tau) = BN_0 \text{ sinc } B\tau \cos 2\pi f_c\tau \qquad\qquad (2\text{-}45)$$

These are illustrated in Fig. 2-9.

Time-Domain Analysis of Linear Systems with Random Inputs

When a sample function of a random process is applied at the input of a linear system, the output is also a sample function of a random process. Consider the situation shown in Fig. 2-10. The mean value of the output $y(t)$ is given by

$$\bar{y} = \bar{x} \int_0^\infty h(t) \, dt \qquad\qquad (2\text{-}46)$$

The mean square value of the output is

$$\overline{y^2} = \int_0^\infty \int_0^\infty R_x(\lambda_2 - \lambda_1) h(\lambda_1) h(\lambda_2) \, d\lambda_2 \, d\lambda_1 \qquad\qquad (2\text{-}47)$$

For the important case where the input is white noise:

$$R_x(\tau) = \frac{N_0}{2} \delta(\tau) \qquad\qquad (2\text{-}48)$$

$$\bar{y} = 0 \qquad\qquad (2\text{-}49)$$

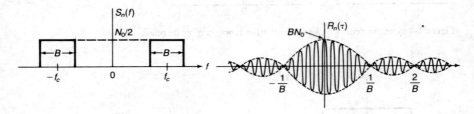

Figure 2-9 Spectral density and autocorrelation function of bandpass white noise.

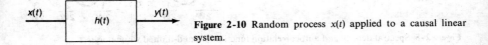

$x(t)$ $h(t)$ $y(t)$

Figure 2-10 Random process $x(t)$ applied to a causal linear system.

$$\overline{y^2} = \int_0^\infty \int_0^\infty \frac{N_0}{2} \delta(\lambda_2 - \lambda_1) h(\lambda_1) h(\lambda_2) \, d\lambda_1 \, d\lambda_2$$

$$= \frac{N_0}{2} \int_0^\infty h^2(\lambda) \, d\lambda \tag{2-50}$$

As an example consider the low-pass RC filter previously discussed. For this filter

$$h(t) = \alpha e^{-\alpha t} u(t) \tag{2-51}$$

where

$$\alpha = \frac{1}{RC}$$

$$\overline{y} = \overline{x} \int_0^\infty \alpha e^{-\alpha \lambda} \, d\lambda = \overline{x} \tag{2-52}$$

Assuming $x(t)$ is white noise with spectral density $N_0/2$,

$$\overline{y^2} = \frac{N_0}{2} \int_0^\infty \alpha^2 e^{-2\alpha \lambda} \, d\lambda = \frac{\alpha N_0}{4} \tag{2-53}$$

Note that $\alpha = 1/RC$ is related to the half-power bandwidth of the filter in the following way.

$$B_{1/2} = \frac{1}{2\pi RC} = \frac{\alpha}{2\pi} \tag{2-54}$$

$$\overline{y^2} = \frac{\pi B_{1/2} N_0}{2} \tag{2-55}$$

From Eq. (2.55) it is evident that the output noise power is directly proportional to the filter bandwidth.

The autocorrelation function of the output of a linear system is given by

$$R_y(\tau) = \int_0^\infty \int_0^\infty R_x(\lambda_2 - \lambda_1 - \tau) h(\lambda_1) h(\lambda_2) \, d\lambda_2 \, d\lambda_1 \tag{2-56}$$

For the special case of a white noise input this becomes

$$R_y(\tau) = \int_0^\infty \int_0^\infty \frac{N_0}{2} \delta(\lambda_2 - \lambda_1 - \tau) h(\lambda_1) h(\lambda_2) \, d\lambda_2 \, d\lambda_1$$

$$= \frac{N_0}{2} \int_0^\infty h(\lambda_1) h(\lambda_1 + \tau) \, d\lambda_1 \tag{2-57}$$

Using the low-pass RC example again,

$$R_y(\tau) = \frac{N_0}{2} \int_0^\infty \alpha e^{-\alpha\lambda} \alpha e^{-\alpha(\lambda+\tau)}\, d\lambda$$

$$= \frac{\alpha N_0}{4} e^{-\alpha\tau} \qquad \tau \geq 0 \tag{2-58}$$

For the case of $\tau < 0$, the range of integration must be altered. However, because $R(\tau)$ is always an even function—that is, $R(\tau) = R(-\tau)$—the final result may be stated immediately as

$$R_y(\tau) = \frac{\alpha N_0}{4} e^{-\alpha|\tau|} \tag{2-59}$$

This is sketched in Fig. 2-11.

Frequency-Domain Analysis of Linear Systems with Random Inputs

Just as with deterministic signals, it is often easier to analyze the performance of linear systems with random inputs by using frequency-domain techniques. Consider the system shown in Fig. 2-12. The spectral density of the output signal is determined directly from that of the input by the following expression.

$$S_y(f) = S_x(f)|H(f)|^2 \tag{2-60}$$

or

$$S_y(\omega) = S_x(\omega)|H(\omega)|^2 \tag{2-61}$$

where $H(f)$ is the system transfer function. The quantity $|H(f)|^2$ is called the power transfer function.

Again using the low-pass RC filter with white noise as the input, the frequency

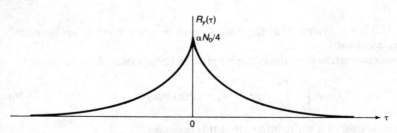

Figure 2-11 Autocorrelation function of output of RC low-pass filter with white noise input.

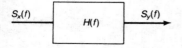

Figure 2-12 Frequency-domain representation of a linear system with a random input.

domain analysis would be

$$S_x(\omega) = \frac{N_0}{2}$$

$$H(\omega) = \frac{\alpha}{j\omega + \alpha}$$

$$S_y(\omega) = \frac{N_0}{2}\left(\frac{\alpha}{j\omega + \alpha}\right)\left(\frac{\alpha}{-j\omega + \alpha}\right) = \frac{N_0}{2}\frac{\alpha^2}{\omega^2 + \alpha^2}$$

$$R_y(\tau) = \mathscr{F}^{-1}\left\{\frac{N_0\alpha}{4}\cdot\frac{2\alpha}{\omega^2 + \alpha^2}\right\} = \frac{N_0\alpha}{4}e^{\alpha|\tau|} \tag{2-62}$$

This is the same result previously obtained for the autocorrelation function of the output.

System and Signal Bandwidths

A convenient way to describe the performance of a network or system with a white noise input is by means of the *equivalent noise bandwidth*. For a system having transfer function $H(f)$, this is defined as

$$\text{Equivalent noise bandwidth} = \frac{1}{|H_{max}|^2}\int_0^\infty |H(f)|^2\,df \tag{2-63}$$

where H_{max} is the maximum magnitude of $H(f)$. This quantity is often just referred to as the *noise bandwidth*. This bandwidth corresponds to the width of an ideal bandpass filter having a gain equal to the maximum gain of the system, and which delivers the same average power from a white noise source as does the actual system. It is convenient to distinguish low-pass bandwidths from bandpass bandwidths. This is done by using W for low-pass bandwidths and B for bandpass bandwidths. For the RC low-pass filter, the equivalent noise bandwidth can be found as follows:

$$|H(f)|^2 = \frac{\alpha^2}{4\pi^2 f^2 + \alpha^2} \tag{2-64}$$

$$|H_{max}|^2 = 1 \qquad (\text{occurs at } f = 0) \tag{2-65}$$

$$W_n = \int_0^\infty \frac{\alpha^2}{4\pi^2 f^2 + \alpha^2}\,df \tag{2-66}$$

Since the area under a Fourier transform is equal to the value of the inverse transform at the origin, W_n can be evaluated as

$$W_n = \frac{\alpha}{4}e^{-\alpha|\tau|}\bigg|_{\tau=0} = \frac{\alpha}{4} = \frac{1}{4RC} \tag{2-67}$$

For a tuned RLC circuit, the equivalent noise bandwidth is

$$B_n = \frac{\pi f_0}{2Q} \quad \text{Hz} \tag{2-68}$$

where f_0 is the center frequency and Q is the quality factor of the circuit.

Other bandwidth definitions are also used to characterize system performance. The *half-power bandwidth* $B_{1/2}$ or $W_{1/2}$ is defined as follows:

1. For a low-pass filter, the half-power bandwidth $W_{1/2}$ corresponds to that frequency at which the magnitude of the power transfer function falls to one half of its value at the origin.
2. For a bandpass filter, the half-power bandwidth $B_{1/2}$ is the width of the frequency span between half-amplitude points of the power transfer function around the center frequency of the pass band.

For a simple RC low-pass filter, the half-power bandwidth is readily found as follows:

$$|H(f)|^2 = \frac{\alpha^2}{4\pi^2 f_{1/2}^2 + \alpha^2} = \frac{1}{2}$$

$$W_{1/2} = f_{1/2} = \frac{\alpha}{2\pi} = \frac{1}{2\pi RC} \tag{2-69}$$

The ratio of the noise bandwidth to the half-power bandwidth for the RC filter is

$$\frac{W_n}{W_{1/2}} = \frac{1}{4RC} \cdot 2\pi RC = \frac{\pi}{2} \tag{2-70}$$

The more rapid the rolloff of the transfer function, the closer the noise bandwidth and the half-power bandwidth become.

Another bandwidth that is frequently encountered is the *rms bandwidth*, W_{rms} or B_{rms}. This bandwidth is defined as the radius of gyration of the power transfer function around the origin for a low-pass filter and the radius of gyration of the positive frequency portion of the power transfer function around its center frequency for a bandpass filter. Mathematically these expressions become

$$W_{\mathrm{rms}}^2 = \frac{\displaystyle\int_{-\infty}^{\infty} f^2 |H(f)|^2 \, df}{\displaystyle\int_{-\infty}^{\infty} |H(f)|^2 \, df} \tag{2-71}$$

$$B_{\mathrm{rms}}^2 = \frac{\displaystyle\int_{0}^{\infty} (f - f_0)^2 |H(f)|^2 \, df}{\displaystyle\int_{0}^{\infty} |H(f)|^2 \, df} \tag{2-72}$$

Only filters having power transfer functions that fall off faster than $1/f^2$ have finite rms bandwidths.

In cases where the rms bandwidth does not exist because the spectrum does not fall off rapidly enough, it is often convenient to use the e-energy bandwidths W_e and B_e defined as follows:

$$\int_{-W_e}^{W_e} |H(f)|^2 \, df = (1-e) \int_{-\infty}^{\infty} |H(f)|^2 \, df \tag{2-73}$$

or

$$\int_{f_0 - B_e/2}^{f_0 + B_e/2} |H(f)|^2 \, df = (1-e) \int_0^{\infty} |H(f)|^2 \, df \tag{2-74}$$

These bandwidths correspond to that segment of the spectrum that contain the fraction $(1-e)$ of the total energy.

In all cases except noise bandwidth, the definitions can be used to calculate signal bandwidths by replacing $H(f)$ by the transform of the signal $X(f)$, or for a random signal, by replacing $|H(f)|^2$ by the spectral density $S_x(f)$. For example, consider a signal that has a Gaussian-shaped spectral density, that is,

$$S_x(f) = A e^{-f^2/\alpha^2} \tag{2-75}$$

The half-power bandwidth is obtained by solving

$$A e^{-(W_{1/2})^2/\alpha^2} = \frac{1}{2} A e^0$$

$$W_{1/2} = 0.833\alpha \tag{2-76}$$

The rms bandwidth is

$$W_{\text{rms}}^2 = \frac{\displaystyle\int_{-\infty}^{\infty} A f^2 e^{-f^2/\alpha^2} \, df}{\displaystyle\int_{-\infty}^{\infty} A e^{-f^2/\alpha^2} \, df}$$

Putting the integrals into the form of PDFs gives

$$W_{\text{rms}}^2 = \frac{\alpha\sqrt{\pi} \displaystyle\int_{-\infty}^{\infty} \frac{f^2}{\sqrt{2\pi}\,\dfrac{\alpha}{\sqrt{2}}} \exp\left[\frac{-f^2}{2\left(\dfrac{\alpha}{\sqrt{2}}\right)^2}\right] df}{\alpha\sqrt{\pi} \displaystyle\int_{-\infty}^{\infty} \frac{1}{\sqrt{2\pi}\,\dfrac{\alpha}{\sqrt{2}}} \exp\left[\frac{-f^2}{2\left(\dfrac{\alpha}{\sqrt{2}}\right)^2}\right] df}$$

$$= \left(\frac{\alpha}{\sqrt{2}}\right)^2$$

$$W_{\text{rms}} = \frac{\alpha}{\sqrt{2}} \tag{2-77}$$

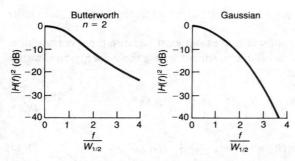

Figure 2-13 Butterworth and Gaussian transfer functions.

Table 2-1 Bandwidths of Butterworth and Gaussian filters

Parameter	Butterworth	Gaussian
Low-pass $\lvert H(f) \rvert^2$	$\dfrac{1}{1 + \left(\dfrac{f}{W_{1/2}}\right)^{2n}}$	$\exp(-f^2/\alpha^2)$
$W_{1/2}$	$W_{1/2}$	0.833α
W_{rms}	$W_{1/2}\left[\dfrac{\sin(\pi/2n)}{\sin(3\pi/2n)}\right]^{1/2}$	0.707α
W_n	$W_{1/2}\left[\dfrac{\pi/2n}{\sin(\pi/2n)}\right]$	0.887α
Bandpass $\lvert H(f) \rvert^2$	$\dfrac{1}{1+\left[\dfrac{f-f_0}{\frac{1}{2}B_{1/2}}\right]^{2n}} + \dfrac{1}{1+\left[\dfrac{f+f_0}{\frac{1}{2}B_{1/2}}\right]^{2n}}$	$\exp[-(f-f_0)^2/\alpha^2] + \exp[-(f+f_0)^2/\alpha^2]$
$B_{1/2}$	$B_{1/2}$	1.665α
B_{rms}	$B_{1/2}\left[\dfrac{\sin(\pi/2n)}{\sin(3\pi/2n)}\right]^{1/2}$	1.414α
B_n	$B_{1/2}\left[\dfrac{\pi/2n}{\sin \pi/2n}\right]$	1.477α

The two filter functions most frequently used for analysis and simulation are the Butterworth and the Gaussian. These transfer functions for the low-pass case are shown in Fig. 2-13.

The various bandwidths for these two transfer functions are given in Table 2-1.

2-3 NARROW-BAND GAUSSIAN NOISE[3]

In most communication systems, the receiver itself is a major contributor of noise. This noise occurs at the input stages of the receiver and can usually be modeled as

Gaussian with a spectral density determined by the bandpass characteristics of the RF and IF stages in the receiver. The bandwidth of such noise is generally small compared with the center frequency of the spectrum, and in such cases it is referred to as narrow-band noise.

The statistical properties of this type of noise can be found from the Fourier series expansion of a finite time interval and then letting the interval go to infinity. Let $x(t)$ be a zero mean Gaussian noise process band-limited to $\pm B/2$ around center frequency f_c. The Fourier series expansion over the interval

$$-\frac{T}{2} \le t \le \frac{T}{2}$$

is given by

$$x(t) = \sum_{n=1}^{\infty} x_{cn} \cos n\omega_0 t + x_{sn} \sin n\omega_0 t \tag{2-78}$$

where

$$\omega_0 = \frac{2\pi}{T}$$

$$x_{cn} = \frac{2}{T} \int_{-T/2}^{T/2} x(t) \cos n\omega_0 t \, dt \qquad n = 1, 2, \dots$$

$$x_{sn} = \frac{2}{T} \int_{-T/2}^{T/2} x(t) \sin n\omega_0 t \, dt \qquad n = 1, 2, \dots$$

Since $x(t)$ is a Gaussian random variable, it follows that x_{cn} and x_{sn} are also Gaussian random variables. The center frequency ω_c can be introduced into the representation by making the substitution $n\omega_0 = (n\omega_0 - \omega_c) + \omega_c$, leading to

$$x(t) = \sum_{n=1}^{\infty} x_{cn} \cos[(n\omega_0 - \omega_c)t + \omega_c t] + x_{sn} \sin[(n\omega_0 - \omega_c)t + \omega_c t] \tag{2-79}$$

By using trigonometric identities, this can be put into the form

$$x(t) = x_c(t) \cos \omega_c t - x_s(t) \sin \omega_c t \tag{2-80}$$

where

$$x_c(t) = \sum_{n=1}^{\infty} x_{cn} \cos(n\omega_0 - \omega_c)t + x_{sn} \sin(n\omega_0 - \omega_c)t$$

$$x_s(t) = \sum_{n=1}^{\infty} x_{cn} \sin(n\omega_0 - \omega_c)t - x_{sn} \cos(n\omega_0 - \omega_c)t$$

This representation can be converted into the form of an envelope and phase representation using the relationships

$$x(t) = r(t) \cos[\omega_c t + \phi(t)] \tag{2-81}$$

where

$$r(t) = \sqrt{x_c^2(t) + x_s^2(t)} \qquad 0 \le r(t) < \infty$$

$$\phi(t) = \tan^{-1}\left[\frac{x_s(t)}{x_c(t)}\right] \qquad 0 \le \phi(t) \le 2\pi$$

$$x_c(t) = r(t) \cos \phi(t)$$

$$x_s(t) = r(t) \sin \phi(t)$$

The only terms that do not vanish in the expansion are those for which nf_0 is in the frequency range $f_c \pm B/2$; that is, components within the narrow band around f_c. Therefore, it follows that the components of $x_c(t)$ and $x_s(t)$ are located in a similar band around zero frequency.

Let

$$x_{ct} = x_c(t)$$

$$x_{st} = x_s(t)$$

These quantities are sums of Gaussian random variables and so are themselves Gaussian random variables. Their means are zero since the original process was zero mean, that is,

$$E\{x_{ct}\} = E\{x_{st}\} = 0 \tag{2-82}$$

The variance and covariance of these quantities can be found as follows:

$$E\{x_{ct}^2\} = E\left\{ \sum_{n=1}^{\infty} \sum_{m=1}^{\infty} [x_{cn} \cos(n\omega_0 - \omega_c)t + x_{sn} \sin(n\omega_0 - \omega_c)t] \right.$$

$$\times \left. [x_{cm} \cos(m\omega_0 - \omega_c)t + x_{sm} \sin(m\omega_0 - \omega_c)t] \right\}$$

$$= \sum_{n=1}^{\infty} \sum_{m=1}^{\infty} [\overline{x_{cn}x_{cm}} \cos(n\omega_0 - \omega_c)t \cos(m\omega_0 - \omega_c)t$$

$$+ \overline{x_{cn}x_{sm}} \cos(n\omega_0 - \omega_c)t \sin(m\omega_0 - \omega_c)t$$

$$+ \overline{x_{sn}x_{cm}} \sin(n\omega_0 - \omega_c)t \cos(m\omega_0 - \omega_c)t$$

$$+ \overline{x_{sn}x_{sm}} \sin(n\omega_0 - \omega_c)t \sin(m\omega_0 - \omega_c)t] \tag{2-83}$$

where

$$\overline{x_{cn}x_{cm}} = E\left\{ \frac{2}{T} \int_{-T/2}^{T/2} x(t) \cos n\omega_0 t \, dt \cdot \frac{2}{T} \int_{-T/2}^{T/2} x(t) \cos m\omega_0 t \, dt \right\}$$

$$= \frac{4}{T^2} \int_{-T/2}^{T/2} \int_{-T/2}^{T/2} \overline{x(t_1)x(t_2)} \cos n\omega_0 t_1 \cos m\omega_0 t_2 \, dt_1 \, dt_2$$

$$= \frac{4}{T^2} \int_{-T/2}^{T/2} \int_{-T/2}^{T/2} R_x(t_2 - t_1) \cos n\omega_0 t_1 \cos m\omega_0 t_2 \, dt_1 \, dt_2 \tag{2-84}$$

Letting $t_2 - t_1 = u$ and $t_1/T = v$, this can be put into the form

$$\overline{x_{cn}x_{cm}} = \frac{4}{T} \int_{-\frac{1}{2}}^{\frac{1}{2}} \cos n\omega_0 Tv \int_{-T(\frac{1}{2}+v)}^{T(\frac{1}{2}-v)} R_x(u) \cos m\omega_0(u+v) \, du \, dv \qquad (2\text{-}85)$$

Noting that $\omega_0 = 2\pi f_0 = 2\pi/T$ and rearranging leads to

$$\overline{x_{cn}x_{cm}} = \frac{4}{T} \int_{-\frac{1}{2}}^{\frac{1}{2}} \cos 2\pi nv \left[\cos 2\pi mv \int_{-T(\frac{1}{2}+v)}^{T(\frac{1}{2}-v)} R_x(u) \cos \frac{2\pi mu}{T} \, du \right.$$

$$\left. - \sin 2\pi mv \int_{-T(\frac{1}{2}+v)}^{T(\frac{1}{2}-v)} R_x(u) \sin \frac{2\pi mu}{T} \, du \right] dv$$

For $(v) \neq 1/2$ and $T \to \infty$ with $f_m = m/T$, the inner integrals are, respectively,

$$\lim_{T \to \infty} \int_{-T(\frac{1}{2}+v)}^{T(\frac{1}{2}-v)} R_x(u) \cos 2\pi f_m u \, du = S_x(f_m)$$

$$\lim_{T \to \infty} \int_{-T(\frac{1}{2}+v)}^{T(\frac{1}{2}-v)} R_x(u) \sin 2\pi f_m u \, du = 0$$

The variable $f_m \to 0$ unless m is chosen so that as T increases, m has a value such that

$$\lim_{m, T \to \infty} \left(\frac{m}{T} = f_m \right) = f$$

With this choice, Eq. (2-85) becomes

$$\lim_{T \to \infty} T \cdot \overline{x_{cn}x_{cm}} = 4S_x(f) \int_{-\frac{1}{2}}^{\frac{1}{2}} \cos 2\pi nv \cos 2\pi mv \, dv$$

$$= 2S_x(f) \qquad m = n$$

$$= 0 \qquad m \neq n \qquad (2\text{-}86)$$

It can be shown similarly that

$$\lim_{T \to \infty} T \overline{x_{sn}x_{sm}} = 2S_x(f) \qquad m = n$$

$$= 0 \qquad m \neq n$$

$$\lim_{T \to \infty} T \overline{x_{cn}x_{sm}} = 0 \qquad \text{all } m, n \qquad (2\text{-}87)$$

Therefore, the coefficients are uncorrelated as $T \to \infty$! By using this result, it is readily shown that

$$E\{x_{ct}^2\} = \lim_{T \to \infty} \sum_{n=1}^{\infty} x_{cn}^2 [\cos^2(n\omega_0 - \omega_c)t + \sin^2(n\omega_0 - \omega_c)t]$$

Recognizing that the trigonometric terms sum to unity and using the result given in Eq. (2-86), this can be evaluated as

$$\lim_{T \to \infty} \sum_{n=1}^{\infty} S_x(f)\left(\frac{2}{T}\right) \to 2 \int_0^{\infty} S_x(f) \, df = \overline{x_t^2}$$

Similarly it can be shown that

$$E\{\overline{x_{st}^2}\} = E\{\overline{x_{ct}^2}\}$$

and since $\bar{x}_{st} = \bar{x}_{ct} = 0$,

$$\sigma_{st}^2 = \sigma_{ct}^2 = \sigma_x^2 \tag{2-88}$$

where σ_x^2 is the variance of $x(t)$. The covariance of x_{ct} and x_{st} can be shown to go to zero as $T \to \infty$. Thus it follows that samples of the in-phase and quadrature components x_{ct} and x_{st} are uncorrelated Gaussian random variables and hence are statistically independent. They have zero means and each has a variance equal to the variance of the original signal.

Spectral Density of the In-Phase and Quadrature Components

Using the Fourier series expansion and following the same procedure as employed to compute the mean square values, it is possible to calculate the autocorrelation functions of $x_c(t)$ and $x_s(t)$ and the cross-correlation function of $x_c(t)$ and $x_s(t)$. The results are as follows:

$$R_{x_c}(\tau) = 2 \int_0^\infty S_x(f) \cos 2\pi(f - f_c)\tau \, df$$

$$R_{x_s}(\tau) = 2 \int_0^\infty S_x(f) \cos 2\pi(f - f_c)\tau \, df$$

$$R_{x_c x_s}(\tau) = 2 \int_0^\infty S_x(f) \sin 2\pi(f - f_c)\tau \, df$$

$$R_{x_s x_c}(\tau) = -R_{x_c x_s}(\tau) \tag{2-89}$$

These expressions can be put into a more easily interpreted form with the following transformations of variables. Consider the expression for $R_{x_c}(\tau)$.

$$R_{x_c}(\tau) = \int_0^\infty S_x(f) \cos 2\pi(f - f_c)\tau \, df + \int_0^{-\infty} S_x(-f) \cos 2\pi(-f - f_c)\tau(-df)$$

$$= \int_{-f_c}^\infty S_x(f + f_c) \cos 2\pi f\tau \, df + \int_{-\infty}^{f_c} S_x(f - f_c) \cos 2\pi f\tau \, df$$

$$= \int_{-f_c}^{f_c} [S_x(f + f_c) + S_x(f - f_c)] \cos 2\pi f\tau \, df$$

Since the autocorrelation function is the Fourier transform of the spectral density, it follows that

$$S_{x_c}(f) = S_x(f + f_c) + S_x(f - f_c) \tag{2-90}$$

Thus $x_c(t)$ has a spectral density that is the sum of the negative and positive frequency components of $S_x(t)$ after their translation to the origin. From Eq. (2-89) it is evident that

$$S_{x_s}(f) = S_{x_c}(f) \tag{2-91}$$

For the case of the cross-spectral density, there is a sine factor under the integral sign and this leads to an imaginary expression of the following form.

$$S_{cs}(f) = -S_{sc}(f) = -j[S_x(f + f_c) - S_x(f - f_c)] \qquad |f| \leq f_c \tag{2-92}$$

It is seen that the cross-spectral density will be zero if the original noise spectrum is symmetrical around f_c.

Fig. 2-14 illustrates the spectral characteristics that result when the spectral density of the original noise is not symmetrical around the carrier frequency f_c.

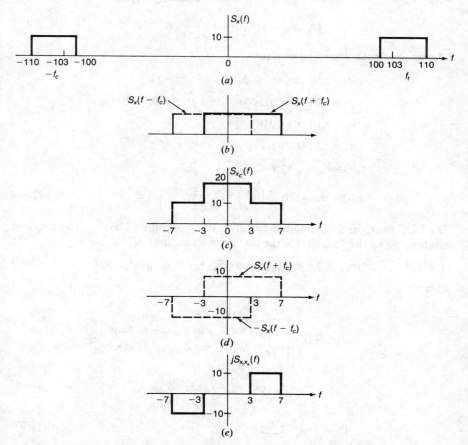

Figure 2-14 Spectral density of in-phase and quadrature components of narrow-band Gaussian random process $x_c(t)$ with carrier frequency $f_c = 103$. (a) $S_x(f)$; (b) components of $S_{x_c}(t)$; (c) $S_{x_c}(f)$; (d) components of $S_{x_c x_s}(f)$; (e) $S_{x_c x_s}(f)$.

Envelope of Narrow-Band Noise

The envelope of narrow-band Gaussian noise can be found by a transformation of variables from the in-phase and quadrature representation for which the statistics are known. The time samples of the in-phase and quadrature components have Gaussian PDFs of the following form:

$$p(n_{ct}) = \frac{1}{\sigma_n \sqrt{2\pi}} e^{-n_{ct}^2/2\sigma_n^2}$$

$$p(n_{st}) = \frac{1}{\sigma_n \sqrt{2\pi}} e^{-n_{st}^2/2\sigma_n^2}$$

where $\sigma_{ct}^2 = \sigma_{st}^2 = \sigma_n^2$. Because n_{ct} and n_{st} are statistically independent, it follows that

$$p(n_{ct}, n_{st}) = p(n_{ct})p(n_{st}) = \frac{1}{2\pi\sigma_n^2} e^{-(n_{ct}^2 + n_{st}^2)/2\sigma_n^2} \tag{2-93}$$

Changing to polar coordinates gives

$$n(t) = r_n(t) \cos[\omega_0 t + \phi_n(t)]$$

$$= r_n(t) \cos \phi_n(t) \cos \omega_0 t - r_n(t) \sin \phi_n(t) \sin \omega_0 t$$

$$n_{ct} = r_n(t) \cos \phi_n(t)$$

$$n_{st} = r_n(t) \sin \phi_n(t)$$

$$r_n(t) = r_{nt} = \sqrt{n_{ct}^2 + n_s^2}$$

$$\phi_n(t) = \phi_{nt} = \tan^{-1}\left(\frac{n_{st}}{n_{ct}}\right) \tag{2-94}$$

The PDF for r_{nt} and ϕ_{nt} is found by substituting into the PDF for n_{ct} and n_{st} and multiplying by the Jacobian of the transformation; that is,

$$p(r_{nt}, \phi_{nt}) = p_{x_{ct}, x_{st}}[r_{nt} \cos \phi_{nt}, r_{nt} \sin \phi_{nt}] |J(r_{nt}, \phi_{nt})| \tag{2-95}$$

where

$$|J(r_n, F_n)| = \begin{vmatrix} \dfrac{\partial n_{ct}}{\partial r_{nt}} & \dfrac{\partial n_{ct}}{\partial \phi_{nt}} \\ \dfrac{\partial n_{st}}{\partial r_{nt}} & \dfrac{\partial n_{st}}{\partial \phi_{nt}} \end{vmatrix} = \begin{vmatrix} \cos \phi_{nt} & -r_{nt} \sin \phi_{nt} \\ \sin \phi_{nt} & r_{nt} \cos \phi_{nt} \end{vmatrix}$$

$$= r_{nt} \cos^2 \phi_{nt} + r_{nt} \sin^2 \phi_{nt} = r_{nt} \tag{2-96}$$

$$p(r_{nt}, \phi_{nt}) = \frac{r_{nt}}{2\pi\sigma_n^2} e^{-r_{nt}^2/2\sigma_n^2} \tag{2-97}$$

This is the joint PDF of the envelope and phase of narrow-band noise. The marginal

PDFs of r_{nt} and ϕ_{nt} are found by integrating out the other variables and lead to

$$p(r_{nt}) = \frac{r_{nt}}{\sigma_n^2} e^{-r_{nt}^2/2\sigma_n^2} \qquad 0 \le r_{nt} \le \infty \tag{2-98}$$

$$p(\phi_{nt}) = \frac{1}{2\pi} \qquad 0 \le \phi_{nt} \le 2\pi \tag{2-99}$$

Thus the envelope has a Rayleigh PDF and the phase has a uniform PDF. The mean and variance of the envelope are related to the variance of the original noise waveform by

$$\overline{r_{nt}} = \sigma_n \sqrt{\frac{\pi}{2}} \tag{2-100}$$

$$\sigma_r^2 = \left(2 - \frac{\pi}{2}\right)\sigma_n^2 \tag{2-101}$$

Since $p(r_{nt}, \phi_{nt}) = p(r_{nt})p(\phi_{nt})$, it follows that the envelope and phase are statistically independent of each other. Note that this has only been shown to be true at the same instant of time.

PDF of the Power of Narrow-Band Noise

The power of a waveform is generally taken to be the square of the amplitude. For narrow-band noise, the instantaneous power of the envelope is

$$P_{nt} = r_{nt}^2 \tag{2-102}$$

The PDF of P_n is given by

$$p(P_{nt}) = p_{r_{nt}}(\sqrt{P_{nt}}) \left|\frac{dr_{nt}}{dP_{nt}}\right|$$

$$= \frac{1}{2\sigma_n^2} e^{-P_{nt}/2\sigma_n^2} \qquad 0 \le P_{nt} < \infty \tag{2-103}$$

This is called an exponential PDF. The mean and variance of the power are

$$\bar{P}_{nt} = 2\sigma_n^2 \tag{2-104}$$

$$\sigma_{P_{nt}}^2 = 4\sigma_n^4 \tag{2-105}$$

The exponential and Rayleigh PDFs are illustrated in Fig. 2-15.

Sine Wave Plus Narrow-Band Noise

When a signal is present at the same time as the noise, the PDF of the envelope is significantly changed. For a sinusoidal signal, the signal plus noise waveform

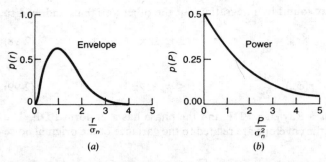

Figure 2-15 Probability density functions of narrow-band noise. (a) Envelope-Rayleigh; (b) power-exponential.

can be written as

$$x(t) = s(t) + n(t) = A \cos \omega_0 t + n_c(t) \cos \omega_0 t - n_s(t) \sin \omega_0 t$$
$$= [A + n_c(t)] \cos \omega_0 t - n_s(t) \sin \omega_0 t \quad (2\text{-}106)$$

The probability density function for $x(t)$ can be found from that of noise alone by subtracting the deterministic component of the signal from the in-phase component. Thus

$$p(x_c, x_s) = \frac{1}{2\pi\sigma_n^2} e^{-[(x_c - A)^2 + x_s^2]/2\sigma_n^2} \quad (2\text{-}107)$$

where

$$x_c(t) = r_n(t) \cos \phi_n(t)$$
$$x_s(t) = r_n(t) \sin \phi_n(t)$$

Converting to polar coordinates:

$$r = \sqrt{x_c^2 + x_s^2}$$
$$\phi = \tan^{-1}\left(\frac{x_s}{x_c}\right)$$

and multiplying by the Jacobian of the transformation leads to

$$p(r, \phi) = \frac{r}{2\pi\sigma_n^2} \exp\left[-\frac{r^2 - 2Ar \cos \phi + A^2}{2\sigma_n^2}\right] \quad (2\text{-}108)$$

The marginal PDFs of r and ϕ are found to be

$$p(r) = \frac{r}{\sigma_n^2} \exp\left[-\frac{r^2 + A^2}{2\sigma_n^2}\right] I_0\left(\frac{Ar}{\sigma_n^2}\right) \qquad 0 \le r < \infty \quad (2\text{-}109)$$

$$p(\phi) = \frac{1}{2\pi} \qquad 0 \le \phi \le 2\pi \quad (2\text{-}110)$$

where $I_0(\cdot)$ is the modified Bessel function of zero order, and first kind.

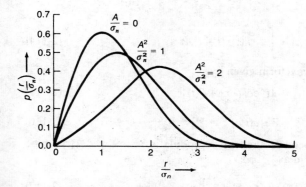

Figure 2-16 Rician probability density function.

The SNR is the ratio of signal power to noise power and is given by

$$\text{SNR} = \frac{A^2}{2\sigma_n^2} \tag{2-111}$$

As $\text{SNR} \to 0$ the modified Bessel function goes to $I(0) = 1$ and the PDF of the envelope becomes Rayleigh. As $\text{SNR} \to \infty$, which is equivalent to $\sigma_n \to 0$, this leads to an impulse at $r = A$. For very large but not infinite SNR, the distribution approaches Gaussian with mean A and variance σ_n^2. For large argument

$$I_0(z) \doteq e^z / \sqrt{2\pi z} \tag{2-112}$$

and for $rA \gg \sigma_n^2$,

$$p(r) \doteq \sqrt{\frac{r}{2\pi A \sigma_n^2}} \cdot \exp\left[-\frac{(r-A)^2}{2\sigma_n^2} \right] \tag{2-113}$$

The PDF of the envelope of a sine wave plus noise is called a Rician PDF. Fig. 2-16 shows several examples of the Rician PDF.

2-4 FILTERING OF MODULATED SIGNALS

In the processing of communication signals, the information waveform is often attached to a carrier by an appropriate modulation process and the combined waveform is passed through a filter. Consider the situation as shown in Fig. 2-17.

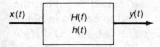

Figure 2-17 Filtering a modulated carrier.

The filter output is

$$y(t) = \int_{-\infty}^{\infty} h(\tau)x(t-\tau)\,d\tau \tag{2-114}$$

where $x(t)$ is the modulated waveform given by

$$x(t) = a(t)\cos[\omega_c t + \theta(t)] \tag{2-115}$$

$$= \text{Re}\{a(t)e^{j\theta(t)}e^{j\omega_c t}\} \tag{2-116}$$

The forms of $m(t)$ and $\theta(t)$ depend on the type of modulation employed. $y(t)$ can be expressed in more general form as

$$y(t) = \int_{-\infty}^{\infty} h(\tau)\text{Re}\{a(t-\tau)e^{j\theta(t-\tau)}e^{j\omega_c(t-\tau)}\}\,d\tau \tag{2-117}$$

$$= \text{Re}\left\{e^{j\omega_c t}\int_{-\infty}^{\infty} h(\tau)a(t-\tau)e^{j\theta(t-\tau)}e^{-j\omega_c \tau}\,d\tau\right\} \tag{2-118}$$

For the linear modulated waveform in which $a(t) = m(t)$, $\theta(t) = \psi$, and $x(t) = m(t)\cos(\omega_c t + \psi)$, this becomes

$$y(t) = \text{Re}\{m_0(t)e^{j(\omega_c t + \psi)}\} \tag{2-119}$$

where

$$m_0(t) = \int_{-\infty}^{\infty} m(t-\tau)h(\tau)e^{-j\omega_c \tau}\,d\tau \tag{2-120}$$

The output envelope function $m_0(t)$ is seen to be a complex time function obtained by filtering $m(t)$ with $h(t)e^{-j\omega_c t}$. This impulse response has an equivalent transfer function of the form

$$H_1(\omega) = \mathscr{F}\{h(t)e^{-j\omega_c t}\} \tag{2-121}$$

$$= H(\omega + \omega_c) \tag{2-122}$$

Thus $H_1(\omega)$ corresponds to the original bandpass function shifted to the left by an amount ω_c so that its center is now at the origin. There is another component of this (nonreal) complex transfer function centered at $-2\omega_c$ but in normal operation there are no components of $m(t)$ at this frequency, so it can be neglected. This is shown in Fig. 2-18. The Fourier transform of the output envelope function is then

$$M_0(\omega) = M(\omega)H_1(\omega) \tag{2-123}$$

If $M_0(\omega)$ can be made an even function of frequency in magnitude and an odd function of frequency in phase, then $m_0(t)$ will be real. This would leave the carrier factor in $p(t)$ unperturbed in phase and frequency. Since $m(t)$ is real, $M(\omega)$ has the required symmetry. Therefore, all that is required is for $H_1(\omega)$ to be odd in phase and even in magnitude in the vicinity of the origin. This result is obtained if $H(\omega)$ has this type of symmetry about the original carrier ω_c. When this condition is

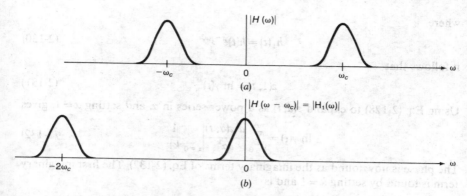

Figure 2-18 Original bandpass filter; (*b*) equivalent complex envelope filter.

met, which is usually the case, the effect on the modulation of the bandpass filtering operation is the same as passing the modulation waveform through the equivalent low-pass filter before attaching it to the carrier.

The filtering of angle-modulated carriers, in which $a(t) = A$, cannot be handled in as simple a manner as in the case of linear modulation. Proceeding as in the previous case for an angle-modulated carrier of the form $A \cos[\omega_c t + \theta(t)]$ gives

$$y(t) = A \operatorname{Re}\left\{ e^{j\omega_c t} \int_{-\infty}^{\infty} h(\tau) e^{-j\omega_c \tau} e^{j\theta(t-\tau)} \, d\tau \right\} \qquad (2\text{-}124)$$

It is seen that the same equivalent low-pass filter function is convolved with the information-bearing signal. In this instance, however, the information signal is contained in an exponential function. Some type of approximation is required to obtain a solution. One method, proposed by Rice and Bedrosian[4], is to obtain a series expansion for the phase of the filtered carrier directly. Let

$$\gamma(t) = \int_{-\infty}^{\infty} h(\tau) e^{-j\omega_c \tau} e^{j\theta(t-\tau)} \, d\tau \qquad (2\text{-}125)$$

Then the phase of $y(t)$ is the same as the phase of $\gamma(t)$; that is,

$$\text{Phase } \{y(t)\} = \text{phase } \{\gamma(t)\} \qquad (2\text{-}126)$$

$$= \operatorname{Im} \{\ln \gamma(t)\} \qquad (2\text{-}127)$$

Using a Taylor's series expansion for $\ln(\cdot)$ leads to one form of solution.

$$\ln(u) = \sum_{k=0}^{\infty} \frac{d^k \ln(u)}{du^k} \bigg|_{u=0} \frac{u^k}{k!} \qquad (2\text{-}128)$$

Letting

$$z(\alpha, t) = \ln \int_{-\infty}^{\infty} h_1(\tau) e^{j\alpha\theta(t-\tau)} \, d\tau \qquad (2\text{-}129)$$

where

$$h_1(t) = h(t)e^{-j\omega_c t} \tag{2-130}$$

it follows that

$$z(1, t) = \ln \gamma(t) \tag{2-131}$$

Using Eq. (2-128) to expand $z(d, t)$ as a power series in α, and setting $\alpha = 1$, gives

$$\ln \gamma(t) = \sum_{k=0}^{\infty} \frac{d^k z(\alpha, t)}{d\alpha^k}\bigg|_{\alpha=0} \frac{1}{k!} \tag{2-132}$$

The phase is now found as the imaginary terms of Eq. (2-132). The first imaginary term is found by setting $k = 1$ and is

$$\frac{\int h_1(\tau)[\theta(t - \tau)]e^{j\alpha\theta(t-\tau)} \, d\tau}{\int h_1(\tau)e^{j\alpha\theta(t-\tau)} \, d\tau}\bigg|_{\alpha=0} \cdot \frac{1}{1} = \frac{1}{H_1(0)} \int_{-\infty}^{\infty} h_1(\tau)\theta(t-\tau) \, d\tau \tag{2-133}$$

This corresponds to the linear filtering of the modulation by the equivalent low-pass filter $H_1(\omega) = H(\omega + \omega_c)$. Define this linear term to be $\phi(t)$ and a new quantity

$$q_n(t) = \frac{1}{n!} \int h_1(\tau)[\theta(t - \tau) - \phi(t)]^n \, d\tau \tag{2-134}$$

Then the first distortion term—that is, the first term not linearly related to the input—is given by q_3 and the second distortion term is

$$q_5 - \frac{10}{5!} q_3(t)q_2(t) \tag{2-135}$$

Several terms are required to predict the distortion accurately. Other methods of calculating this type of distortion have been described in the literature.[5]

2-5 NONLINEAR PROCESSING

Frequency Translation

Frequency translation is a widely used nonlinear operation performed on communication signals. This operation is shown in block diagram form in Fig. 2-19 as employed in converting an RF signal to an IF signal. Assuming a multiplier gain (or loss) factor of G_m, the multiplier output is given by

$$G_m[s(t) + n(t)]A_{LO} \cos[\omega_{LO}t + \theta_{LO}(t)] \tag{2-136}$$

where ω_{LO} is the local oscillator frequency and $\theta_{LO}(t)$ represents the phase instability of the local oscillator. Substituting for $s(t)$ and dropping the sum frequency terms, the mixer output is

$$K_m a(t) \cos[(\omega_c - \omega_{LO})t + \theta(t) - \theta_{LO}(t)] + n_m(t) \tag{2-137}$$

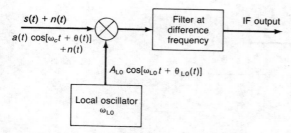

Figure 2-19 Frequency translation.

where $n_m(t)$ is the noise signal translated to the IF frequency and

$$K_m = \frac{G_m A_{LO}}{2} \tag{2-138}$$

In dealing with bandpass noise, it is convenient to represent it in a form that relates it to the carrier frequency. Such a representation is the in-phase and quadrature form discussed in Sect. 2-3 and given by

$$n(t) = n_c(t) \cos \omega_c t - n_s(t) \sin \omega_c t \tag{2-139}$$

$n_c(t)$ and $n_s(t)$ are low-pass random processes, and for Gaussian noise having a spectral density symmetrical about the carrier, they are uncorrelated with identical correlation functions given by

$$R_c(\tau) = R_s(\tau) = 2\mathscr{F}^{-1}\{S_x(f + f_c)\} \tag{2-140}$$

Using the in-phase and quadrature representation for the input noise, the mixer noise output is found to be

$$n_m(t) = K_m\{n_c(t) \cos[(\omega_c - \omega_{LO})t - \theta_{LO}(t) + \psi] - n_s(t) \sin[(\omega_c - \omega_{LO})t - \theta_{LO}(t) + \psi]\} \tag{2-141}$$

If $\theta_{LO}(t)$ is constant, then $n_m(t)$ is just a shifted version of the input noise and this is the usual approximation that is made. This approximation is valid as long as $\theta_{LO}(t)$ is small compared with the phase variations produced by the noise, which are

$$\theta_n(t) = \tan^{-1}\left[\frac{n_s(t)}{n_c(t)}\right] \tag{2-142}$$

Limiting

Nonlinear elements are often employed at the front end of a receiving system to prevent the occurrence of large signals such as impulsive noise from overloading (saturating) the following stages. Such elements are called *limiters*. An ideal hard limiter has an input/output function as shown in Fig. 2-20. Such an element converts a sinusoid into a square wave of the same frequency but with an amplitude of

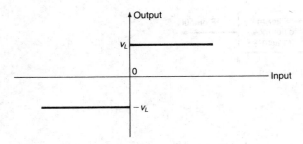

Figure 2-20 Hard limiter.

$\pm V_L$. When followed by a bandpass filter tuned to the carrier frequency, it is called a bandpass hard limiter. Such an element removes the amplitude modulation of a signal. For an input signal of the form

$$s(t) = a(t) \cos[\omega_c t + \theta(t)]$$

the output of a bandpass hard limiter would be

$$\frac{4V_L}{\pi} \cos[\omega_c t + \theta(t)]$$

When the input is the sum of an angle-modulated carrier and noise, changes occur in the carrier-to-noise power ratio (CNR). For a large input CNR, there is a further increase at the output. However, for a low input CNR, a slight degradation ($\simeq 2$ dB) is produced by the bandpass hard limiter.

Square-Law Detector

Detectors are used to extract information from modulated signals. The two most common types of detectors are the square-law detector and the envelope detector. The spectral density and autocorrelation function at the detector output are of great importance in determining the effects of postdetection filtering on the detected waveforms. The performances of these two types of detectors are quite similar, however, the analysis of the envelope detector is very involved and will not be carried out here. Analysis of the square-law detector is relatively straightforward and is considered in some detail. Figure 2-21 is a block diagram of this type of detector. Let the signal $s(t)$ and the noise $n(t)$ be sample functions from

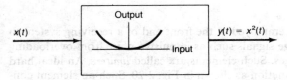

Figure 2-21 Square-law detector.

independent stationary random processes.

$$x(t) = s(t) + n(t)$$

$$s_1 = s(t_1) \qquad s_2 = s(t_2) = s(t_1 + \tau)$$

$$n_1 = n(t_1) \qquad n_2 = n(t_2) = n(t_1 + \tau)$$

The autocorrelation function of the detector output is

$$
\begin{aligned}
R_y(\tau) &= E\{(s_1 + n_1)^2 (s_2 + n_2)^2\} \\
&= E\{(s_1^2 + 2s_1 n_1 + n_1^2)(s_2^2 + 2s_2 n_2 + n_2^2)\} \\
&= E\{s_1^2 s_2^2 + 2s_1^2 s_2 n_2 + s_1^2 n_2^2 + 2s_2^2 s_1 n_1 + 4s_1 s_2 n_1 n_2 + 2n_2^2 s_1 n_1 + n_1^2 s_2^2 \\
&\quad + 2n_1^2 s_2 n_2 + n_1^2 n_2^2\} \\
&= \overline{s_1^2 s_2^2} + \overline{s^2, n_2^2} + 4\overline{s_1 s_2 n_1 n_2} + \overline{n_1^2 s_2^2} + \overline{n_1^2 n_2^2}
\end{aligned}
\tag{2-143}
$$

The first term results from the interaction of the signal with itself; the second, third and fourth from interaction of the signal and noise; and the last from interaction of the noise with itself. Using the assumption that the noise is Gaussian and independent of the sign leads to the following expressions:

$$R_{s \times s} = E\{(s_1 s_2)^2\} \tag{2-144}$$

$$R_{s \times n} = 4R_s(\tau)R_n(\tau) + 2\sigma_s^2 \sigma_n^2 \tag{2-145}$$

$$
\begin{aligned}
R_{n \times n} &= E\{n_1 n_2 n_1 n_2\} \\
&= \overline{n_1 n_2}\ \overline{n_1 n_2} + \overline{n_1 n_1}\ \overline{n_2 n_2} + \overline{n_1 n_2}\ \overline{n_2 n_1} \\
&= 2R_n^2(\tau) + \sigma_n^4
\end{aligned}
\tag{2-146}
$$

The corresponding spectral density is

$$S_y(f) = S_{s \times s}(f) + 4S_s(f) * S_n(f) + 2\sigma_s^2 \sigma_n^2 \delta(f) + \sigma_n^4 \delta(f) + 2S_n(f) * S_n(f) \tag{2-147}$$

Consider the case where $s(t)$ is a sinusoid

$$s(t) = A \cos(\omega_c t + \phi) \tag{2-148}$$

where A is a constant and ϕ is a random variable uniformly distributed over $(0, 2\pi)$.

$$R_s(\tau) = \frac{A^2}{2} \cos \omega_c \tau \tag{2-149}$$

$$S_s(f) = \frac{A^2}{4} [\delta(f + f_c) + \delta(f - f_c)] \tag{2-150}$$

$$
\begin{aligned}
s_1^2 s_2^2 &= A^4 \cos^2(\omega_c t + \phi) \cos^2[\omega_c(t + \tau) + \phi] \\
&= \frac{A^4}{4} \{\cos^2[\omega_c(t + \tau) + 2\phi] + 2 \cos[\omega_c(2t + \tau + 2\phi)] \cos \omega_c \tau + \cos^2 \omega_c \tau\}
\end{aligned}
\tag{2-151}
$$

$$R_{s \times s}(\tau) = \frac{A^4}{4} \left[1 + \frac{1}{2} \cos 2\omega_c \tau \right] \tag{2-152}$$

$$S_{s \times s}(f) = \frac{A^4}{4} \left[\delta(f) + \frac{1}{4} \delta(f + 2f_c) + \frac{1}{4} \delta(f - 2f_c) \right] \tag{2-153}$$

$$S_n(f) * S_n(f) = \frac{A^2}{4} \left[S_n(f + f_c) + S_n(f - f_c) \right] \tag{2-154}$$

The spectral density of the output, therefore, has the following form:

$$S_y(f) = \underset{\substack{\text{Around dc} \\ \text{and} \pm f_c}}{S_{s \times s}} + \underset{\substack{\text{around dc} \\ \text{and} \pm f_c}}{4 S_s * S_n} + \underset{dc}{(2\sigma_s^2 \sigma_n^2 + \sigma_n^4)\delta(f)} + \underset{\substack{\text{Around dc} \\ \text{and} \pm 2f_c}}{2 S_n * S_n} \tag{2-155}$$

$$S_y(f) = \left[\frac{A^4}{4} + A^2 \sigma_n^2 + \sigma_n^4 \right] \delta(f) + \frac{A^4}{16} \left[\delta(f + 2f_c) + \delta(f - 2f_c) \right]$$

$$+ A^2 [S_n(f + f_c) + S_n(f - f_c)] + 2 S_n(f) * S_n(f) \tag{2-156}$$

As an example assume band-limited white noise with bandwidth B centered at f_c. The resulting spectral components are as shown in Fig. 2-22.

The effect of the nonlinear behavior of the detector is readily seen by computing the output SNR as a function of the input SNR as follows:

$$\text{SNR}_i = \frac{\text{signal power}}{\text{noise power}} = \frac{A^2}{2BN_o} \tag{2-157}$$

$$\text{SNR}_o = \frac{(\text{change in dc level produced by signal})^2}{(\text{variance in bandwidth} \pm B \text{ around origin})}$$

$$= \frac{\left[\frac{A^2}{2} + N_0 B - B N_0 \right]^2}{\frac{1}{2}(2 B N_0^2 B) + N_0 B A^2}$$

$$= \frac{\left(\frac{A^2}{2 N_0 B} \right)^2}{1 + 2 \left(\frac{A^2}{2 N_0 B} \right)}$$

$$= \frac{(\text{SNR}_i)^2}{1 + 2 \text{SNR}_i} \tag{2-158}$$

It is seen from Eq. (2-158) that $\text{SNR}_0 = \frac{1}{2}\text{SNR}_i$ for large SNR_i and $\text{SNR}_0 = (\text{SNR}_i)^2$ for small SNR_i. This relationship is sketched in Fig. 2-23.

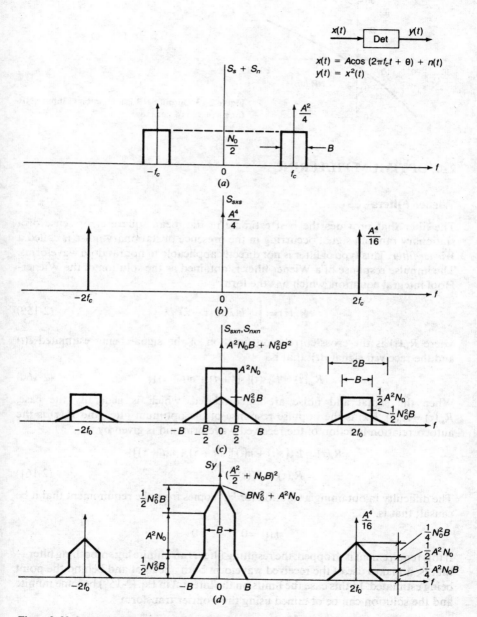

Figure 2-22 Square-law detector spectral density. (a) Input; (b) signal–signal component; (c) signal–noise and noise–noise components; (d) output.

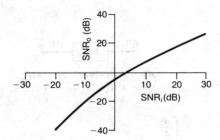

Figure 2-23 Output SNR is a function of input SNR for a square-law detector.

2-6 OPTIMUM FILTERING

Wiener Filters

The filter that provides the best estimate in the mean square error sense of a stationary random signal occurring in the presence of stationary noise is called a *Wiener filter*. This type of filter is not directly applicable to nonrandom waveforms. The impulse response of a Wiener filter is obtained as the solution of the Wiener-Hopf integral equation, which has the form[3]

$$R_{sr}(\tau) = \int_0^\infty h(\lambda) R_r(\tau - \lambda) \, d\lambda \qquad (2\text{-}159)$$

where $R_{sr}(\tau)$ is the cross-correlation function of the signal being estimated $s(t)$ and the received signal $r(t)$; that is,

$$R_{sr}(\tau) = E\{s(t)[s(t+\tau) + n(t+\tau)]\} \qquad (2\text{-}160)$$

When the signal and noise are uncorrelated, which is generally the case, $R_{sr}(\tau) = R_s(\tau)$. $h(\lambda)$ is the impulse response of the optimum filter and $R_r(\tau)$ is the autocorrelation function of the received waveform and is given by

$$R_r(\tau) = E\{[s(t) + n(t)][s(t+\tau) + n(t+\tau)]\}$$
$$= R_s(\tau) + R_n(\tau) \qquad (2\text{-}161)$$

The difficulty in obtaining a solution for $h(t)$ comes from the requirement that it be causal; that is,

$$h(t) = 0 \qquad t < 0$$

If this requirement is dropped, the resulting filter is an infinite lag smoothing filter—that is, it makes use of the received waveform both ahead of and behind the point being estimated. In this case the limits on the integral in Eq. (2-159) become infinite and the solution can be obtained using the Fourier transform

$$R_s(\tau) = \int_{-\infty}^{\infty} h(\lambda)[R_s(\tau - \lambda) + R_n(\tau - \lambda)] \, d\lambda$$

$$S_s(\omega) = H(\omega)[S_s(\omega) + S_n(\omega)]$$

$$H(\omega) = \frac{S_s(\omega)}{S_s(\omega) + S_n(\omega)} \tag{2-162}$$

The filter transfer function in Eq. (2-162) is seen to be an even function of frequency with no imaginary components. Therefore, its inverse transform $h(t)$ is symmetrical about the time origin and consequently is noncausal. This filter function is useful in two ways. First, it is the filter that gives the best performance possible—better than the physically realizable causal filter. It, therefore, can be used to compute a bound on how good an estimate is possible and for evaluation of various suboptimum filters compared with this bound. Second, the impulse response of this filter can be truncated and used as an approximation to the optimum filter when a delayed output can be utilized. The basis of the filtering operation is readily seen from Eq. (2-162), where it is observed that the output is increased when the signal spectral density exceeds the noise spectral density and decreased when the opposite is true.

As an example of the design of a noncausal Wiener filter, consider the case of additive signal plus noise in which the statistical properties are as follows:

$$R_s(\tau) = S_0 e^{-\alpha|\tau|} \longleftrightarrow \frac{2S_0\alpha}{\alpha^2 + \omega^2} \tag{2-163}$$

$$R_n(\tau) = \frac{N_0}{2}\delta(\tau) \longleftrightarrow \frac{N_0}{2} \tag{2-164}$$

The optimum filter is then

$$H(\omega) = \frac{\dfrac{2S_0\alpha}{\alpha^2 + \omega^2}}{\dfrac{2S_0\alpha}{\alpha^2 + \omega^2} + \dfrac{N_0}{2}}$$

$$= \frac{4S_0\alpha/N_0}{\omega^2 + \left(\dfrac{4S_0\alpha}{N_0} + \alpha^2\right)} \tag{2-165}$$

$$h(t) = \frac{2S_0\alpha}{N_0\sqrt{\dfrac{4S_0\alpha}{N_0} + \alpha^2}}\, e^{-\sqrt{(4S_0\alpha/N_0) + \alpha^2}\,|t|} \tag{2-166}$$

This impulse response is a double exponential and is clearly noncausal.

To make a filter that is capable of processing data in real time, it is necessary to incorporate the causality constraint into the design. There is one situation that leads to a simple solution to the Wiener-Hopf equation under the causality constraint. It corresponds to the case when the received waveform has a white spectrum; that is,

$$S_r(\omega) = 1 \tag{2-167}$$

$$R_r(\tau) = \delta(\tau) \tag{2-168}$$

In this case the Wiener-Hopf equation becomes

$$R_s(\tau) = \int_0^\infty h_0(\lambda)\delta(\tau - \lambda)\, d\lambda = h_0(\tau) \qquad t \geq 0 \tag{2-169}$$

therefore,

$$h_0(t) = R_s(t) \qquad t \geq 0$$
$$= 0 \qquad t < 0 \tag{2-170}$$

This solution is not directly applicable in most practical situations but it suggests a solution procedure:

1. Preprocess the incoming waveform so that its spectral density is white.
2. Find the optimum filter for the whitened data and process the waveform with this filter.
3. Compensate the filtered data with a filter designed to remove the effects of the original whitening operation.

For rational spectra—that is, spectral densities that can be expressed as the ratio of polynomials in ω^2—the design procedure can be carried out in the following manner. Let $w(t)$ be the impulse response of the whitening filter. Then

$$W(\omega) = F\{w(t)\} \tag{2-171}$$

$$|W(\omega)|^2 S_r(\omega) = 1$$

$$|W(\omega)|^2 = \frac{1}{S_r(\omega)} \tag{2-172}$$

Since $|W(\omega)|^2 = W(\omega)W^*(\omega)$, the transfer function $W(\omega)$ can be found by factoring $1/S_r(\omega)$. Consider an example in which the spectral density is

$$S_r(\omega) = \frac{1}{\omega^2 + 1} \tag{2-173}$$

Often it is easier to use $s = j\omega$ in the factoring operations. In the present case, this would give

$$S_r(s) = \frac{1}{-s^2 + 1} = \frac{1}{s+1} \cdot \frac{1}{-s+1} \tag{2-174}$$

When the variable s is used, the power transfer function equivalent to $|W(\omega)|^2$ is $W(s)W(-s)$. Therefore, from Eqs. (2-172) and (2-174):

$$W(s)W(-s) = -s^2 + 1 = (s+1)(-s+1) \tag{2-175}$$

General practice is to assign the factor with poles and zeros in the left half of the s-plane (LHP) to $W(s)$. Therefore,

$$W(s) = 1 + s \qquad (2\text{-}176)$$

$$w(t) = \delta(t) + \dot{\delta}(t) \qquad (2\text{-}177)$$

The inverse whitening filter $w^{-1}(t)$ undoes the whitening operation and is specified by

$$\int_{-\infty}^{\infty} w(\lambda) w^{-1}(t - \lambda)\, d\lambda = \delta(t) \qquad (2\text{-}178)$$

$$F\{w^{-1}(t)\} = \frac{1}{W(\omega)} = W^{-1}(\omega) \qquad (2\text{-}179)$$

For the example considered previously,

$$W^{-1}(s) = \frac{1}{W(s)} = \frac{1}{s+1} \qquad (2\text{-}180)$$

$$w^{-1}(t) = e^{-t} u(t) \qquad (2\text{-}181)$$

The poles and zeros of rational spectral densities expressed in terms of the variable $s = j\omega$ always occur in complex conjugate pairs and are located symmetrically in the right and left halves of the s-plane. The factors of $S_r(s)$ are then

$$S_r(s) = G(s)G(-s)$$

$$\text{Poles in LHP} \cdot \text{poles in RHP} \qquad (2\text{-}184)$$

Any rational spectral density can be factored in this manner. The filter design now can be represented as shown in Fig. 2-24 where the whitening filter is followed by the optimum filter $H'_0(s)$, which estimates $\hat{s}(t)$ from $z(t)$, which is the whitened version of $r(t)$. In this filter the desired signal is still $s(t)$ but it now must use the whitened $s(t)$ as the reference. For this filter the Wiener-Hopf equation is

$$R_{sz}(\tau) = \int_{0}^{\infty} h'_0(\lambda) R_z(\tau - \lambda)\, d\lambda \qquad (2\text{-}183)$$

But since $z(t)$ is now white, it follows that

$$h'_0(t) = R_{sz}(t) \qquad t \geq 0$$
$$= 0 \qquad t < 0 \qquad (2\text{-}184)$$

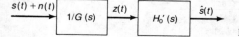

Figure 2-24 Wiener filter.

The quantity $R_{sz}(\tau)$ is given by

$$R_{sz}(\tau) = E\{s(t)z(t-\tau)\} = E\left\{s(t)\int_{-\infty}^{\infty} w(v)r(t-\tau-v)\,dv\right\} \tag{2-185}$$

$$= \int_{-\infty}^{\infty} w(v)R_{sr}(\tau+v)\,dv$$

$$= \int_{-\infty}^{\infty} w(-x)R_{sr}(\tau-x)\,dx \tag{2-186}$$

In the frequency domain, this can be expressed as

$$S_{sz}(s) = W(-s)S_{sr}(s) = \frac{S_{sr}(s)}{G(-s)} \tag{2-187}$$

With the variable ω, it would be

$$S_{sz}(\omega) = W(\omega)S_{sr}(\omega) = \frac{S_{sr}(\omega)}{[G(\omega)]^*} \tag{2-188}$$

The design procedure can be summarized as follows:

1. Find $S_{sr}(s)$.
2. Find $G(s)$ and $\dfrac{1}{G(s)}$, the whitening filter.
3. Compute $\dfrac{S_{sr}(s)}{G(-s)}$.
4. Find the inverse transform of item 3, retaining only the positive time terms to give $h'_0(t)$.
5. Cascade the whitening filter and the filter found in item 4 to give the complete filter.

Sometimes considerable simplification results by combining the separate parts of the filter, that is, the whitening filter, the optimum filter, and the compensating filter. The overall composite filter has a transfer function of the form

$$H_0(s) = W(s)H'_0(s) \tag{2-189}$$

$$= \frac{1}{G(s)}\left[\frac{S_{sr}(s)}{G(-s)}\right]_+ \tag{2-190}$$

where $|_+$ signifies that only the part of this transform corresponding to positive time is used.

As an example of this filter design procedure, consider the following.

$$r(t) = a(t) + n(t) \tag{2-191}$$

$$S_a(\omega) = \frac{A_0}{\omega^2+1} \rightarrow S_a(s) = \frac{A_0}{-s^2+1} \tag{2-192}$$

$$S_n(\omega) = \frac{N_0}{2} \rightarrow S_n(s) = \frac{N_0}{2} \tag{2-193}$$

$$S_r(s) = S_a(s) + S_n(s) = \frac{A_0}{-s^2+1} + \frac{N_0}{2} \tag{2-194}$$

$$= \frac{A_0 + \frac{N_0}{2}(-s^2+1)}{-s^2+1}$$

$$= \frac{\left(-\sqrt{\frac{N_0}{2}}s + \sqrt{A_0 + \frac{N_0}{2}}\right)\left(\sqrt{\frac{N_0}{2}}s + \sqrt{A_0 + \frac{N_0}{2}}\right)}{(-s+1)(s+1)}$$

$$G(s) = \frac{\sqrt{\frac{N_0}{2}}s + \sqrt{A_0 + \frac{N_0}{2}}}{s+1} \tag{2-195}$$

$$R_{sr}(\tau) = E\{s(t)[a(t-\tau) + n(t-\tau)]\} \tag{2-196}$$

and if $s(t) = a(t)$, it follows that

$$R_{sr}(\tau) = R_a(\tau) \tag{2-197}$$

$$S_{sr}(s) = S_a(s) = \frac{A_0}{-s^2+1} \tag{2-198}$$

$$H_0'(s) = \left[\frac{\dfrac{A_0}{-s^2+1}}{-\sqrt{\dfrac{N_0}{2}}s + \sqrt{A_0 + \dfrac{N_0}{2}}}\right]_+ \tag{2-199}$$

$$= \left[\frac{A_0}{(s+1)\left(-\sqrt{\dfrac{N_0}{2}}s + \sqrt{A_0 + \dfrac{N_0}{2}}\right)}\right]_+$$

$$= \left[\frac{\dfrac{A_0}{\sqrt{\dfrac{N_0}{2}} + \sqrt{A_0 + \dfrac{N_0}{2}}}}{s+1} + \frac{K}{-\sqrt{\dfrac{N_0}{2}}s + \sqrt{A_0 + \dfrac{N_0}{2}}}\right]_+$$

$$= \frac{\dfrac{A_0}{\sqrt{\dfrac{N_0}{2}} + \sqrt{A_0 + \dfrac{N_0}{2}}}}{s+1} \tag{2-200}$$

$$H_0(s) = \frac{1}{G(s)} H_0'(s)$$

$$= \frac{s+1}{\sqrt{\dfrac{N_0}{2}} s + \sqrt{A_0 + \dfrac{N_0}{2}}} \cdot \frac{A_0}{\dfrac{\sqrt{\dfrac{N_0}{2}} + \sqrt{A_0 + \dfrac{N_0}{2}}}{s+1}} \tag{2-201}$$

$$= \frac{\dfrac{A_0}{1 + \sqrt{1 + \dfrac{2A_0}{N_0}}}}{s + \sqrt{1 + \dfrac{2A_0}{N_0}}} \tag{2-202}$$

$$h_0(t) = \frac{A_0}{1 + \sqrt{1 + \dfrac{2A_0}{N_0}}} e^{-\sqrt{1 + (2A_0/N_0)}\,t}\, u(t) \tag{2-203}$$

Thus the optimum filter in this case corresponds to a simple low-pass RC circuit.

Matched Filters

In many cases it is not necessary to recover the signal waveform from the noise, but only to determine whether or not it is present. An example of this situation is in binary signaling where it is only necessary to determine which of two possible signals is present in a given segment of received signal. For such situations a filter that maximizes the ratio of output signal power to output noise power is required. If the filter impulse response is designated $h(t)$, then the output is

$$y(t) = [s(t) + n(t)] * h(t) \tag{2-204}$$

The signal component of this output is

$$s_0(t) = s(t) * h(t) = F^{-1}\{S(f)H(f)\} \tag{2-205}$$

$$= \int_{-\infty}^{\infty} S(f)H(f)e^{j2\pi ft}\, df \tag{2-206}$$

At the time t_0 this output is

$$s_0(t_0) = \int_{-\infty}^{\infty} S(f)H(f)e^{j2\pi f t_0}\, df \tag{2-207}$$

The spectral density of the noise at the output is

$$S_{n_0}(f) = S_n(f)|H(f)|^2 \tag{2-208}$$

The mean square value is the area under the spectral density and is given by

$$\overline{n_0^2} = \int_{-\infty}^{\infty} S_n(f)|H(f)|^2 \, df \tag{2-209}$$

The signal-to-noise ratio at time t_0 is then

$$SNR = \frac{\left| \int_{-\infty}^{\infty} S(f)H(f)e^{j2\pi f t_0} \, df \right|^2}{\int_{-\infty}^{\infty} S_n(f)|H(f)|^2 \, df} \tag{2-210}$$

It is desired to choose the $H(f)$ that will maximize Eq. (2-210). This can be done by use of the Schwartz inequality, which states the following property about the integral of the product of two functions.

$$\left| \int_{-\infty}^{\infty} X(f)Y(f) \, df \right|^2 \le \int_{-\infty}^{\infty} |X(f)|^2 \, df \int_{-\infty}^{\infty} |Y(f)|^2 \, df \tag{2-211}$$

Rearranging the numerator of Eq. (2-210) and applying Eq. (2-211) gives

$$SNR = \frac{\left| \int_{-\infty}^{\infty} \frac{S(f)e^{j2\pi f t_0}}{\sqrt{S_n(f)}} \cdot \sqrt{S_n(f)}H(f) \, df \right|^2}{\int_{-\infty}^{\infty} S_n(f)|H(f)|^2 \, df} \tag{2-212}$$

$$\le \frac{\int_{-\infty}^{\infty} \frac{|S(f)|^2}{S_{n(f)}} \, df \int_{-\infty}^{\infty} S_n(f)|H(f)|^2 \, df}{\int_{-\infty}^{\infty} S_n(f)|H(f)|^2 \, df} \tag{2-213}$$

$$SNR \le \int_{-\infty}^{\infty} \frac{|S(f)|^2}{S_n(f)} \, df \tag{2-214}$$

The equality holds when the two functions in the integrand of Eq. (2-212) are such that one is equal to a constant times the complex conjugate of the other. Thus the condition for a maximum SNR is

$$K \left[\frac{S(f)e^{j2\pi t_0 f}}{\sqrt{S_n(f)}} \right]^* = \sqrt{S_n(f)}H(f) \tag{2-215}$$

$$H(f) = \frac{S^*(f)e^{-j2\pi t_0 f}}{S_n(f)} \tag{2-216}$$

where the constant is taken as unity since it affects only the gain and not the SNR. For white noise,

$$S_n(f) = \frac{N_0}{2} \tag{2-217}$$

$$H(f) = \frac{S^*(f)e^{-j2\pi t_0 f}}{N_0/2} \tag{2-218}$$

Neglecting the gain factor $2/N_0$, this becomes

$$H(f) = S^*(f)e^{-j2\pi t_0 f} \tag{2-219}$$

$$h(t) = s(t_0 - t) \qquad t \geq 0 \tag{2-220}$$

It is seen from Eq. (2-220) that in white noise the matched filter has an impulse response that corresponds to the signal reflected around the origin and translated to the right by an amount t_0. This is illustrated in Fig. 2-25.

The waveform out of a matched filter can be computed as follows:

$$s_0(t) = s(t) * h(t) = \int_0^\infty s(\lambda)s(t_0 + \lambda - t) \, d\lambda \tag{2-221}$$

$$= R_s(t - t_0) \tag{2-222}$$

This corresponds to the time-correlation function of $s(t)$ translated to the right by an amount t_0. For this case (white noise), the maximum SNR is the value at t_0 and is

$$\text{SNR} = \int_{-\infty}^\infty \frac{|S(f)|^2}{N_0/2} \, df = \frac{2E_s}{N_0} \tag{2-223}$$

where E_s is the signal energy in the observation interval.

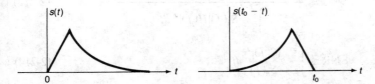

Figure 2-25 Matched filter impulse response.

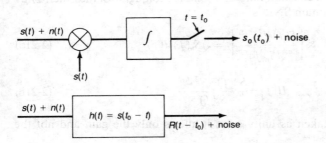

Figure 2-26 Equivalence of matched filtering and signal correlating.

It is evident that in white noise a matched filter and a signal correlator accomplish the same function at time $t = t_0$. This is illustrated in Fig. 2-26. However, the outputs of the two devices are *not* the same at times other than $t = t_0$.

2-7 MESSAGE DIGITIZATION

Sampling

The conversion of signals from analog to digital representation involves two operations: sampling and quantization. With regard to sampling, one of the principal requirements is that it be sufficiently fast to represent the waveform adequately. This requirement is spelled out in the *sampling theorem*, which may be stated as follows:

A signal whose Fourier transform is zero outside the frequency interval $|f| < W$ can be uniquely represented by a set of samples of the waveform taken at intervals of $1/2W$ seconds.

Uniquely represented means that the original waveform can be completely reconstructed from its samples. The sampling frequency $f = 2W$ is twice the highest frequency present in the waveform and is usually referred to as the *Nyquist frequency*. The sampling and reconstruction operations are illustrated in Fig. 2-27. The sampling operation consists of multiplying the incoming signal $x(t)$ by a narrow pulse sequence $q(t)$ giving an output

$$y(t) = x(t)q(t) \tag{2-224}$$

In the frequency domain, this becomes

$$Y(f) = X(f) * Q(f) \tag{2-225}$$

where

$$Q(f) = \frac{t_0}{T} \sum_{n=-\infty}^{\infty} \text{sinc}\left(\frac{nt_0}{T}\right) \delta\left(f - \frac{n}{T}\right) \qquad T = \frac{1}{2W} \tag{2-226}$$

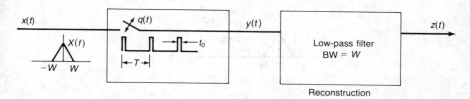

Figure 2-27 Sampling and reconstruction of a band-limited waveform.

Therefore,

$$Y(f) = \frac{t_0}{T} \sum_{n=-\infty}^{\infty} \text{sinc}\left(\frac{nt_0}{T}\right) X\left(f - \frac{n}{T}\right) \tag{2-227}$$

The spectra of these waveforms are shown in Fig. 2-28. It is seen from Fig. 2-28c that the spectrum of the sampled signal consists of the original signal spectrum replicated around each harmonic of the sampling frequency. The original wave-shape can be recovered by passing the sampled signal pulse train through a low-pass filter. The output of such a filter would be the component in Eq. (2-227) corresponding to $n = 0$; that is,

$$Z(f) = 2Wt_0 X(f) \tag{2-228}$$

$$z(t) = 2Wt_0 x(t) \tag{2-229}$$

By incorporating a gain of $1/2Wt_0$ into the LPF, $x(t)$ would be recovered exactly. The function $x(t)$ can also be expressed in terms of the sample values by

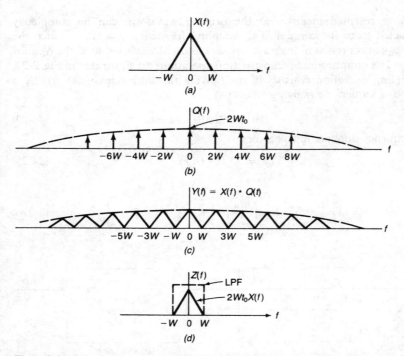

Figure 2-28 Spectra of sampling process. (*a*) Original signal; (*b*) sampling pulse train; (*c*) sampled signal; (*d*) LPF sampled signal.

means of the formula

$$x(t) = \sum_{n=-\infty}^{\infty} x\left(\frac{n}{2W}\right) \text{sinc}(2Wt - n) \tag{2-230}$$

The function $x(t)$ is obtained as a weighted sum of basis functions of the form $\text{sinc}(2Wt - n)$ with the sample values providing the weights.

If the original time function is not band-limited, or if the sampling frequency is not twice W, then recovery of the signal from its samples will not be exact. In this situation the replicated spectra overlap as shown in Fig. 2-29, where f_s is the sampling frequency.

In the recovery process, part of the signal is lost because it falls outside the LPF bandwidth, and equally seriously, there are components that overlap from adjacent replications of the spectra that are added into the reconstructed waveform producing distortion of the reproduced wave. The distortion produced by the overlapping of components from adjacent bands is called *aliasing error* or aliasing distortion.

If $S_x(f)$ is the spectral density of $x(t)$, then the power in the aliasing distortion is

$$P_D = 2 \int_{f_s/2}^{\infty} S_x(f) \, df \tag{2-231}$$

The signal-to-distortion ratio (SDR) at the output then becomes (assuming an LPF with cutoff at $1/2 f_s$)

$$\text{SDR} = \frac{\displaystyle\int_0^{f_s/2} S_x(f) \, df}{\displaystyle\int_{f_s/2}^{\infty} S_x(f) \, df} \tag{2-232}$$

Aliasing often results from sampling data, which, although low-pass filtered, still have frequency components above one half of the sampling frequency. The potential seriousness of this problem can be illustrated by considering the SDR that results from sampling at various rates an initially flat spectrum signal after

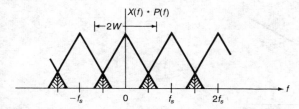

Figure 2-29 Effect of sampling below the Nyquist rate.

it has been filtered by a Butterworth filter. For this case the SDR is given by

$$\text{SDR} = \frac{\int_0^{f_s/2} \dfrac{df}{1+(f/W)^{2n}}}{\int_{f_s/2}^{\infty} \dfrac{df}{1+(f/W)^{2n}}} \tag{2-233}$$

where W is the half-power bandwidth of the filter and n is the order of the filter. The SDRs for various sampling frequencies and orders of filters are shown in Fig. 2-30. It is seen from the figure that either high-order filters must be used or sampling rates many times the half-power bandwidth must be used to obtain SDRs of 30 dB or more.

Bandpass signals can also be sampled at rates determined by their bandwidths, although the relationship of the bandwidth to center frequency also has an effect. Consider the bandpass signal shown in Fig. 2-31. It is seen that difficulties can arise because of the presence of both positive and negative frequency bands. The spacing of the replicated spectra must be such that overlapping does not occur. One solution can be obtained by relating the carrier and sampling frequencies according to the expression

$$f_c = \left(\frac{n}{2} + \frac{1}{4}\right) f_s \qquad f_s \geq 2B \tag{2-234}$$

The relationship shown in Fig. 2-31b corresponds to $f_c = 3B$ and $f_s = \frac{2}{3} f_c = 2B$.

Quantizing

For digital signal processing or digital communication, the amplitude of each signal sample must be quantized to one of a set of previously established values. This process is illustrated in Fig. 2-32 for a sine wave. Figure 2-32b shows the error signal that results from employing uniform quantizing steps. It is seen that the error signal has a periodic structure with sharp discontinuities. This type of error

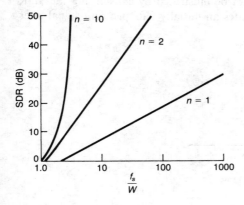

Figure 2-30 SDR for nth-order Butterworth filter with half-power frequency W.

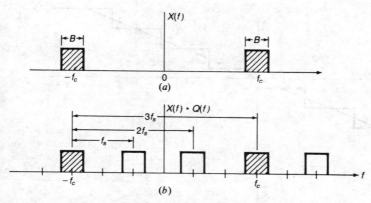

Figure 2-31 Sampling a bandpass signal. (*a*) Source spectrum; (*b*) sample spectrum.

signal has a very broad frequency spectrum and has been found to be reasonably well represented as having a uniform spectral density extending to a frequency several times that of the sampling frequency.[6] As discussed in Chap. 1, the mean square quantizing error for a uniform step size Δ, when the number of discrete steps and the rate of signal change are sufficiently high to make adjacent errors independent, is

$$\text{MSQE} = \frac{\Delta^2}{12} \tag{2-235}$$

For a more general type of signal, the MSQE can be written as

$$\text{MSQE} = \sum_{i=1}^{L} (\text{MSQE})_i P_i \tag{2-236}$$

where

$$L = \text{number of discrete quantization levels}$$

$$(\text{MSQE})_i = \text{mean square quantization error in } i\text{th interval}$$

$$P_i = \text{probability that the signal falls in the } i\text{th interval}$$

The $(\text{MSQE})_i$ is given by

$$(\text{MSQE})_i = \int_{I_i} (x - \xi_i)^2 p(x|i) \, dx \tag{2-237}$$

where I_i is the ith quantization interval, ξ_i is the quantized value for the ith interval, and $p(x|i)$ is the conditional probability density of x given that its amplitude falls in the ith interval. It follows that

$$p(x|i) = \frac{p(x)}{P_i} \tag{2-238}$$

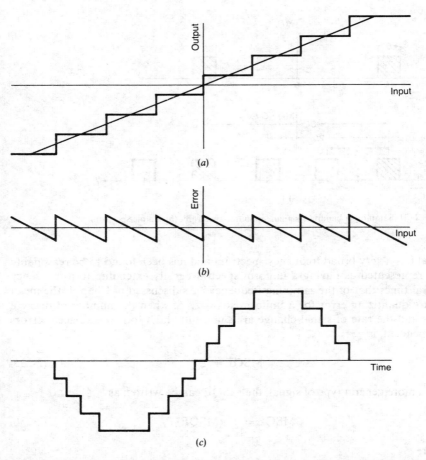

Figure 2-32 Quantization of a sine wave. (a) Transfer characteristics of quantizer; (b) error characteristic; (c) quantized full-load sine wave.

$$P_i = \int_{I_i} p(x)\, dx \tag{2-239}$$

Once the MSQE is determined, the signal-to-quantization-noise ratio (SQNR) can be expressed as

$$\text{SQNR} = \frac{E\{x^2\}}{\text{MSQE}} \tag{2-240}$$

The MSQE can be written in a more convenient form in the following manner.

$$\text{MSQE} = \sum_{i=1}^{L} \left[\int_{I_i} (x - \xi_i)^2 p(x|i)\, dx \right] P_i$$

$$= \sum_{i=1}^{L} \int_{I_i} (x - \xi_i)^2 \frac{p(x)}{P_i} P_i \, dx$$

$$= \sum_{i=1}^{L} \int_{I_i} (x - \xi_i)^2 p(x) \, dx \qquad (2\text{-}241)$$

As an example consider the case of quantizing a zero mean signal having a Gaussian PDF. Assuming L to be even, as is usually the case because of the widespread use of binary encoding, and making use of the symmetry of $p(x)$, the MSQE can be written as

$$\text{MSQE} = \frac{2}{\sigma\sqrt{2\pi}} \left\{ \sum_{i=0}^{(L/2)-2} \int_{i\Delta}^{(i+1)\Delta} (x - i\Delta - \Delta/2)^2 e^{-x^2/2\sigma^2} \, dx \right.$$

$$\left. + \int_{[(L/2)-1]\Delta}^{\infty} \left(x - \frac{L}{2}\Delta + \Delta/2 \right)^2 e^{-(x^2/2\sigma^2)} \, dx \right\} \qquad (2\text{-}242)$$

The infinite limit on the second term is required to account for the errors produced by the signal amplitudes that exceed the range of the quantizer. For the Gaussian case, the SQNR is given by

$$\text{SQNR} = \frac{\sigma^2}{\text{MSQE}} \qquad (2\text{-}243)$$

The SQNR for 4-bit, 6-bit, and 8-bit quantization corresponding to 16, 64, and 256 levels is shown in Fig. 2-33 as a function of the ratio of the quantizer range to the signal standard deviation, v/σ. It is seen from the figure that the SQNR increases with the number of levels employed in the quantizer. Also there is an optimum v/σ ratio that varies with the number of bits. The optimum v/σ varies from about 2.9 for a 4-bit quantizer to about 4 for an 8-bit quantizer.

The SQNR can be increased by using unequal step sizes. The idea is to use small step sizes in regions where the signal is most likely to occur and large step sizes where it is less likely to occur. The step sizes are selected as follows. Let the end points of the quantization levels be specified by $l_1, l_2, \ldots, l_{L+1}$. The quantizer

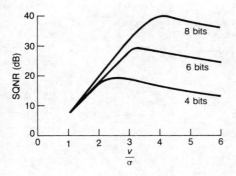

Figure 2-33 SQNR for a Gaussian signal quantized to 4, 6, and 8 bits.

output in the ith interval is ξ_i. The MSQE is then given by

$$\text{MSQE} = \sum_{i=1}^{L} \int_{l_i}^{l_{i+1}} (x - \xi_i)^2 p(x) \, dx \tag{2-244}$$

The optimum l_i and ξ_i are chosen to minimize the MSQE and are found as solutions to the following set of simultaneous equations.

$$\frac{\partial \text{MSQE}}{\partial \xi_i} = 2 \int_{l_i}^{l_{i+1}} (x - \xi_i) p(x) \, dx = 0 \qquad i = 1, 2, \ldots, L \tag{2-245}$$

$$\frac{\partial \text{MSQE}}{\partial l_i} = [(l_i - \xi_{i-1})^2 - (l_i - \xi_i)^2] p(l_i) = 0 \qquad i = 2, 3, \ldots, L \tag{2-246}$$

$$\frac{\partial \text{MSQE}}{\partial l_1} = -(l_L - \xi_1)^2 p(l_L) = 0 \tag{2-247}$$

$$\frac{\partial \text{MSQE}}{\partial l_{L+1}} = (l_{L+1} - \xi_L)^2 p(l_{L+1}) = 0 \tag{2-248}$$

The first equation leads to

$$P_i \xi_i = \int_{l_i}^{l_{i+1}} x p(x) \, dx$$

$$\xi_i = \frac{1}{P_i} \int_{l_i}^{l_{i+1}} x p(x) \, dx \tag{2-249}$$

Thus the optimum ξ_i is the conditional mean in the ith interval. In general the equations are not easy to solve. Solutions for a Gaussian PDF have been obtained for values up to $L = 36$.[8] The optimum step sizes for a 4-bit quantizer are given in Table 2-2. The SQNR obtained with this quantizer is 20.2 dB. This compares to a SQNR of 19.4 dB for the optimum uniform-step-size quantizer. In this case the

Table 2-2 Optimum step sizes for 4-bit quantization of a signal having a Gaussian PDF

Level	Step size	Output
1	0	0.128
2	0.258	0.388
3	0.522	0.657
4	0.800	0.942
5	1.099	1.256
6	1.437	1.618
7	1.844	2.069
8	2.401	2.733

nonuniform quantization does not appear to provide much improvement, however, consider the case of speech signals. Such signals typically have a dynamic range of 40 dB or more but certain speech sequences may be at low amplitude for an extended time. Such signals would have a much lower SQNR than would normal or loud speech. By utilizing nonuniform quantization steps, considerable improvement in the quality of reproduction of such signals can be obtained. Figure 2-34 shows the transfer characteristic and error characteristic of a nonuniform quantizer. In this case many more steps are assigned to the smaller amplitudes, thus reducing the error for such signals. The errors for larger signals are clearly much greater.

This same result can be obtained in a different way. If the waveform to be quantized is passed through a nonlinear element that compresses large amplitudes but has little effect on small amplitudes, it can then be processed with a uniform quantizer, giving an equivalent nonuniform-step-size performance. The signal is regenerated by passing the samples through a different nonlinear element called an expandor that restores the proper amplitude levels. The combined compression/expansion is called *companding* and a typical operating characteristic for such a system is shown in Fig. 2-35. The mean square error for a companded uniform quantizer is

$$\text{MSQE} = \frac{1}{12} \sum_{j=1}^{L} \frac{\Delta^2}{[F'(x)]^2} (x_{j+1} - x_j) P_j \qquad (2\text{-}250)$$

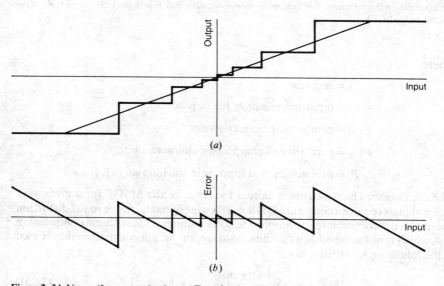

Figure 2-34 Nonuniform quantization. (a) Transfer characteristic; (b) error characteristic. (Reproduced with permission from *Transmission Systems for Communication*, Bell Telephone Laboratories, Winston-Salem, N.C., 1971.)

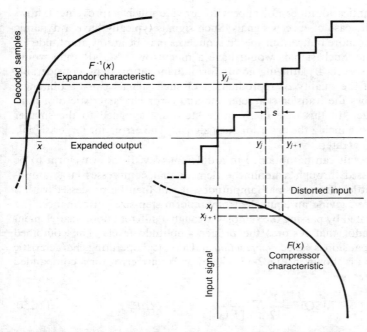

Figure 2-35 Nonuniform quantization by companding with a uniform step quantizer. (Reproduced with permission from *Transmission Systems for Communication*, Bell Telephone Laboratories, Winston-Salem, N.C., 1971.)

where

$$\Delta = step\ size$$

$$(x_{j+1} - x_j) = \text{input increment of } j\text{th step}$$

$$F(x) = \text{compressor characteristic}$$

$$F'(x) = \text{derivative of compressor characteristic}$$

$$P_j = \text{probability that input is in the interval } (x_{j+1} - x_j)$$

The compressor characteristic is chosen to minimize the MSQE for a given $p(x)$. Voice signals require a constant SQNR over a wide dynamic range so the distortion should be proportional to the amplitude. This requires a logarithmic compression law, which is not achievable with finite word length. An approximation that is used is the following so-called μ law.

$$F(x) = \text{sgn } x \, \frac{\ln(1 + \mu|x|)}{\ln(1 + \mu)} \qquad -1 < x < 1 \qquad (2\text{-}251)$$

Compression curves for several values of μ are shown in Fig. 2-36. The performances for sinusoids and speech are shown in Figs. 2-37 and 2-38.[9]

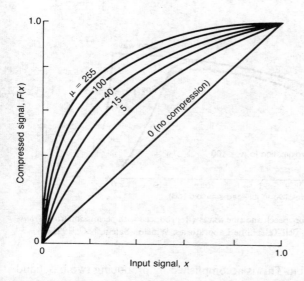

Figure 2-36 Compression curves for μ-law companding. (Reproduced with permission from *Transmission Systems for Communication*, Bell Telephone Laboratories, Winston-Salem, N.C., 1971.)

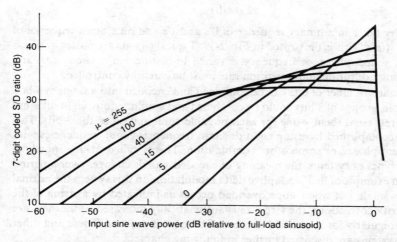

Figure 2-37 Signal-to-distortion performance of logarithmic compandor. (Reproduced with permission from *Transmission Systems for Communication*, Bell Telephone Laboratories, Winston-Salem, N.C., 1971.)

Channel Encoding

To be modulated onto a carrier, the numbers corresponding to quantized samples must be put into a baseband signal. For certain types of sources, the encoding is quite simple. For example, in PCM and delta modulation it is required that binary

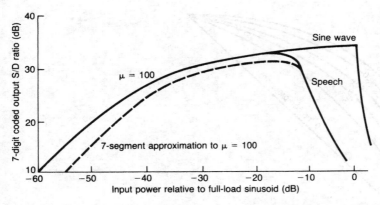

Figure 2-38 Comparison of SDRs for speech and sine waves. (Reproduced with permission from *Transmission Systems for Communication*, Bell Telephone Laboratories, Winston-Salem, N.C., 1971.)

digits be transmitted bit by bit. This is accomplished by producing two baseband signals—one for each symbol.

$$0 \to s_0(t)$$

$$1 \to s_1(t)$$

The encoder takes in a binary sequence of 0's and 1's and puts out a sequence of $s_0(t)$'s and $s_1(t)$'s. This is illustrated in Fig. 2-39. Typical signals are pulses in which amplitude, shape, phase, or frequency is varied to distinguish between $s_0(t)$ and $s_1(t)$. The pulse duration and repetition rate must be carefully controlled.

A specified number of bits can be combined and encoded into a single symbol. For example, groups of 3 bits could be combined and would require eight different signals to represent them—one for each possible combination of the 3 bits. This could be accomplished by using eight different frequencies, four frequencies with two different phases or some other combination. This is called *M*-ary encoding. In certain types of systems, the meaning of encoded signals changes over a period of time. An example of this is adaptive delta modulation. In this system the normal step size is small, but when slope overload occurs as indicated by a string of the same polarity corrections, the step size is automatically increased. If the step size alternates regularly for a specified number of times, it is reduced. These and other encoding systems are discussed further in following chapters.

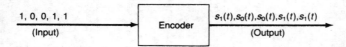

Figure 2-39 Channel encoding.

2-8 REFERENCES

1. McGillem, C. D., and G. R. Cooper: *Continuous and Discrete Signal and System Analysis, 2nd Ed.*, Holt, Rinehart and Winston, Inc., New York, 1984.
2. Cooper, G. R., and C. D. McGillem: *Probabilistic Methods of Signal and System Analysis*, Holt, Rinehart and Winston, Inc., New York, 1971.
3. Davenport, W. B. Jr., and W. L. Root: *Introduction to Random Signals and Noise*, McGraw-Hill Book Co., New York, 1958, 1965.
4. Bedrosian, E., and S. Rice: Distortion and Crosstalk of Filtered Angle Modulated Signals, *Proc. IEEE*, vol. 56, January 1968.
5. Williams, O.: A Comparison of the Carson-Fry and Rice-Bedrosian Methods of FM Analysis, *IEEE Trans. Comm.*, vol. 52, August 1973.
6. Schwartz, M., W. R. Bennett, and S. Stein: *Communication Systems and Techniques*, McGraw-Hill Book Company, New York, 1966.
7. Gagliardi, R. M.: *Introduction to Communications Engineering*, Wiley-Interscience, New York, 1978.
8. Max, J.: Quantizing for Minimum Distortion, *IRE Trans. Infor. Theory*, vol. IT-C, March 1960.
9. Bell Telephone Laboratories: *Transmission Systems for Communications*, Bell Telephone Laboratories, Winston-Salem, N.C., 1971.

2-9 PROBLEMS

2.1-1 The rise time of a low-pass system is defined as the time required for the output to go from 10 to 90 percent of its final value when the input is a unit step. Find the rise time for (a) a single time constant RC low-pass filter with half-power bandwidth $W_{1/2}$ and (b) an ideal low-pass filter with bandwidth W.

2.1-2 Show that for the same signal bandwidth the rise time of the envelope of a bandpass function is twice that of the corresponding low-pass function. Assume that the bandpass function is obtained by frequency translation of a low-pass function to an appropriate carrier.

2.1-3 A system is said to be distortion-free if it introduces the same time delay and attenuation to all spectral components of the input. Write an expression for the output of such a system given that the input is $x(t)$, the output $y(t)$, the attenuation 3 dB, and the time delay of each spectral component is τ. What is the system transfer function? What special properties must the transfer function of such a system have?

2.1-4 The duration of a positive pulse $x(t)$ having a single maximum can be approximated by a rectangular pulse of the same peak amplitude and area the same as $x(t)$. A similar procedure can be used to obtain an equivalent rectangular spectrum for the pulse. Show that if T is the time duration of the equivalent rectangular pulse and W is the bandwidth of the equivalent rectangular spectrum, then $W = 1/2T$ or $T = 1/2W$.

2.2-1 A random process has a spectral density given by

$$S(f) = 10^{-4} \operatorname{sinc}^2 10^4 f \quad \text{V}^2/\text{Hz}$$

(a) Find the mean value of the process.
(b) Find the mean square value of the process.
(c) Find the autocorrelation function of the process.

2.2-2 A stationary white noise process having a spectral density of $N_0/2 = 0.01$ V^2/Hz is passed through an RC low-pass filter having a time constant of 1 ms. Find the rms value of the filter output.

2.2-3 A random process has an autocorrelation function of the form

$$R(\tau) = 1 - |10^4 \tau| \qquad |\tau| \leq 10^{-4}$$
$$= 0 \qquad\qquad |\tau| > 10^{-4}$$

(a) Find the half-power bandwidth.

(b) Find the rms bandwidth.

2.2-4 A low-pass filter has a power transfer function shaped like a Hanning window; that is,

$$|H(f)|^2 = 0.5 + 0.5 \cos(\pi f/3500) \qquad |f| \le 3500$$
$$= 0 \qquad\qquad |f| > 3500$$

(a) Find the half-power bandwidth.

(b) Find the equivalent noise bandwidth.

2.2-5 A finite time integrator has an impulse response of the form

$$h(t) = \text{rect}\left(\frac{t}{T} - \frac{1}{2}\right)$$

Find the variance of the output of such a component when the input is white noise with a one-sided spectral density of N_0.

2.2-6 Find the half-power and equivalent noise bandwidths of the finite time integrator in Problem 2.2-5.

2.2-7 The input filter of an AM receiver has a transfer function of the form:

$$|H(f)|^2 = \frac{1}{1 + \left(\dfrac{f - f_c}{5 \times 10^3}\right)^2} + \frac{1}{1 + \left(\dfrac{f + f_c}{5 \times 10^3}\right)^2}$$

(a) Determine the power in the output when the input is of the form:

$$x(t) = A[1 + 0.5 \cos 2\pi(1000)t] \cos 2\pi f_c t$$

(b) Determine the power in the output when the input is a random modulated AM signal of the form:

$$x(t) = A[1 + m(t)] \cos 2\pi f_c t$$

where $m(t)$ is a low-pass white noise with spectral density 5×10^{-6} V^2/Hz and the bandwidth is 5 kHz.

2.2-8 Show that the following expression for the rms bandwidth of a low-pass random process in terms of its autocorrelation function is valid.

$$W_{\text{rms}} = \frac{1}{2\pi}\left[\frac{-\ddot{R}(0)}{R(0)}\right]^{1/2}$$

2.3-1 A narrow-band Gaussian noise $n(t)$ has a spectral density as shown. Sketch the spectral density of the in-phase and quadrature components for cases when the carrier frequency is (a) $f_c = f_a$, (b) $f_c = f_b$, and (c) $f_c = \frac{1}{2}(f_a + f_b)$. For which of these cases are the in-phase and quadrature components uncorrelated?

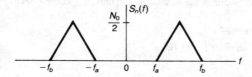

2.3-2 Repeat Problem 2.3-1 for a noise having the following spectral density.

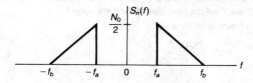

2.3-3 For the narrow-band Gaussian noise whose spectral density is shown find the autocorrelation function $R_{n_c n_c}(\tau)$ and the cross-correlation function $R_{n_c n_s}(\tau)$ for: (a) $f_c = f_a$ and (b) $f_c = \frac{1}{2}(f_a + f_b)$.

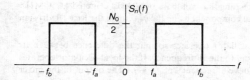

2.3-4 A stationary Gaussian random process can be expressed in the form

$$x(t) = x_c(t)\cos(\omega_c t + \psi) - x_s(t)\sin(\omega_c t + \psi)$$

where ψ is a random variable uniformly distributed $0-2\pi$, and $x_c(t)$ and $x_s(t)$ are low-pass functions having identical autocorrelation functions $R_s(\tau) = R_c(\tau)$ and cross-correlation functions having the property that $R_{cs}(\tau) = -R_{sc}(\tau)$. Find the autocorrelation function and spectral density of $x(t)$ in terms of the corresponding quantities for $x_c(t)$ and $x_s(t)$.

2.3-5 Assume that a white Gaussian noise process has been bandpass filtered to 20-Hz bandwidth. Assume that $\overline{n^2} = .02 \text{ V}^2$.

    (a) What is the probability that a sample of the noise has an amplitude greater than .4 volts?
    (b) Find the average value of the envelope of this noise.
    (c) Find the variance of the envelope.
    (d) What is the probability that the envelope has an amplitude greater than .4 volts?

2.3-6 Assume that a 4-volt unmodulated carrier is added to the noise of Problem 2.3-5.

    (a) Find the average value of the envelope of this waveform.
    (b) Find the variance of the envelope of this waveform.

2.4-1 A bandpass filter can be obtained from a low-pass prototype by proper scaling and frequency translation. If the low-pass filter has a half-power frequency of 1 rad/sec and a transfer function $H(\omega)$, then the equivalent bandpass filter transfer function is obtained by replacing the variable in the low-pass filter by the new variable $(\omega^2 - \omega_U \omega_L)/\omega(\omega_L - \omega_U)$ where ω_U is the upper half-power frequency and ω_L is the lower half-power frequency. For the second-order prototype filter shown, find the transfer function of the equivalent bandpass filter having a center frequency of 1 MHz and a bandwidth of 10 kHz.

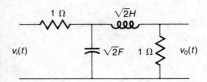

2.4-2 A bandpass filter has an impulse response of the form

$$h(t) = e^{-\alpha t}\cos \omega_0 t u(t)$$

Find the low-pass equivalent filter $h_1(t)$. For an input signal of the form $x(t) = A \cos \omega_m t \cos \omega_0 t$, write a simple expression for the output signal.

2.4-3 A bandpass filter has a transfer function given by

$$H(\omega) = \frac{\omega_0}{(j\omega + \alpha)^2 + \omega_0^2}$$

A linear modulated waveform $x(t) = A \cos \omega_m t \cos \omega_0 t$ is applied to this filter. Find the filter output using Eq. (2-119). Why does this filter not satisfy the requirements of a real low-pass function when the segment around ω_0 is translated to the origin?

2.4-4 An FM signal has the following form:

$$x(t) = A \cos [2\pi(5 \times 10^7)t + 5 \cos 2\pi(10^4)t]$$

This signal is passed through a filter having a rectangular bandpass of 50 kHz centered at 50 MHz. Determine the distortion as a fraction of the signal component using the q_3 term in the Rice–Bedrosian expansion to estimate the distortion component.

2.5-1 In a superhetrodyne receiver, the local oscillator operates so that the difference between its frequency and the incoming (desired) signal is equal to the IF frequency. If the local oscillator operates above the desired carrier frequency, then it will have a frequency of $f_{IF} + f_c$. The output of the mixer will have frequency components at $|f_c + f_{IF} - f_{in}|$ and, therefore, will produce a component at f_{IF} when $f_{in} = f_c$ and also when $f_{in} = f_c + 2f_{IF}$. The component at $f_c + 2f_{IF}$, called the image, is undesirable and is generally eliminated by filtering ahead of the mixer. Assume that a superhetrodyne receiver has a 10-kHz rectangular IF passband centered at 455 kHz with no image suppression. Sketch the receiver passbands for $f_c = 1$ MHz and (a) the LO above the carrier and (b) the LO below the carrier. Which arrangement would be most practical for an AM broadcast receiver (550–1600 kHz)?

2.5-2 A signal of the following form is passed through a bandpass limiter whose center frequency is f_c.

$$x(t) = 4 \cos 2\pi(f_c + f_1)t + 2 \cos 2\pi(f_c - f_1)t$$

Find the output of the limiter.

2.5-3 A band-limited white noise having bandwidth B centered at f_c is passed through an ideal square-law detector and then through an ideal low-pass filter having bandwidth $W = B/2$. Assuming the rms value of the input noise was 1 volt, find the rms value of the output noise.

2.5-4 A square-law detector is preceded by a low-pass RC filter with a time constant of $1/\alpha$. Assuming an input of white noise with one-sided spectral density N_0 V^2/Hz, find the autocorrelation function and spectral density of the output of the square-law detector.

2.5-5 Assume that the square-law detector of Problem 2.5-4 is followed by an ideal low-pass filter. Sketch the spectrum of the output and find the variance of the output in terms of the system parameters.

2.6-1 A random signal $a(t)$ is mixed with noise $n(t)$ such that the received signal is

$$s(t) = a(t) + n(t)$$

Assuming that $a(t)$ is a stationary random process with spectral density

$$S_a(f) = \frac{A}{1 + \left(\dfrac{f}{2000}\right)^2}$$

and $n(t)$ is a white noise process with spectral density $N_0/2$, find the (noncausal) Wiener filter transfer function for estimating $a(t)$ and determine the mean square error of the estimate.

2.6-2 Consider a signal having a spectral density

$$S_x(\omega) = \frac{1}{\omega^2 + 1}$$

and a noise having a spectral density of unity. Find the transfer function of the physically realizable Wiener filter that minimizes the output error and evaluate the error. Compare this error with that of the noncausal Wiener filter.

2.6-3 Find the transfer function of the matched filter for a 1-μs-long pulse having unit amplitude. Assume that additive white noise is present having a spectral density $N_0/2 = 10^{-7}$ V^2/Hz. Sketch the output waveform when only the signal is present. What is the SNR at the filter output?

2.6-4 Find the transfer function of the matched filter for the signal $e^{-10^7 t} u(t)$ when the noise has a spectral density given by

$$s_n(f) = \frac{10^7}{10^{14} + (2\pi f)^2}$$

Sketch the filter impulse response neglecting the time delay required for causality.

2.6-5 Design an optimum RC filter for detection of the waveform of Problem 2.6-3. Sketch the output of the filter and compute the maximum SNR at the output. (*Hint*: Use a low-pass filter and select the time constant to maximize the SNR.)

2.7-1 Find the signal-to-distortion ratio for a signal whose spectral density corresponds to a first-order Butterworth power transfer function and which is regenerated from samples taken at twice the half-power frequency and at four times the half-power frequency.

2.7-2 What is the lowest frequency at which a 10-kHz signal centered at 50 kHz can be sampled and unambiguously reconstructed?

2.7-3 Find the SQNR for a random phase sinusoid that covers only the fraction α of the available number of steps in an N-bit quantizer.

2.7-4 Find the optimum equal-amplitude steps for quantizing a zero-mean Gaussian random process with unity variance and (*a*) a 4-bit quantizer; (*b*) a 6-bit quantizer.

2.7-5 When a compandor is used, noise is often introduced in the transmission channel. Show that if the magnitude of the noise is small compared with the magnitude of the compressed signal, the output of the expandor is given by

$$y(\ ) \doteq s(t) + \frac{n(t)}{dF[s(t)]/ds(t)}$$

where $s(t)$ is the input signal, $n(t)$ is the noise signal, and $F[x]$ is the compressor characteristic. (*Hint*: Use a Taylor's series expansion of the expandor output.)

2.7-6 A compandor has compression and expansion characeristics given by

$$F(x) = \sqrt{|x|}\ \text{sgn}\ x$$

$$G(x) = x^2\ \text{sgn}\ x$$

Assume that a sinusoidal signal $A \cos \omega_1 t$ is passed through the compandor but that a spurious signal given by $B \cos \omega_2 t$ is added within the channel (i.e., after the compressor). Find the SDR at the channel output using the compandor and compare with the result when the compandor is not used. Assume that $B \ll A$ and use the expression in Problem 2.7-5 for the expandor output.

THREE
ANALOG SIGNAL DETECTION

3-1 DEMODULATION OF AM, DSB/SC, AND SSB SIGNALS

Receiver Noise

Because of their proven performance and well-established theoretical basis, analog communication systems are often used as the standards of comparison to evaluate other systems. The basic analog receiving system configuration is shown in Fig. 3-1.

It is convenient to convert the RF noise process to an equivalent bandpass noise at the IF frequency, which, as discussed in Chap. 2, can be represented by its in-phase and quadrature components as

$$n(t) = n_c(t)\cos(\omega_{IF}t + \psi) - n_s(t)\sin(\omega_{IF}t + \psi) \tag{3-1}$$

The spectral densities of $n_c(t)$ and $n_s(t)$ are

$$S_{n_c}(\omega) = S_{n_s}(\omega) = S_n(\omega + \omega_{IF}) + S_n(\omega - \omega_{IF}) \qquad |\omega| < \omega_{IF} \tag{3-2}$$

Assuming that the IF input noise is white with spectral density $N_0/2$, then $S_{n_c}(\omega)$ is given by

$$S_{n_c}(\omega) = N_0 G_1 |\tilde{H}_{IF}(\omega)|^2 \tag{3-3}$$

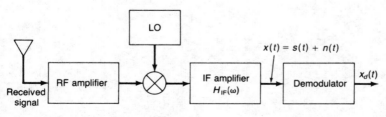

Figure 3-1 Analog receiving system.

where

$$N_0 = kT_{eq}$$

T_{eq} = equivalent noise temperature of receiver (K)

k = Boltzmann's constant (1.38×10^{-23} erg/K)

$\tilde{H}_{IF}(\omega)$ = positive frequency portion of $H_{IF}(\omega)$ translated to the origin

G_1 = total gain from antenna to demodulator input.

These spectral densities are shown in Fig. 3-2.

Product Demodulator

All linear modulation signals (AM, DSB/SC, SSB/SC) can be demodulated using a product demodulator. Such a processor has the form shown in Fig. 3-3.

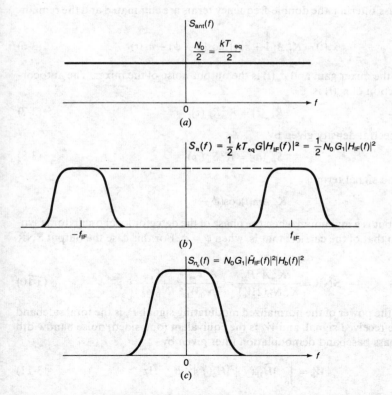

Figure 3-2 Receiver noise spectra. (a) RF spectral density; (b) IF spectral density; (c) equivalent low-pass spectrum.

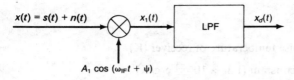

Figure 3-3 Product demodulator.

The input signal plus noise is assumed to be at the IF and for an AM signal has the form

$$x(t) = s(t) + n(t) = A[1 + m(t)]\cos(\omega_{IF}t + \phi) + n(t) \qquad (3-4)$$

This signal is mixed with a local oscillator also at the IF. The output of the mixer is given by

$$x_1(t) = \{A[(1 + m(t)]\cos(\omega_{IF}t + \phi) + n(t)\}A_1 \cos(\omega_{IF}t + \psi) \qquad (3-5)$$

After low-pass filtering, the double-frequency terms are eliminated and the remainder is

$$x_d(t) = K_m A[1 + m(t)]\cos(\phi - \psi) + n_m(t) \qquad (3-6)$$

where K_m is the mixer gain and $n_m(t)$ is the output noise of the mixer. The autocorrelation function of $n_m(t)$ is

$$R_{n_m}(\tau) = K_m^2 R_{n_c}(\tau) \qquad (3-7)$$

and has a spectral density given by

$$S_{n_m}(\omega) = K_m^2 S_{n_c}(\omega) \qquad (3-8)$$

The baseband signal term is

$$K_m A m(t)\cos(\phi - \psi)$$

The output is a maximum when the phase of the detector local oscillator is synchronized to that of the carrier; that is, when $\psi = \phi$. For this case the output SNR is

$$\text{SNR} = \frac{K_m^2 A^2 P_m}{K_m^2 N_0(2W_b)} = \frac{A^2 P_m}{2N_0 W_b} = \frac{P_r}{N_0 W_b} \qquad (3-10)$$

where P_m is the power of the normalized modulating signal, P_r is the total sideband power in the received signal, and W_b is the equivalent (one-sided) noise bandwidth of the low-pass baseband demodulation filter given by

$$W_b = \int_0^\infty |\tilde{H}_{IF}(f)|^2 |H_b(f)|^2 \, df \quad \text{Hz} \qquad (3-11)$$

In Eq. (3-11) $\tilde{H}_{IF}(f)$ and $H_b(f)$ are assumed to have unity (maximum) gain at the origin. For computation purposes W_b is usually set equal to the message signal bandwidth W_m. The same result is obtained when the modulation is DSB/SC. Because of the requirement for a frequency-coherent local oscillator signal, this technique is only used when such a signal is readily available through transmission of a pilot tone or in instrumentation systems employing a known carrier. Some examples are given in the problems.

To analyze product demodulation of SSB signals properly, it is necessary to obtain an expression for the time function corresponding to the modulated waveform. Unfortunately this cannot be done in a simple manner. The most useful approach employs the analytic signal representation of a waveform and the closely related Hilbert transform. The analytic signal is a means of representing a real signal in terms of a complex signal having only positive frequency components. In particular, consider the signal $x(t)$ with spectrum as shown in Fig. 3-4a.

The analytic signal representation of $x(t)$ is designated as $\psi_x(t)$ and is defined such that

$$x(t) = \text{Re}\{\psi_x(t)\} \tag{3-12}$$

$$\mathscr{F}\{\psi_x(t)\} = \Psi_x(f) = 2X(f)u(f) \tag{3-13}$$

It is evident from Eq. (3-13) that $\psi_x(t)$ has a one-sided spectrum with components twice as large as those of $x(t)$. The spectrum of $\psi_x(t)$ is illustrated in Fig. 3-4b. The factor of 2 is included in the definition so that the real part of $\psi_x(t)$ will have the same energy as $x(t)$. By taking the inverse transform of Eq. (3-13) it is found that

$$\psi_x(t) = x(t) + j\left(x(t) * \frac{1}{\pi t}\right)$$

$$= x(t) + j\hat{x}(t) \tag{3-14}$$

where

$$\hat{x}(t) = x(t) * \frac{1}{\pi t} = \frac{1}{\pi}\int_{-\infty}^{\infty}\frac{x(\tau)\,d\tau}{t - \tau} \tag{3-15}$$

is the Hilbert transform of $x(t)$. The Hilbert transform is most easily evaluated in the frequency domain by the relationship

$$\mathscr{F}\{\hat{x}(t)\} = \mathscr{F}\left\{x(t) * \frac{1}{\pi t}\right\} = -j\,\text{sgn}\,f\,X(f) \tag{3-16}$$

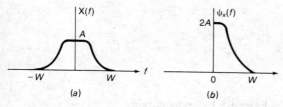

Figure 3-4 (a) Spectra of $x(t)$ and (b) its analytic signal representation $\psi_x(t)$.

The function $-j\,\text{sgn}\,f$ can be written as

$$-j\,\text{sgn}\,f = -j \qquad f > 0$$
$$= j \qquad f < 0 \qquad (3\text{-}17)$$

and, therefore, corresponds to a -90-degree phase shift at all frequencies. A Hilbert transformer is often shown as a phase shifter as in Fig. 3-5. Using this representation it is evident that

$$\text{HT}\{\hat{x}(t)\} = \overset{\approx}{x}(t) = \mathscr{F}^{-1}\{-j\,\text{sgn}\,f)^2 X(f)\} = -x(t) \qquad (3\text{-}18)$$

Three other useful properties of signals and their Hilbert transforms are as follows:

1. The energy or power in a signal and in its Hilbert transform are equal; that is,

$$\int_{-\infty}^{\infty} x^2(t)\,dt = \int_{-\infty}^{\infty} \hat{x}^2(t)\,dt \qquad (3\text{-}19)$$

2. A signal and its Hilbert transform are orthogonal; that is,

$$\int_{-\infty}^{\infty} x(t)\hat{x}(t)\,dt = 0 \qquad (3\text{-}20)$$

3. If two signals $x(t)$ and $y(t)$ are band-limited with nonoverlapping spectra, and if the spectrum of $x(t)$ is below that of $y(t)$, then the Hilbert transform of the product is

$$\text{HT}\{x(t)y(t)\} = x(t)\hat{y}(t) \qquad (3\text{-}21)$$

This last result is useful in cases of modulated signals where $x(t)$ is a low-pass signal and $y(t)$ is a carrier waveform.

As an example of computation of the Hilbert transform of a signal, let $x(t) = \cos 2\pi f_c t$.

$$X(f) = \frac{1}{2}\delta(f + f_c) + \frac{1}{2}\delta(f - f_c) \qquad (3\text{-}22)$$

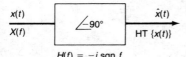

$$H(f) = -j\,\text{sgn}\,f$$

Figure 3-5 Hilbert transformer.

$$\mathscr{F}\{\hat{x}(t)\} = -j \, \text{sgn} \, f \left[\frac{1}{2} \delta(f + f_c) + \frac{1}{2} \delta(f - f_c) \right]$$

$$= \frac{-j}{2} \, \text{sgn}(-f_c) \, \delta(f + f_c) - \frac{j}{2} \, \text{sgn}(f_c) \, \delta(f - f_c)$$

$$= \frac{j}{2} \delta(f + f_c) - \frac{j}{2} \delta(f - f_c)$$

$$= \mathscr{F}\{\sin 2\pi f_c t\} \tag{3-23}$$

Therefore,

$$\text{HT}\{\cos 2\pi f_c t\} = \sin 2\pi f_c t \tag{3-24}$$

A table of signals and their Hilbert transforms is included in Appendix A.

The analytic signal contains spectral components only at positive frequencies. It is also possible to generate a signal that has only negative frequency components. Such a signal can be expressed as follows:

$$2X(f)u(-f) = 2X^*(-f)u(-f) = \Psi_x^*(-f) \tag{3-25}$$

$$\mathscr{F}^{-1}\{2X(f)u(-f)\} = \mathscr{F}^{-1}\{\Psi_x^*(-f)\} = \psi_x^*(t) = x(t) - j\hat{x}(t) \tag{3-26}$$

Figure 3-6 shows the amplitude spectrum of an arbitrary signal $x(t)$ and the analytic signal representations corresponding to the positive-frequency-only and negative-frequency-only positions.

By using these relationships, it is now possible to write expressions for SSB-modulated waveforms. Consider the upper sideband signal (USB) first. The spectrum of this signal is obtained by translating the positive-frequency portion of the message signal spectrum to the right by f_c and the negative portion of the spectrum to the left by the same amount, and multiplying by 1/2 to keep the energy the same.

$$X_{\text{USB}}(f) = \frac{A}{2} \Psi^*(f + f_c) + \frac{A}{2} \Psi(f - f_c) \tag{3-27}$$

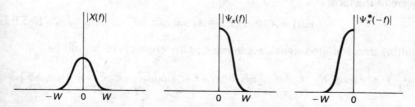

Figure 3-6 Representation of the positive-frequency-only and negative-frequency-only portions of the spectrum of $x(t)$.

where A is the factor necessary to relate the modulated signal power to the message power. The corresponding time function is

$$x_{USB}(t) = \frac{A}{2} \{[m(t) - j\hat{m}(t)]e^{-j2\pi f_c t} + [m(t) + j\hat{m}(t)]e^{j2\pi f_c t}\}$$

$$= \frac{A}{2} m(t)[e^{j2\pi f_c t} + e^{-j2\pi f_c}] + \frac{j}{2} \hat{m}(t)[e^{j2\pi f_c t} - e^{-j2\pi f_c t}]$$

$$= A[m(t)\cos 2\pi f_c t - \hat{m}(t)\sin 2\pi f_c t] \tag{3-28}$$

Similarly, the lower sideband signal can be expressed as

$$x_{LSB}(t) = A [m(t)\cos 2\pi f_c t + \hat{m}(t)\sin 2\pi f_c t] \tag{3-29}$$

The detection of SSB signals is carried out using a product demodulator in the same manner as for DSB/SC signals. Assuming that the locally generated carrier has a frequency error of $\Delta\omega$ and a phase error of $\Delta\phi$, the detector output would be

$$x_d(t) = LPF\{A [m(t)\cos \omega_c t \mp \hat{m}(t)\sin \omega_c t]\cos[(\omega_c + \Delta\omega)t + \Delta\phi]\} \tag{3-30}$$

$$= \frac{A}{2} [m(t)\cos(\Delta\omega t + \Delta\phi) \pm \hat{m}(t)\sin(\Delta\omega t + \Delta\phi)] \tag{3-31}$$

where the upper of the $\pm$ or $\mp$ signs refers to the USB and the lower to the LSB. From Eq. (3-31) it is possible to compute the distortion that results when the local carrier has frequency or phase errors present. As an example suppose there were no frequency error but there was a phase error of $\Delta\phi$. The SDR for this case would be

$$SDR = \frac{(A^2/4)\overline{m^2(t)} \cos^2 \Delta\phi}{(A^2/4)\overline{\hat{m}^2(t)} \sin^2 \Delta\phi} = \cot^2 \Delta\phi$$

A phase error of 30 degrees would result in an SDR of 4.7 dB. At first look this appears to be very poor. However, the error in this case is in the phases of the spectral components, and for certain types of communication this is not a serious problem. For example, in the case of speech signals, it is found that a phase variation does not affect intelligibility of the speech. Suppose that the modulating signal were of the form

$$m(t) = A_1 \cos \omega_1 t + A_2 \sin \omega_2 t \tag{3-32}$$

The resulting demodulated signal, assuming a phase error of $\Delta\phi$, would be

$$x_d(t) = \frac{A}{2} [(A_1 \cos \omega_1 t + A_2 \sin \omega_2 t)\cos \Delta\phi + (A_1 \sin \omega_1 t - A_2 \cos \omega_2 t)\sin \Delta\phi]$$

$$= \frac{A}{2} [A_1 \cos(\omega_1 t - \Delta\phi) + A_2 \sin(\omega_1 t - \Delta\phi)] \tag{3-33}$$

It is seen that the modulated signal is correctly reproduced except for a phase error of $\Delta\phi$ in each component. For a more general case, it is readily shown that when only a phase error is present in the locally generated carrier, the demodulated signal is given by

$$x_d(t) = \frac{A}{2} \operatorname{Re}\{\psi_m(t)e^{j\Delta\phi}\} \tag{3-34}$$

When receiver noise is present, the demodulator input can be expressed as

$$x(t) = s(t) + n(t)$$

$$= A[m(t)\cos \omega_c t \mp \hat{m}(t)\sin \omega_c t] + n_c(t)\cos \omega_c t - n_s(t)\sin \omega_c t$$

$$= [Am(t) + n_c(t)]\cos \omega_c t \mp [A\hat{m}(t) \pm n_s(t)]\sin \omega_c t \tag{3-35}$$

Assuming that the frequency and phase of the locally generated carrier are correct, the demodulator output will be

$$x_d(t) = K_m[Am(t) + n_c(t)] \tag{3-36}$$

where K_m is the mixer gain constant as before. In this case the bandwidth of the input noise is one half of that for the AM or DSB/SC waveforms, which means that the SNR ahead of the demodulator is twice as large for the SSB waveform. The output SNR for the SSB demodulator can be obtained from Eq. (3-36) as

$$(\text{SNR})_0 = \frac{K_m^2 A^2 P_m}{K_m^2 N_0 W} = \frac{A^2 P_m}{N_0 W} \tag{3-37}$$

The received power can be obtained from

$$s(t) = Am(t)\cos \omega_c t - A\hat{m}(t)\sin \omega_c t \tag{3-38}$$

and is given by

$$P_r = \frac{1}{2} A^2 \overline{m^2(t)} + \frac{1}{2} A^2 \overline{\hat{m}^2(t)} \tag{3-39}$$

Therefore, the output SNR can be expressed as

$$(\text{SNR})_0 = \frac{P_T}{N_0 W} \tag{3-40}$$

which is the same as Eq. (3-10) for product demodulation of DSB/SC or AM if P_r is taken to be the power in the message component only.

Envelope Detection

For AM signals the presence of a large carrier component permits a much simpler demodulation procedure to be employed. This procedure is called *envelope detection* and is the almost universally used method for detecting AM waveforms. The AM signal plus noise, after conversion to the IF frequency, can be written as

$$x(t) = A[1 + m(t)]\cos(\omega_{\text{IF}}t + \psi) + n_c(t)\cos(\omega_{\text{IF}}t + \psi) - n_s(t)\sin(\omega_{\text{IF}}t + \psi) \tag{3-41}$$

Combining the sine and cosine terms, this can be written in terms of an envelope and carrier as follows:

$$x(t) = w(t)\cos[\omega_{IF}t + \psi + \gamma(t)] \tag{3-42}$$

where

$$w(t) = \sqrt{[A(1 + m(t)) + n_c(t)]^2 + n_s^2(t)} \tag{3-43}$$

is the envelope function, and

$$\gamma(t) = \tan^{-1}\left[\frac{n_s(t)}{A[1 + m(t)] + n_c(t)}\right] \tag{3-44}$$

is the phase noise term, and ψ is the carrier phase. When this signal is passed through an ideal envelope detector as shown in Fig. 3-7, the output is

$$x_1(t) = K_d\{[A(1 + m(t)) + n_c(t)]^2 + n_s^2(t)\}^{1/2} \tag{3-45}$$

It is evident from Eq. (3-45) that, in general, the baseband signal is not distinct from the noise. However, for the important case in which the amplitude of the modulated carrier is greater than the noise amplitude; that is,

$$A[1 + m(t)] \gg |n_c(t)|, |n_s(t)| \tag{3-46}$$

the output is closely approximated by

$$x_1(t) \simeq K_d[A(1 + m(t)) + n_c(t)] \tag{3-47}$$

$$= K_d A + K_d Am(t) + K_d n_c(t) \tag{3-48}$$

The first term in Eq. (3-48) is a dc signal, the second term represents the baseband signal, and the third term is the in-phase component of the noise. The noise power at the output of the lowpass filter is

$$P_n = K_d^2 \int_{-\infty}^{\infty} \tilde{S}_n(f)|H_d(f)|^2 \, df \tag{3-49}$$

$$= K_d^2 N_0 G_1 \int_{-\infty}^{\infty} |\tilde{H}_{IF}(f)|^2 |H_d(f)|^2 \, df \tag{3-50}$$

The integral in Eq. (3-50) corresponds to the area under the baseband power transfer function and in a properly designed system would correspond to twice the message bandwidth W_m of the baseband signal [see Eq. (3-11)]. Under this condition the noise power referred to the input is given by

$$P_n = 2K_d^2 N_0 W_m$$

Assuming that the lowpass filter passes $m(t)$ without distortion, the signal-to-noise

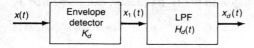

Figure 3-7 Envelope detector.

ratio at the filter output is then

$$(\text{SNR})_0 = \frac{K_d^2 A^2 P_m}{K_d^2 N_0 (2W_m)} = \frac{A^2 P_m}{2N_0 W_m} = \frac{P_r}{N_0 W_m} \tag{3-51}$$

where P_r is the message component of the received power. It is seen that this result is the same as for the coherent (product) detector. It is evident from Eq. (3-45) that this result only applies to the case of a large SNR. For SNRs below which Eq. (3-46) is no longer valid, the performance deteriorates much more rapidly than linearly with decreasing SNR. For such situations envelope detection is not practical and coherent detection must be used.

The nature of the deterioration in performance of an envelope detector can be illustrated by considering a square-law device instead of a linear envelope detector. Analysis of the envelope detector is very involved, whereas analysis of the square-law detector is quite straightforward and the results have been shown to be very comparable for small SNRs.[1] The square-law detector is illustrated in Fig. 3-8. For the case of no noise, the demodulator output would be

$$x_d(t) = \text{LPF}\{A(1 + \alpha m(t))\cos \omega_0 t\}^2$$

$$= \text{LPF}\{A^2[1 + 2\alpha m(t) + \alpha^2 m^2(t)]\cos^2 \omega_0 t\}$$

$$= \frac{A^2}{2}[1 + 2\alpha m(t) + \alpha^2 m^2(t)] \tag{3-52}$$

Neglecting the dc term, it is seen that there is a signal component $A^2 \alpha m(t)$ and a distortion component $A^2 \alpha^2 m^2(t)/2$. The signal-to-distortion ratio (SDR) depends on the waveform of $m(t)$ and for a sinusoid is given by

$$\text{SDR} = \frac{16}{3\alpha^2} \tag{3-53}$$

This type of detector clearly operates best for small modulation indices. For a modulation index of one and a sinusoidal modulating signal, the SDR would be 5.3 or 7.3 dB. If the modulation index were reduced to 0.2, the SDR would increase to 21 dB.

When noise is present, it is convenient to write the input in terms of an envelope and phase function. Thus

$$x(t) = s(t) + n(t)$$

$$= A[1 + \alpha m(t)]\cos \omega_c t + n_c(t)\cos \omega_c t - n_s(t)\sin \omega_c t$$

$$= \sqrt{[A[1 + \alpha m(t)] + n_c(t)]^2 + n_s^2(t)} \cos[\omega_c t + \theta(t)] \tag{3-54}$$

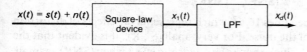

Figure 3-8 Square-law demodulator.

where

$$\theta(t) = \tan^{-1} \frac{n_s(t)}{A[1 + \alpha m(t)] + n_c(t)} \tag{3-55}$$

The output of the detector is obtained by squaring Eq. (3-54) and retaining only the baseband terms. The result is

$$x_d(t) = \frac{1}{2} \{[A[1 + \alpha m(t)] + n_c(t)]^2 + n_s^2(t)\} \tag{3-56}$$

Dropping the factor of $1/2$ since it is only a gain factor and does not affect the SNR, the detector output becomes

$$x_d(t) = A^2 + 2A^2\alpha m(t) + A^2\alpha^2 m^2(t) + 2An_c(t) + 2A\alpha m(t)n_c(t) + n_c^2(t) + n_s^2(t) \tag{3-57}$$

The first term in Eq. (3-57) represents direct current and can be omitted. The second term is the signal component given by $2A^2\alpha m(t)$. The third term is a distortion term but it can be lumped with the remaining noise terms to determine the output SNR. For purposes of illustration, assume that $m(t) = \cos(\omega_m t + \theta)$. The power in the signal term is then

$$P_s = 2A^4\alpha^2 \tag{3-58}$$

The power in the distortion and noise terms is

$$P_n = E\{[A^2\alpha^2 \cos^2 \omega_m t + 2An_c(t) + 2A\alpha n_c(t)\cos \omega_m t + n_c^2(t) + n_s^2(t)]^2\}$$

$$= \frac{A^4\alpha^4}{8} + 4A^2\sigma_n^2 + 2A^2\alpha^2\sigma_n^2 + 8\sigma_n^4 \tag{3-59}$$

The output SNR is, therefore,

$$(\text{SNR})_0 = \frac{2A^4\alpha^2}{(A^4\alpha^4/8) + 2A^2\sigma_n^2(2 + \alpha^2) + 8\sigma_n^4} \tag{3-60}$$

Defining the input signal-to-noise ratio as

$$(\text{SNR})_i = \frac{A^2\alpha^2}{2\sigma_n^2} \tag{3-61}$$

the output SNR can be written as

$$(\text{SNR})_0 = \frac{8(\text{SNR})_i^2}{(\alpha^2/2)(\text{SNR})_i^2 + 4(2 + \alpha^2)(\text{SNR})_i + 8\alpha^2} \tag{3-62}$$

$$(\text{SNR})_0 = \frac{16/\alpha^2}{1 + \frac{8(2 + \alpha^2)}{\alpha^2(\text{SNR})_i} + \frac{16}{(\text{SNR})_i^2}} \tag{3-63}$$

For very large $(\text{SNR})_i$, this becomes $16/\alpha^2$, which represents the effects of distortion in the output independent of the noise. For very small $(\text{SNR})_i$, it is evident that the output $(\text{SNR})_0 = (\text{SNR})_i^2/\alpha^2$, which shows the quadratic effect when $(\text{SNR})_i$ is small.

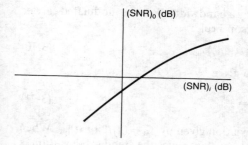

Figure 3-9 Output SNR versus input SNR for a square-law detector with a tone-modulated AM waveform input.

Figure 3-9 is a plot of the output $(SNR)_0$ versus the input $(SNR)_i$ for a square-law detector. The very poor performance at low $(SNR)_i$ is clearly evident. This same effect occurs in an envelope detector.

An improvement in $(SNR)_0$ from IF (or RF) to baseband occurs in DSB and AM as a result of the coherent addition of the signal components in the sidebands while the noise components add incoherently. For an AM or DSB/SC signal, this improvement is

$$\frac{(SNR)_0}{(SNR)_i} = 2 \tag{3-64}$$

For equal power in the sidebands, all of the output SNRs are identical. For the same average power,

$$\frac{(SNR)_0 DSB}{(SNR)_0 AM} = 1 + \left(\frac{P_m \text{ peak}}{P_m \text{ avg}}\right) \tag{3-65}$$

$$\frac{(SNR)_0 DSB}{(SNR)_0 SSB} = 1 \tag{3-66}$$

For equal peak power,

$$\frac{(SNR)_0 DSB}{(SNR)_0 AM} = 4 \tag{3-67}$$

3-2 DEMODULATION OF ANGLE-MODULATED SIGNALS

FM Demodulation

An FM signal has the form

$$s(t) = A \cos[\omega_{IF} t + \theta(t)] \tag{3-68}$$

where

$$\theta(t) = k_f \int_{t_0}^{t} m(\xi) \, d\xi \tag{3-69}$$

The modulation $m(t)$ is assumed to occupy a bandwidth W_m. The modulated carrier occupies an IF bandwidth given by Carson's rule:

$$B_{IF} = 2(D + 1)W_m \qquad (3\text{-}70)$$

where D is the frequency deviation ratio given by

$$D = \frac{f_{pk}}{W_m} \qquad (3\text{-}71)$$

where f_{pk} is the maximum frequency deviation given by $f_{pk} = (k_f/2\pi)m(t)|_{pk}$. When $m(t)$ is a random process, an appropriate constant times the standard deviation is sometimes used for $|m(t)|_{pk}$. The case of $D \gg 1$ is of most interest for practical communication systems; that is, FM waveforms having peak frequency deviations substantially greater than the bandwidth of the modulation.

The waveform to be demodulated is the sum of the carrier and a narrow-band Gaussian noise whose bandwidth corresponds to the IF bandwidth. This waveform can be represented as

$$x(t) = s(t) + n(t)$$
$$= A \cos[\omega_{IF}t + \theta(t)] + n_c(t)\cos \omega_{IF}t - n_s(t)\sin \omega_{IF}t \qquad (3\text{-}72)$$

The noise term can be transformed in the following manner to permit its combination with the signal.

$$n_c(t)\cos \omega_{IF}t = n_c(t)\cos[\omega_{IF}t + \theta(t) - \theta(t)]$$
$$= n_c(t)\cos \theta(t)\cos[\omega_{IF}(t) + \theta(t)]$$
$$\quad + n_c(t)\sin \theta(t)\sin[\omega_{IF}t + \theta(t)] \qquad (3\text{-}73)$$

Similarly,

$$n_s(t)\sin \omega_{IF}t = - n_s(t)\sin \theta(t)\cos[\omega_{IF}t + \theta(t)] + n_s(t)\cos \theta(t)\sin[\omega_{IF}t + \theta(t)] \qquad (3\text{-}74)$$

Calling

$$n_c(t)\cos \theta(t) - n_s(t)\sin \theta(t) = n'_c(t) \qquad (3\text{-}75)$$

$$n_s(t)\cos \theta(t) + n_c(t)\sin \theta(t) = n'_s(t) \qquad (3\text{-}76)$$

the received signal can be written as

$$x(t) = A \cos[\omega_{IF}t + \theta(t)] + n'_c(t)\cos[\omega_{IF}t + \theta(t)]$$
$$\quad - n'_s(t)\sin[\omega_{IF}t + \theta(t)] \qquad (3\text{-}77)$$

This is similar to the expression found for linear modulation signals except that the noise components are modified to $n'_c(t)$ and $n'_s(t)$. Assuming $\theta(t)$ is nonrandom, it can be readily shown that

$$R_{n'_c}(\tau, t) = R_{n'_s}(\tau, t) = R_{n_c}(\tau)\cos[\theta(t) - \theta(t + \tau)] - R_{n_c n_s}(\tau)\sin[\theta(t) - \theta(t + \tau)] \qquad (3\text{-}78)$$

$$R_{n'_c n'_s}(\tau, t) = R_{n_c n_s}(\tau)\cos[\theta(t) - \theta(t + \tau)] + R_{n_c}(\tau)\sin[\theta(t) - \theta(t + \tau)] \qquad (3\text{-}79)$$

It is seen from Eqs. (3-78) and (3-79) that the varying carrier phase modifies the correlation functions of the noise. However, if the carrier phase does not change significantly over the time the noise remains correlated—that is, the noise is wide band compared with the signal—the $nn\theta(t) - \theta(t + \tau) \simeq 0$ and

$$R_{n'_c}(\tau) = R_{n'_s}(\tau) = R_{n_c}(\tau) \tag{3-80}$$

$$R_{n'_c n'_s}(\tau) = R_{n_c n_s}(\tau) \tag{3-81}$$

Furthermore, if the noise spectrum is symmetrical about the IF frequency, as is usually the case,

$$R_{n'_c n'_s}(\tau) = R_{n_c n_s}(\tau) = 0 \tag{3-82}$$

Under these assumptions the modified noise is essentially the same as the unmodified noise and the analysis can be carried out using the unmodified noise with little error. By using this procedure, the received waveform can be represented in terms of envelope and phase functions as

$$x(t) = A \cos[\omega_{IF}t + \theta(t)] + n_c(t)\cos[\omega_{IF}t + \theta(t)] - n_s(t)\sin[\omega_{IF}t + \theta(t)]$$

$$= r(t)\cos[\omega_{IF}t + \theta(t) + \gamma(t)] \tag{3-83}$$

where $r(t)$ is the envelope function given by

$$r(t) = \{[A + n_c(t)]^2 + [n_s(t)]^2\}^{1/2} \tag{3-84}$$

and $\gamma(t)$ is the phase noise from $n(t)$ given by

$$\gamma(t) = \tan^{-1}\left\{\frac{n_s(t)}{A + n_c(t)}\right\} \tag{3-85}$$

The expression for the envelope noise is not needed since this noise is removed by a limiter before the detection processing is carried out.

One type of FM demodulator uses a frequency discriminator as shown in Fig. 3-10. The frequency discriminator is built from resonant circuit elements and has an output that varies linearly in amplitude with the frequency deviation of the input from the center frequency. The output of an ideal frequency demodulator with detector constant K_d is

$$x_1(t) = K_d[\omega(t) - \omega_{IF}]$$

$$= K_d \frac{d}{dt}[\theta(t) - \gamma(t)] \tag{3-86}$$

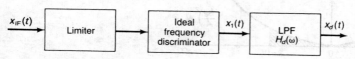

Figure 3-10 FM demodulation.

Under the generally valid assumption that $A \gg |n_c(t)|, |n_s(t)|$, the $\tan^{-1}(\cdot)$ expression for $\gamma(t)$ can be replaced by its argument and the derivative expanded to give

$$x_1(t) \doteq K_d \left\{ k_f m(t) + \frac{[A + n_c(t)]\dot{n}_s(t) - n_s(t)\dot{n}_c(t)}{[A + n_c(t)]^2} \right\} \tag{3-87}$$

Making further use of the fact that A is large relative to the noise amplitude, this can be put into the following form:

$$x_d(t) \doteq K_d \left\{ k_f m(t) + \frac{\dot{n}_s(t)}{A} - \frac{n_s(t)\dot{n}_c(t)}{A^2} \right\} \tag{3-88}$$

Neglecting the term of second order in $1/A$ gives

$$x_1(t) \simeq K_d k_f m(t) + K_d \frac{\dot{n}_s(t)}{A} \tag{3-89}$$

For this large SNR condition, the noise is the second term of Eq. (3-89); that is,

$$n_1(t) = \frac{K_d \dot{n}_s(t)}{A} = \left(\frac{K_d}{A} \right) \frac{dn_s(t)}{dt} \tag{3-90}$$

The spectral density of the derivative of a stationary random process is ω^2 times the spectral density of the original process. Therefore,

$$S_{n1}(\omega) = \frac{K_d^2}{A^2} \omega^2 S_{n_c}(\omega) = \left(\frac{K_d}{A} \right)^2 \omega^2 N_0 |H_{IF}(\omega)|^2 \tag{3-91}$$

This is shown in Fig. 3-11.

The bandwidth of the LPF at the detector output is much smaller than the IF bandwidth so there is a substantial reduction in the noise in going from IF to baseband. The output signal power is

$$P_s = K_d^2 k_f^2 P_m \tag{3-92}$$

The noise power for a rectangular LPF function $H_d(\omega)$ having a bandwidth equal to the modulation bandwidth W_m is

$$P_n = \left(\frac{K_d}{A} \right)^2 \frac{N_0}{2\pi} \int_{-2\pi W_m}^{2\pi W_m} \omega^2 \, d\omega = \frac{2K_d^2 N_0 (2\pi)^2 W_m^3}{3A^2} \tag{3-93}$$

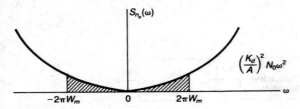

Figure 3-11 FM discriminator output noise spectral density.

The output SNR is then

$$(\text{SNR})_0 = \frac{P_s}{P_n} = \frac{3A^2 K_d^2 k_f^2 P_m}{2K_d^2 N_0 (2\pi)^2 W_m^3} = \frac{3A^2 k_f^2 P_m}{8\pi^2 N_0 W_m^3} \qquad (3\text{-}94)$$

Assuming $P_m = 1$, this can be written as

$$(\text{SNR})_0 = \frac{3D^2}{|m(t)|_{pk}^2} \left(\frac{A^2/D}{N_0 W_m} \right) = \frac{3D^2}{|m(t)|_{pk}^2} (\text{CNR})_{W_m} \qquad (3\text{-}95)$$

where D is the deviation ratio f_{pk}/W_m and $(\text{CNR})_{W_m}$ is the ratio of carrier power to noise power in the *baseband bandwidth*. It is evident from Eq. (3-95) that by increasing the frequency deviation (i.e., increasing D), an improvement in $(\text{SNR})_0$ can be obtained for the same carrier power. Thus FM provides a trade-off between SNR and bandwidth. To achieve this performance, it is necessary that the carrier-to-noise ratio $(\text{CNR})_{B_{IF}}$ in the IF passband exceed a threshold sufficient that the approximations in Eqs. (3-87) and (3-88) are valid; that is, it is necessary that

$$(\text{CNR})_{B_{IF}} = \frac{A^2/2}{N_0 B_{IF}} > \xi_0 \qquad (3\text{-}96)$$

This threshold value corresponds to about 12 dB. Figure 3-12 shows typical FM

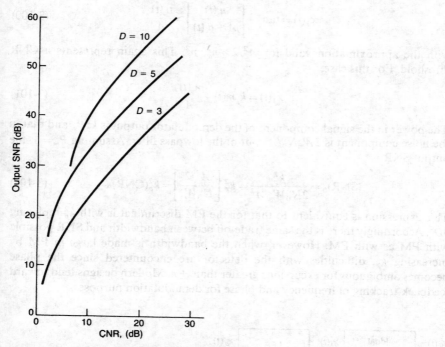

Figure 3-12 FM detector performance $(|m(t)|_{pk} = 1)$

performance curves. A precise analysis of the FM threshold problem is very complex but a number of studies have been carried out.[2-4]

PM Demodulation

An ideal phase detector can be used to recover the baseband signal from a phase-modulated signal. The PM signal has the form

$$s(t) = A \cos[\omega_{IF}t + k_p m(t)] \tag{3-97}$$

where as before $A^2/2$ is the carrier power and k_p is the phase-modulation index. The IF signal is

$$\begin{aligned} x_{IF}(t) &= s(t) + n(t) \\ &= r(t)\cos[\omega_{IF}t + k_p m(t) + \gamma(t)] \end{aligned} \tag{3-98}$$

as in the case of the FM waveform. The processor is illustrated in Fig. 3-13. The output of the ideal phase detector is

$$x_1(t) = k_p m(t) + \gamma(t) \tag{3-99}$$

where, as before,

$$x_1(t) = \tan^{-1}\left[\frac{n_s(t)}{A + n_c(t)}\right] \simeq \frac{n_s(t)}{A} \tag{3-100}$$

with the approximation valid for $A^2/2 \gg \overline{n_s^2}, \overline{n_c^2}$. This again represents a CNR threshold. For this case,

$$x_1(t) = k_p m(t) + \frac{n_s(t)}{A} \tag{3-101}$$

The power in the signal component of the demodulator output is $k_p^2 P_m$ and that in the noise component is $2W_m N_0/A^2$ out of the lowpass filter. Assuming $P_m = 1$, the output SNR is

$$(\text{SNR})_0 = \frac{k_p^2}{2N_0 W_m/A^2} = k_p^2\left[\frac{A^2/2}{N_0 W_m}\right] = k_p^2(\text{CNR})_{W_m} \tag{3-102}$$

This expression is equivalent to that for the FM discriminator with k_p^2 replacing $3D^2$. Accordingly there is the same trade-off between bandwidth and SNR possible with PM as with FM. However, when the bandwidth is made large in PM by increasing k_p, difficulties with the detector are encountered since the phase becomes ambiguous for excursions greater than $\pm\pi$. Modern designs tend toward feedback tracking of frequency and phase for demodulation purposes.

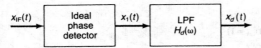

Figure 3-13 Phase demodulator.

FM SNR Improvement Using Preemphasis/Deemphasis

Because the spectral density of the noise at the output of an FM demodulator increases with f^2, it has its greatest effect on the high-frequency components of the message waveform. This effect can be greatly reduced by passing the modulating signal through a preemphasis filter that amplifies its high-frequency components more than its low-frequency components before using it to modulate the carrier. After demodulation at the receiver, the signal is passed through a deemphasis filter that attenuates the high frequencies and restores the proper spectral shape. The preemphasis and deemphasis filters are chosen such that over the band of interest

$$H_e(f) = \frac{1}{H_d(f)} \qquad -W_m \le f \le W_m \qquad (3\text{-}103)$$

where $H_e(f)$ and $H_d(f)$ are the transfer functions of the preemphasis and deemphasis filters respectively. A typical system arrangement is shown in Fig. 3-14.

This noise-reduction technique is used in commercial FM broadcasting. Typical preemphasis and deemphasis networks are shown in Fig. 3-15, along with

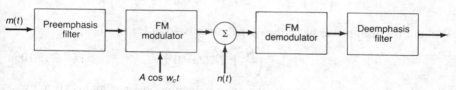

Figure 3-14 FM preemphasis and deemphasis.

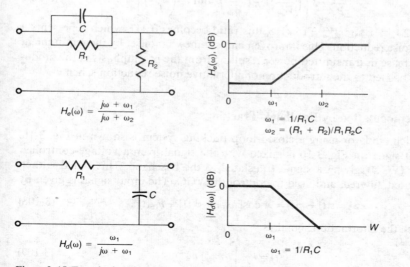

Figure 3-15 Transfer functions of preemphasis and deemphasis networks.

asymptotic plots of their transfer functions. The value of ω_1 is selected to be at the point where the spectrum of the modulating signal has fallen about 3 dB. This is taken for purposes of standardization to be 2.1 kHz. The value of ω_2 is chosen to be well above the audio-frequency band.

The improvement in SNR resulting from using preemphasis is found by computing the reduction in noise power out of the demodulator that results from cascading $H_d(\omega)$ with the already present LPF. The net effect of the preemphasis and deemphasis filters on the signal leaves it unchanged. From Eq. (3-91), assuming the IF amplifier has a uniform passband, the demodulator output spectral density is

$$S_{n_d}(\omega) = \left(\frac{K_d}{A}\right)^2 N_0 \omega^2 \tag{3-104}$$

When this signal is passed through the demodulator low-pass filter and deemphasis filter of Fig. 3-15, the output noise power is given by

$$
\begin{aligned}
P'_n &= \left(\frac{K_d}{A}\right)^2 \frac{N_0 \omega_1^2}{2\pi} \int_{-2\pi W_m}^{2\pi W_m} \frac{\omega^2 \, d\omega}{\omega^2 + \omega_1^2} \\
&= \left(\frac{K_d}{A}\right)^2 \frac{N_0 \omega_1^2}{\pi} \left[2\pi W_m - \omega_1 \tan^{-1}\left(\frac{2\pi W_m}{\omega_1}\right) \right]
\end{aligned}
\tag{3-105}
$$

Without the deemphasis filter, the output noise is, from Eq. (3-93),

$$P_n = \frac{2}{3}\left(\frac{K_d}{A}\right)^2 N_0 (2\pi)^2 W_m^3 \tag{3-106}$$

The ratio of P'_n and P_n represents the noise reduction obtained and is given by

$$\frac{P'_n}{P_n} = 3\left[\left(\frac{f_1}{W_m}\right)^2 - \left(\frac{f_1}{W_m}\right)^3 \tan^{-1}\left(\frac{W_m}{f_1}\right) \right] \tag{3-107}$$

For $f_1 = 2.1$ kHz and $W_m = 15$ kHz, this ratio becomes 0.047, which represents a 13.3-dB noise reduction. This improvement is somewhat offset by the requirement for an increase in transmitted power resulting from the preemphasis at the modulator and has led to the introduction of alternative noise reduction schemes.[5]

Angle Demodulation by Feedback Tracking

An FM demodulator using a closed-loop tracking system is shown in Fig. 3-16. The input signal in Fig. 3-16 is mixed with the signal from a voltage-controlled oscillator (VCO), giving a signal translated to the baseband. This signal is then differentiated, filtered, and used to control the VCO. The input signal is given by

$$s(t) + n(t) = A \cos[\omega_{IF} t + \theta(t)] + n_{IF}(t) \tag{3-108}$$

where $\theta(t)$, the information signal, is

$$\theta(t) = k_f \int_{t_0}^{t} m(\xi) \, d\xi \tag{3-109}$$

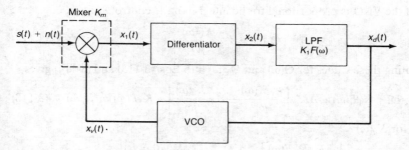

Figure 3-16 FM feedback demodulator.

The input is mixed with the VCO signal given by

$$x_v(t) = \sin[\omega_{\text{IF}}t + \theta_v(t)] \tag{3-110}$$

where $\theta_v(t)$ is the phase variation of the VCO, including any error in its value. The mixer output is

$$x_1(t) = K_m A \sin[\theta_e(t)] + n_m(t) \tag{3-111}$$

where $n_m(t)$ is the mixer noise output and $\theta_e(t)$ is given by

$$\theta_e(t) = \theta(t) - \theta_v(t) \tag{3-112}$$

For small phase errors (i.e., tight tracking), $\sin[\theta_e(t)] \simeq \theta_e(t)$. For this case the output of the differentiator is

$$x_2(t) = \frac{dx_1(t)}{dt} \simeq AK_m \frac{d\theta_e(t)}{dt} + \frac{dn_m(t)}{dt} \tag{3-113}$$

This waveform is low-pass filtered by $K_1 F(\omega)$ and can be described and interpreted more easily in the frequency domain. Strictly speaking the Fourier transform of $n_m(t)$ or its derivative does not exist since it is a sample function of a stationary random process. However, the corresponding time-domain operations do exist, and if the frequency-domain representation is thought of as a shorthand way of writing the time-domain operations, the procedure can be used. The Fourier transform representation of the output is

$$X_d(\omega) = K_1 F(\omega)[j\omega(AK_m \Theta_e(\omega) + N_m(\omega))] \tag{3-114}$$

where

$$\Theta_e(\omega) = \Theta(\omega) - \Theta_v(\omega) \tag{3-115}$$

$$\Theta(\omega) = k_f \frac{M(\omega)}{j\omega} \tag{3-116}$$

The signal $x_d(t)$ is fed back to control the frequency of the VCO. Accordingly the phase of the VCO is proportional to the integral of the control voltage, or

$$\Theta_v(\omega) = K_2 \frac{X_0(\omega)}{j\omega} \tag{3-117}$$

Substituting these values for $\Theta(\omega)$ and $\Theta_1(\omega)$ into Eqs. (3-115) and (3-114) gives

$$X_d(\omega) = K_1 F(\omega) j\omega A K_m \left[\frac{k_f M(\omega)}{j\omega} - K_2 \frac{X_d(\omega)}{j\omega} \right] + K_1 F(\omega) j\omega N_m(\omega) \tag{3-118}$$

Solving for $X_d(\omega)$,

$$X_d(\omega) = \left[\frac{A K_m K_1 F(\omega)}{1 + K A F(\omega)} \right] k_f M(\omega) + \left[\frac{K_1 F(\omega)}{1 + K A F(\omega)} \right] j\omega N_m(\omega) \tag{3-119}$$

where $K = K_m K_1 K_2$ is the gain around the closed loop. Equation (3-119) is of the form

$$\text{Output} = \text{filtered modulation} + \text{filtered noise}$$

Analyzing the performance of a nonlinear loop of this type operating at the carrier frequency can be greatly simplified by representing it in terms of an equivalent linear loop operating at baseband. Such a representation is shown in Fig. 3-17. The input to the linear baseband loop is the phase modulation $\theta(t)$. The mixer is replaced by a phase subtractor and the VCO is replaced by a phase integrator. Loop filtering is the same as in the actual system so that the results of the analysis will be directly applicable to the actual loop.

Consider the case where $F(\omega)$ is an ideal low-pass filter having a bandwidth equal to the baseband frequency range; that is,

$$F(\omega) = 1 \qquad |\omega| \le 2\pi W_m$$
$$= 0 \qquad \text{elsewhere} \tag{3-120}$$

For this case,

$$X_d(\omega) = \left[\frac{A K_m K_1}{1 + A K} \right] \left[k_f M(\omega) + j\omega \frac{N_m(\omega)}{A K_m} \right] \tag{3-121}$$

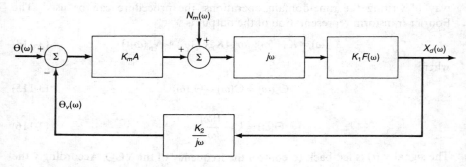

Figure 3-17 Equivalent baseband frequency-tracking loop.

The first term on the right-hand side of Eq. (3-121) corresponds to the baseband signal and the second term is noise. The noise components outside the baseband have been eliminated by $F(\omega)$. The spectral density of the noise out of the mixer is $K_m^2 S_{n_c}(\omega)$ so that the output noise spectral density is

$$S_{n_d}(\omega) = \frac{\omega^2 N_0}{A^2} \left(\frac{AK_mK_1}{1+AK}\right)^2 \qquad |\omega| \geq 2\pi W_m \qquad (3\text{-}122)$$

The noise power is then

$$P_n = \frac{N_0}{A^2} \left(\frac{AK_mK_1}{1+AK}\right)^2 \cdot \frac{2(2\pi W_m)^3}{3(2\pi)} \qquad (3\text{-}123)$$

and, assuming $P_m = 1$, the signal power is

$$P_s = k_f^2 \left(\frac{AK_mK_1}{1+AK}\right)^2 \qquad (3\text{-}124)$$

The SNR is, therefore,

$$(\text{SNR})_0 = \frac{3A^2 k_f^2}{2(2\pi)^2 W_m^3 N_0} \qquad (3\text{-}125)$$

It is seen that this is the same as the SNR for the frequency discriminator given by Eq. (3-95). However, in this case no assumption regarding the CNR was made—the only assumption made was that the tracking error was not large. When the phase error becomes large, the system is "out of lock." In this system the loop gain AK varies with the carrier amplitude. This effect is eliminated by using limiting or by using automatic gain control (AGC) in the receiver to keep A constant.

Phase-Tracking Loop

Instead of using a frequency-tracking loop, it is possible to demodulate angle-modulated signals through use of a phase-tracking loop as shown in Fig. 3-18. Since the mixer output is proportional to the phase error (for small errors), it is the desired signal after low-pass filtering. Because the VCO has a frequency proportional to voltage, it is necessary to convert the phase signal $x_0(t)$ into a

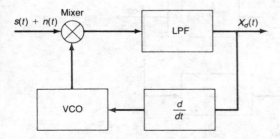

Figure 3-18 Phase-tracking loop.

frequency signal by differentiating. As before the input is

$$s(t) + n(t) = A \cos[\omega_{IF}t + \theta(t)] + n_{IF}(t) \tag{3-126}$$

The mixer output is also the same as in the previous case except that the angle $\theta(t)$ is now a phase modulation; that is,

$$\Theta(\omega) = k_p M(\omega) \tag{3-127}$$

Again assuming a small phase error, the output can be represented as

$$X_d(\omega) = K_1 F(\omega)\{AK_m[\Theta(\omega) - \Theta_1(\omega)]\} + K_1 F(\omega)N_m(\omega) \tag{3-128}$$

This signal is differentiated and used to control the VCO so that

$$\Theta_1(\omega) = j\omega \left(\frac{K_2}{j\omega}\right) X_0(\omega) = K_2 X_d(\omega) \tag{3-129}$$

Substituting into Eq. (3-128) and solving for $X_0(\omega)$ gives

$$X_d(\omega) = \left[\frac{AK_1 K_m F(\omega)}{1 + AKF(\omega)}\right] k_p M(\omega) + \left[\frac{K_1 F(\omega)}{1 + AKF(\omega)}\right] N_m(\omega) \tag{3-130}$$

The equivalent baseband linear loop is shown in Fig. 3-19.

Again assuming $F(\omega)$ to be an ideal low-pass filter extending over the baseband, the output signal power is found to be

$$P_s = \left(\frac{AK_1 K_m k_p}{1 + AK}\right)^2 \tag{3-131}$$

The output noise power spectral density is $K_m^2 S_{n_d}(\omega)$ over the interval $|\omega| < 2\pi W_m$ and the output noise power is

$$P_n = \frac{(K_1 K_m)^2 2N_0 W_m}{(1 + AK)^2} \tag{3-132}$$

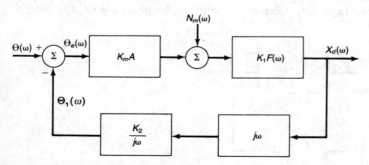

Figure 3-19 Baseband phase-tracking loop.

The SNR is then

$$(\text{SNR})_0 = k_p^2 \left(\frac{A^2/2}{N_0 W_m} \right) \tag{3-133}$$

which is the same as for the ideal phase detector.

A serious problem with this type of loop is the differentiator in the return path. This causes stability problems and also increases noise in the loop. Because of this, tracking phase demodulators are usually designed with the differentiator omitted. Such a demodulator is called a *phase-lock loop* (PLL). This type of loop is shown in Fig. 3-20. By omitting the differentiator, the overall transfer function is no longer correct to provide the demodulated signal at the output and the output must be further processed to give the modulation. This is illustrated in Fig. 3-21.

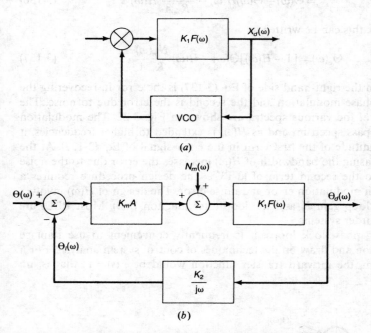

Figure 3-20 Phase-lock loop. (*a*) Basic loop; (*b*) equivalent linear baseband loop.

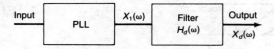

Figure 3-21 Recovery of modulation from PLL output.

The transfer function of the linearized PLL and the output filter in cascade is

$$H_m(\omega) = \left[\frac{A K_m K_1 F(\omega)}{1 + \dfrac{A K F(\omega)}{j\omega}} \right] H_d(\omega) \qquad (3\text{-}134)$$

The PLL can be analyzed in terms of a transfer function of the form

$$H(\omega) = \frac{A K F(\omega)}{j\omega + A K F(\omega)} \qquad (3\text{-}135)$$

This is the closed loop gain function from the baseband input to the VCO output in the equivalent linear baseband system. The error signal is

$$\Theta_e(\omega) = \Theta(\omega) - \Theta_1(\omega)$$

$$= \Theta(\omega) - \Theta(\omega)H(\omega) - \frac{N_m(\omega)}{K_m A} H(\omega) \qquad (3\text{-}136)$$

In terms of $H(\omega)$, this can be written as

$$\Theta_e(\omega) = [1 - H(\omega)]\Theta(\omega) - H(\omega) \frac{N_m(\omega)}{K_m A} \qquad (3\text{-}137)$$

The first term on the right-hand side of Eq. (3-137) is the error in recovering the correct carrier phase modulation and the second is the error due to noise. The general natures of the various spectra are shown in Fig. 3-22. The modulation $\Theta(\omega)$ has a low-pass spectrum and as $H(\omega)$ is extended to higher frequencies, it reduces the magnitude of the first term in the error signal of Eq. (3-137). At the same time, increasing the bandwidth of $H(\omega)$ increases the error due to the noise corresponding to the second term of (3-137). The design procedure requires a trade-off between modulation error and noise error. The design of $H(\omega)$ involves selection of the loop gain K and the loop filter function $F(\omega)$. Most PLLs are first- or second-order systems.

In analyzing phase-lock loops, it is frequently convenient to use Laplace transform notation and draw on the techniques of control system analysis. For a first-order system, the forward transfer function would be $F(s) = 1$; that is, no

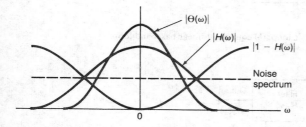

Figure 3-22 Spectra of PLL components.

filter would be present in the loop. The closed-loop transfer function is

$$H(s) = \frac{AK}{s + AK} \tag{3-138}$$

The equivalent noise bandwidth of the loop is

$$W_L = \frac{AK}{4} \text{ Hz} \tag{3-139}$$

As in most feedback systems, the bandwidth is directly related to the loop gain.

A second-order loop results from incorporating a filter in the forward path. Such filters can be either passive or active. Figure 3-23 shows two frequently used filter structures. The passive RC filter is widely used, however, the active filter provides better tracking performance. For the passive filter, the closed-loop transfer function can be written as follows using the Laplace transform variable $s = j\omega$

$$H_1(s) = \frac{s(2\zeta\omega_n - 1/\tau) + \omega_n^2}{s^2 + 2\zeta\omega_n s + \omega_n^2} \tag{3-140}$$

The active filter, after compensating for the phase reversal of the amplifier, provides a closed-loop transfer function of

$$H(s) = \frac{2\zeta\omega_n s + \omega_n^2}{s^2 + 2\zeta\omega_n s + \omega_n^2} \tag{3-141}$$

The parameters in Eqs. (3-140) and (3-141) are as defined in Fig. 3-23. ω_n is called the natural frequency of the system and ζ is called the damping constant. The two transfer functions are essentially the same if $\tau_2 \gg 1/K_m K_1 K_2$. The half-power

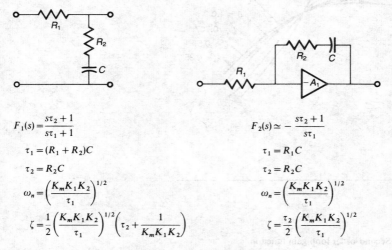

$$F_1(s) = \frac{s\tau_2 + 1}{s\tau_1 + 1}$$

$$\tau_1 = (R_1 + R_2)C$$

$$\tau_2 = R_2 C$$

$$\omega_n = \left(\frac{K_m K_1 K_2}{\tau_1}\right)^{1/2}$$

$$\zeta = \frac{1}{2}\left(\frac{K_m K_1 K_2}{\tau_1}\right)^{1/2}\left(\tau_2 + \frac{1}{K_m K_1 K_2}\right)$$

$$F_2(s) \simeq -\frac{s\tau_2 + 1}{s\tau_1}$$

$$\tau_1 = R_1 C$$

$$\tau_2 = R_2 C$$

$$\omega_n = \left(\frac{K_m K_1 K_2}{\tau_1}\right)^{1/2}$$

$$\zeta = \frac{\tau_2}{2}\left(\frac{K_m K_1 K_2}{\tau_1}\right)^{1/2}$$

Figure 3-23 Filter transfer functions for second-order PLLs.

bandwidth of the closed-loop transfer function is

$$W = \frac{\omega_n}{8\zeta}(1 + 4\zeta^2) \quad \text{Hz} \tag{3-142}$$

In control system terminology, the loop type is determined by the number of integrators contained within the loop. A first-order PLL is a type I servo. A second-order PLL with a high-gain active filter closely approximates a type II servo. A second-order loop with the passive filter is still a type I servo. The type of loop determines the capability of the loop to track changing signals. A type I servo can track a nonchanging signal with zero error but has an error proportional to the rate of change when tracking a changing signal. A type II servo can track nonchanging signals or signals changing at a constant rate with zero error but has an error proportional to the derivative of the rate of change for nonconstant rates of change.

The gain function of a second-order PLL as a function of ω_n and ζ is shown in Fig. 3-24. Typically the damping factor is in the range of $0.5 \leq \zeta \leq 1.5$ with the critical damping value of $\zeta = 0.707$ a common value.

The error signal in the time domain is

$$\theta_e(t) = \theta_{em}(t) + \theta_{en}(t) \tag{3-143}$$

and again consists of a modulation error term $\theta_{em}(t)$ and a noise error term $\theta_{en}(t)$. The mixer output noise has a spectral density $K_m^2 S_{n_c}(\omega)$ and the mean square

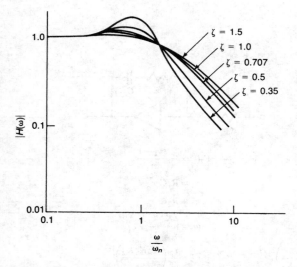

Figure 3-24 Second-order loop gain function.

value of the noise component is given by

$$\sigma_n^2 = \frac{1}{2\pi} \int_{-\infty}^{\infty} |H(\omega)|^2 \frac{S_{n_c}(\omega)}{A^2} \, d\omega \tag{3-144}$$

$$= \frac{N_0}{A^2} \left[\frac{1}{2\pi} \int_{-\infty}^{\infty} |\tilde{H}_{IF}(\omega)|^2 |H(\omega)|^2 \, d\omega \right]$$

$$= \frac{2N_0 W_L}{A^2} \tag{3-145}$$

where W_L is the loop noise bandwidth given by

$$W_L = \frac{1}{2\pi} \int_0^{\infty} |\tilde{H}_{IF}(\omega)|^2 |H(\omega)|^2 \, d\omega \tag{3-146}$$

The CNR in the loop bandwidth is

$$(\text{CNR})_{W_L} = \frac{A^2/2}{N_0 W_L} \tag{3-147}$$

and, therefore,

$$\sigma_n^2 = \frac{1}{(\text{CNR})_{W_L}} \tag{3-148}$$

Since $H(\omega)$ and, therefore, W_L, depend on loop gain, which in turn depends on carrier level, the power level for a given W_L must be specified. In most practical cases, $\tilde{H}_{IF}(\omega)$ is much wider than $H(\omega)$ and, therefore,

$$W_L \doteq \frac{1}{2\pi} \int_0^{\infty} |H(\omega)|^2 \, d\omega \tag{3-149}$$

The modulation-related term $\theta_{em}(t)$ in the phase error depends on the nature of the modulating waveform. If $m(t)$ is a sinusoid, then $\theta_{em}(t)$ will also be a sinusoid with amplitude $k_p|1 - H(\omega)|$, and if $H(\omega) = 1$ for the frequency of the sinusoid, this term will be zero.

If $m(t)$ were modeled as a Gaussian process with spectral density $S_m(\omega)$, then the mean square value of the modulation error term would be

$$\sigma_m^2 = \frac{k_p^2}{2\pi} \int_{-\infty}^{\infty} |1 - H(\omega)|^2 S_m(\omega) \, d\omega \tag{3-150}$$

and the total mean square error would be

$$\sigma_e^2 = \sigma_m^2 + \sigma_n^2 \tag{3-151}$$

By modeling the error as a Gaussian process, an estimate can be made as to how frequently the errors become large enough to cause the loop to depart from the small error assumption used to obtain the linearized equivalent. The probability that the phase error amplitude exceeds a magnitude θ_0 is given by

$$P_r[|\theta_e(t)| > \theta_0] = 1 - \int_{-\theta_0}^{\theta_0} \frac{1}{\sigma_e\sqrt{2\pi}} \exp\left[\frac{-(x-m)^2}{2\sigma_e^2}\right] dx \tag{3-152}$$

$$= Q\left(\frac{\theta_0 - m}{\sigma_e}\right) + Q\left(\frac{\theta_0 + m}{\sigma_e}\right) \tag{3-153}$$

where $Q(\alpha)$ is the area under the normalized probability density function from α to ∞, as discussed previously and m is the mean value of $\Theta_e(t)$. Choosing $\theta_e \simeq \sin \theta_e$ as the criterion for linear operation, then θ_0 can be taken as $\pi/6$ ($= 30$ degrees) and the probabilities computed from Eq. (3-153). The results of this computation are shown in Fig. 3-25 as $Pr[|\Theta_e| < 30^\circ]$. These probabilities can be interpreted as the fraction of the time that the PLL is within the linear range. It is seen from Fig. 3-25 that increasing m, the mean modulation error, increases the probability of exceeding the linear range. If $m = 0$, only noise determines the probability of staying in phase lock. For this case,

$$Pr[|\theta_e(t)| \geq 30^\circ] = 2Q\left(\frac{\pi/6}{\sigma_e}\right) \tag{3-154}$$

If this probability is limited to 0.05, then

$$\frac{\pi/6}{\sigma_e} \geq 2$$

$$\sigma_e^2 \leq \left(\frac{\pi}{12}\right)^2 \tag{3-155}$$

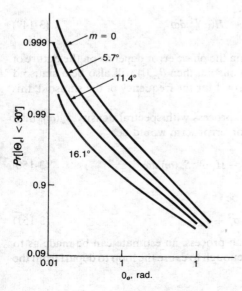

Figure 3-25 Probability of PLL not exceeding filter. (a) Block diagram: (b) binary

For a deterministic modulation with zero modulation tracking error ($m = 0$), the total tracking error variance is $\sigma_e^2 = \sigma_n^2$ and from Eq. (3-148) it follows that

$$(\text{CNR})_{W_L} \geq \frac{1}{\sigma_n^2} = \left(\frac{12}{\pi}\right)^2 \rightarrow 11.6 \text{ dB} \qquad (3\text{-}156)$$

For random modulation this becomes

$$(\text{CNR})_{W_L} \geq \frac{1}{(\pi/12)^2 - \sigma_m^2} \qquad (3\text{-}157)$$

which corresponds to a larger $(\text{CNR})_{W_L}$.

These requirements on $(\text{CNR})_{W_L}$ correspond to a type of threshold condition on the CNR in the loop bandwidth. Tracking demodulators can operate effectively with much less carrier power than standard demodulators for wide-band signals.

Tracking Loop Design

Assume that a sinusoidal modulating signal is employed. Such a signal can be represented as

$$\theta(t) = k_p \sin(\omega_m t + \psi_m) \qquad (3\text{-}158)$$

where ψ_m is a random phase angle uniformly distributed over $0\text{-}2\pi$. The power spectrum of this signal consists of impulses at $\pm \omega_m$. The mean square tracking error will be

$$\sigma_e^2 = \frac{1}{2} |1 - H(\omega_m)|^2 k_p^2 + \sigma_n^2 \qquad (3\text{-}159)$$

For a second-order tracking loop,

$$|1 - H(\omega_m)|^2 = \left| \frac{(\omega_m/\omega_n)^2}{-\left(\dfrac{\omega_m}{\omega_n}\right)^2 + j\dfrac{2\zeta\omega_m}{\omega_n} + 1} \right|^2 \qquad (3\text{-}160)$$

and

$$\sigma_n^2 = \frac{N_0 W_L}{A^2/2} = \frac{N_0}{A^2} \left[\frac{\omega_n}{4\zeta} (1 + 4\zeta^2) \right] \qquad (3\text{-}161)$$

It is seen that the mean square error depends on ω_n and ζ. Assuming that ω_m is well within the loop bandwidth so that $\omega_m \ll \omega_n$, it follows that

$$\sigma_e^2 \simeq \left(\frac{\omega_m}{\omega_n}\right)^4 \frac{k_p^2}{2} + \frac{N_0}{A^2} \left[\frac{\omega_n}{4\zeta} (1 + 4\zeta^2) \right] \qquad (3\text{-}162)$$

The mean square error can now be minimized with respect to ω_n and ζ by setting the partial derivatives of Eq. (3-162) with respect to these variables equal to zero and solving the equations simultaneously. Thus

$$\frac{d\sigma_e^2}{d\zeta} = \frac{N_0 \omega_n}{4}\left(-\frac{1}{\zeta^2}+4\right) = 0 \tag{3-163}$$

$$\frac{d\sigma_e^2}{d\omega_n} = \frac{\omega_m^4 k_p^2}{2}\left(\frac{-4}{\omega_n^5}\right) + \frac{N_0}{A^2 4\zeta}(1+4\zeta^2) = 0 \tag{3-164}$$

The optimum values of ζ and ω_n are found to be

$$\zeta = \frac{1}{2} \tag{3-165}$$

$$\omega_n = \left(\frac{2k_p^2 \omega_m^4 A^2}{N_0}\right)^{1/5} \tag{3-166}$$

The loop damping coefficient $\zeta = 1/2$ is straightforward to implement. However, the loop natural frequency is strongly dependent on the signal parameters ω_m and k_p.

Consider now the case of random modulation. Let $\theta(t)$ be a Gaussian random process with zero mean and power spectrum $S_m(\omega)$. The mean square error is

$$\sigma_e^2 = \frac{1}{2\pi}\int_{-\infty}^{\infty}\left[|1-H(\omega)|^2 S_m(\omega) + |H(\omega)|^2\frac{N_0}{A^2}\right]d\omega \tag{3-167}$$

This cannot be integrated in closed form and must be solved using the calculus of variations. The resulting optimum $H(\omega)$ is

$$H_{op}(\omega) = \frac{S_m(\omega)}{S_m(\omega) + (N_0/A^2)} \tag{3-168}$$

$H_{op}(\omega)$ is seen to be the Wiener filter in unrealizable (noncausal) form and it must be converted to the realizable form for incorporation into the PLL.

The transient performance of phase-lock loops is often of interest in estimating how they behave when signals are suddenly applied or removed. As an example consider the PLL of Fig. 3-20 with the following parameters:

$$K_m = K_2 = 1$$

$$F(\omega) = 1 \qquad |\omega| \le 8000\pi$$

$$= 0 \qquad |\omega| > 8000\pi$$

$$K_1 = 180$$

Let the input signal be $s(t) = A\cos[\omega_0 t + \theta(t)]$ where

$$\theta(t) = 30\pi t + 10\pi \qquad t \ge 0$$

$$= 0 \qquad t < 0$$

It is now desired to determine the value of carrier amplitude A required to keep the final steady-state phase error $\theta_e(\infty)$ less than $\pi/6$ radians. Using the variable s, the phase error can be expressed as

$$\Theta_e(s) = \Theta(s)[1 - H(s)]$$

$$= \Theta(s)\left[1 - \frac{AKF(s)}{s + AKF(s)}\right]$$

$$= \frac{\Theta(s)s}{s + AKF(s)} \tag{3-169}$$

The Laplace transform of the input is

$$\Theta(s) = \mathscr{L}\{(30\pi t + 10\pi)u(t)\}$$

$$= \frac{30\pi}{s^2} + \frac{10\pi}{s}$$

Using the final value theorem of the Laplace transform, $\theta_e(\infty)$ is found to be

$$\theta_e(\infty) = \lim_{s \to 0} s\Theta_e(s) \tag{3-170}$$

$$= \lim_{s \to 0} \frac{s^2}{s + AKF(s)}\left(\frac{30\pi}{s^2} + \frac{10\pi}{s}\right)$$

$$= \frac{30\pi}{AK} \leq \pi/6$$

$$A \geq \frac{30 \times 6}{K} = \frac{180}{180} = 1$$

A smaller tracking error would require a larger carrier or a larger loop gain. A great deal of work has been done on the theoretical analysis of phase-lock loops. However, actual designs are frequently optimized by empirical procedures. For example, it has been found experimentally that the best performance results from having the loop bandwidth larger than the Carson bandwidth while employing an input filter with a bandwidth just sufficient to pass the modulated signal. Much practical information on loop design is available elsewhere.[6]

Zero-Crossing Detector

The rate of occurrence of zero crossings in an oscillatory signal is directly related to the frequency of the signal and can be used as a means of frequency measurement. Consider the FM signal:

$$s(t) = A \cos\left[\omega_c t + k_f \int_{t_0}^{t} m(\lambda)\, d\lambda\right] \tag{3-171}$$

Let $m(t)$ be a periodic ramp such that

$$m(t) = at \qquad 0 < t < T \tag{3-172}$$

$$m(t) = m(t - T) \qquad \text{for all } t \tag{3-173}$$

The total phase angle of the modulated wave is then

$$\theta(t) = \omega_c t + \frac{k_f a}{2} t^2 \tag{3-174}$$

The FM waveform is as shown in Fig. 3-26.
The first zero crossing is at t_1 and the second is at $t_2 = t_1 + \Delta t$. This corresponds to

$$\theta(t_2) - \theta(t_1) = \pi \tag{3-175}$$

Assuming that the carrier frequency is much higher than the bandwidth of the modulation, $m(t)$ will not change appreciably in the interval Δt. Therefore,

$$\theta(t + \Delta t) - \theta(t) = \pi = \omega_c(t + \Delta t - t) + k_f \int_t^{t+\Delta t} m(\lambda)\, d\lambda \tag{3-176}$$

$$\doteq \omega_c \Delta t + k_f at \Delta t$$

$$= [\omega_c + k_f m(t_1)] \Delta t \tag{3-177}$$

The first factor on the right-hand side of Eq. (3-177) corresponds to the instantaneous angular frequency; that is,

$$[\omega_c + k_f m(t)] = \frac{d\theta(t)}{dt} = \omega_i$$

Therefore,

$$\omega_i = \omega_c + k_f m(t) \doteq \frac{\pi}{\Delta t} \tag{3-178}$$

$$f_i \doteq \frac{1}{2\,\Delta t} \tag{3-179}$$

If only positive zero crossings are considered,

$$f_i \doteq \frac{1}{\Delta t}$$

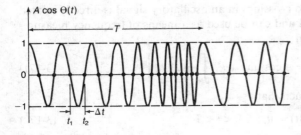

Figure 3-26 FM waveform.

The measurement technique is to count the number of zero crossings in a given interval T_c chosen so that

$$\frac{1}{f_c} < T_c \ll 1/B \tag{3-180}$$

The frequency is then found from n_c, the number of zero crossings in T_c, as

$$f_i = \frac{n_c}{T_c} \tag{3-181}$$

For example, if $f_c = 10$ MHz and $B = 20$ kHz, then a suitable T_c would be 1 μs and the frequency would be

$$f_i = n_c \qquad \text{MHz} \tag{3-182}$$

A procedure for implementing this measurement scheme would be the following.

1. Half-wave rectify $m(t)$.
2. Differentiate to accentuate the zero crossings.
3. Rectify again to eliminate the negative pulses.
4. Pass the resulting sequence of positive pulses into a low-pass filter.

The output of the low-pass filter would be a direct measure of $f_i = f_c + k_f m(t)/2\pi$. By using a balanced device, it would be possible to obtain $m(t)$ directly.

3-3 REFERENCES

1. Davenport, W. B., Jr. and W. L. Root: *Introduction to Random Signals and Noise*, McGraw-Hill Book Company, New York, 1958.
2. Middleton, D.: *An Introduction to Statistical Communication Theory*, McGraw-Hill Book Company, New York, 1960.
3. Rice, S. O.: Noise in FM Receivers, *Proc. Symp. Time Series Analysis*, M. Rosenblatt (ed.), John Wiley & Sons, Inc., New York, 1963.
4. Lawson, J. L., and G. E. Uhlenbeck: *Threshold Signals*, McGraw-Hill Book Company, New York, 1950.
5. Buren, R. S.: A Dynamic Noise Filter, *J. Audio Eng. Soc.*, vol. 19, no. 2, February 1971.
6. Gardner, F. M.: *Phaselock Techniques*, 2nd ed., John Wiley & Sons, Inc., New York, 1979.

3-4 PROBLEMS

3.1-1 A DSB/SC waveform of the following form is received:

$$s(t) = Am(t) \cos \omega_c t$$

where $A = 2 \times 10^{-6}$ volts and $m(t)$ is a sample function of a zero mean Gaussian random process having a spectral density

$$S_m(f) = 5 \times 10^{-3} \text{ rect}(f/10^4) \quad \text{V}^2/\text{Hz}$$

and the receiver noise has a flat spectral density

$$S_n(f) = 5 \times 10^{-14} \quad \text{V}^2/\text{Hz}$$

(a) Find the output SNR for an ideal product demodulator followed by a low-pass filter with an equivalent noise bandwidth of 5 kHz.

(b) Find the output SNR assuming the same product demodulator as in (a) but with a local oscillator phase error of 30 degrees.

3.1-2 Let the reference signal for a coherent AM-DSB product demodulator have a phase error of θ. Derive an expression for the output signal-to-noise ratio in terms of "input" SNR, where the "input bandwidth" is defined to be twice the equivalent noise bandwidth of the output low-pass filter.

3.1-3 Let the reference signal for a coherent SSB/SC product demodulator have a phase error of θ. What is the largest value of θ that is permissible to keep the rms distortion less than 10 percent, if the distortion is assumed to come from the Hilbert transform of the message?

3.1-4 A paging system transmits a carrier that is amplitude modulated with a 1000-Hz sine wave. The degree of modulation is 50 percent. The receiver has a block diagram as shown below:

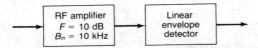

What received signal power (carrier plus sidebands) is required to achieve an output SNR of 20 dB?

3.1-5 Repeat Problem 3.1-4 assuming the detector is a square-law envelope detector.

3.1-6 The RF amplifier in Problem 3.1-4 is assumed to have an infinite-order Butterworth transfer function. The detector is then followed by a bandpass filter (tuned to 1000 Hz) that has a first-order Butterworth transfer function and a half-power bandwidth of 100 Hz. What does the output SNR become for the signal power calculated in Problem 3.1-4?

3.1-7 For the receiver of Problem 3.1-6, sketch (without computing numerical values) a curve of output SNR as a function of B_n, the bandwidth of the RF amplifier. Explain the shape of the curve.

3.1-8 A zero-mean Gaussian message has a two-sided spectral density of

$$S_m(f) = 10^{-4}\left[1 - \frac{|f|}{4000}\right] \quad |f| \le 4000$$

$$= 0 \quad |f| > 4000$$

If this message is to be transmitted by AM with full carrier, what value of P_R/N_0 is required at the receiver to achieve an output SNR of 20 dB with a linear envelope detector?

3.1-9 Repeat Problem 3.1-8 assuming the transmission is SSB/SC and the carrier is reinserted at the receiver with the proper phase.

3.1-10

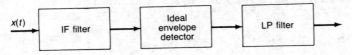

In the above AM demodulator, both filters have infinite-order Butterworth responses and the output filter has a bandwidth of 4000 Hz. The input signal plus noise is

$$x(t) = 4 \times 10^{-4}[1 + m(t)]\cos(3 \times 10^6 t) + n(t)$$

where $n(t)$ is white noise with a one-sided spectral density of 10^{-13} V²/Hz. The message, $m(t)$, is band-limited to 4000 Hz. Given that the output SNR and the SNR into the envelope detector are both to be 100, find the power in $m(t)$ and the bandwidth of the IF filter needed to achieve this condition.

3.1-11 Prove the following theorem regarding Hilbert transforms. If $s_1(t)$ and $s_2(t)$ have band-limited spectra that are nonoverlapping and the spectrum of $s_2(t)$ lies above that of $s_1(t)$, then the Hilbert transform of the product $s_1(t)s_2(t)$ is equal to $s_1(t)\hat{s}_2(t)$.

3.1-12 Show that the following properties of Hilbert transforms are true.

(a) $\displaystyle \int_{-\infty}^{\infty} x^2(t)\, dt = \int_{-\infty}^{\infty} \hat{x}^2(t)\, dt$

(b) $\displaystyle \int_{-\infty}^{\infty} x(t)\hat{x}(t)\, dt = 0$

3.1-13 Derive a mathematical expression for the linear modulation $m(t)$ and its Hilbert transform $\hat{m}(t)$ in terms of $x_{\text{USB}}(t)$ and $\hat{x}_{\text{USB}}(t)$. Repeat for $x_{\text{LSB}}(t)$ and $\hat{x}_{\text{LSB}}(t)$. Using these results, make a block diagram of a receiver for demodulating SSB signals.

3.1-14 Investigate the conditions under which an SSB signal plus a carrier component could be demodulated by an ideal envelope detector. Let the SSB wave be represented by

$$s(t) = A \cos \omega_c t + m(t) \cos \omega_c t - \hat{m}(t) \sin \omega_c t$$

3.1-15 Using the Hilbert transform representation for an SSB wave, show that single-tone modulation results in a single spectral component.

3.1-16 Show how to obtain a single-sideband wave by use of a network that produces a phase retardation of α rad at all positive frequencies. Assume that $0 < \alpha < \pi/2$ and that the *amplitude of the signal input* and the *phase of the carrier oscillator* output can be adjusted as desired.

3.1-17 An upper-sideband SSB signal has a reinserted carrier that has a phase shift relative to the original carrier so that the resulting waveform becomes

$$s(t) + A \cos(\omega_c t + \theta) + m(t) \cos \omega_c t - \hat{m}(t) \sin \omega_c t$$

(a) Write an exact expression for the envelope of $s(t)$.

(b) Assuming that $A \gg |m(t)|$, simplify the expression in part (a) by dropping the insignificant terms.

(c) Find the signal-to-distortion ratio in decibels. (Do not include dc terms.) (*Hint*: For small x, $(1 + x)^{0.5} \simeq 1 + 0.5x$.)

3.2-1 A zero-mean Gaussian message has a two-sided spectral density of

$$S_m(f) = 10^{-4}\left(1 - \frac{|f|}{4000}\right) \qquad |f| \leq 4000$$

$$= 0 \qquad |f| > 4000$$

This message is used to frequency-modulate a carrier with a frequency deviation constant of 10^5 rad/V-s.

(a) Find the bandwidth of the modulated signal using Carson's rule. Assume that the peak value of the Gaussian process is 3.3 times its standard deviation.

(b) Assuming the signal is demodulated in a receiver having an equivalent noise bandwidth equal to that calculated in (a), what output SNR is obtained when the input SNR is 10 dB?

3.2-2 The message of Problem 3.2-1 is used in a narrow-band FM system for which $k_f \sigma_m = 4000$.

(a) Find the half-power bandwidth of the resulting signal using the narrow-band FM spectrum approximation.

(b) Using the results of (a) as the equivalent noise bandwidth for the receiver, compute the output SNR when the input SNR is 10 dB.

3.2-3 A commercial broadcast FM receiver operating at 100 MHz requires a 20-μV signal across its input resistance of 300 ohms to provide an output SNR of 70 dB. Find the receiver noise figure. Assume $P_m = 1$, the input impedance is matched to the source, and all components are at 290K.

3.2-4 An FM wave

$$s(t) = A \cos\left(\omega_c t + \frac{\omega_d}{\omega_1} \sin \omega_1 t\right) \qquad \omega_c \gg \omega_1$$

is added to a delayed replica of itself to form the resultant wave

$$y(t) = s(t) + ks(t - \tau)$$

Assume that the delay τ is sufficiently small to make the approximations $\sin \omega_1 \tau \simeq \omega_1 \tau$ and $\cos \omega_1 \tau \simeq 1$ valid.

    (a) What is the envelope of $y(t)$?
    (b) Evaluate the envelope when $\omega_c \tau = \pi/2$ and $\sin \omega_d \tau \simeq \omega_d \tau$.
    (c) What change would be made in the answer to (b) if $s(t - \tau)$ were *subtracted* from $x(t)$?
    (d) What are the approximate results of (b) and (c) if $2k\tau\omega_d \ll 1$?
    (e) Use the results (a) through (d) to design a frequency detector giving a response proportional to the instantaneous frequency deviation of the input wave.

3.2-5 Show how the system shown in the accompanying figure can operate as an FM demodulator.

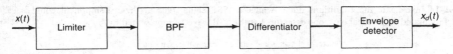

3.2-6 A zero-mean Gaussian message having a spectral density

$$S_m(f) = 10^{-5}\left(1 - \frac{|f|}{4000}\right)\mathrm{rect}\left(\frac{f}{4000}\right)$$

is used to modulate a carrier with a frequency deviation constant of $f_\Delta = k_f/2\pi = 2 \times 10^5$. This signal is demodulated by a receiver having a bandwidth equal to the (modified) Carson's bandwidth of the signal. Find the output SNR assuming an input CNR of 10 dB in the IF bandwidth, an ideal FM discriminator, and a demodulator output filter having a rectangular bandwidth equal to the message bandwidth.

3.2-7 Repeat Problem 3.2-6 assuming phase modulation is used with a phase deviation constant of 10 rad/V and the receiver noise bandwidth equals the rms bandwidth of the signal.

3.2-8 A voice signal normalized to a peak value of 1 volt, having a bandwidth of 4 kHz and variance ($\sigma_m^2 = 0.1$), is to be transmitted by an RF carrier having an allowable bandwidth of 1 MHz. The RF carrier power is sufficient to produce a 15-dB CNR in the 1-MHz bandwidth and correspondingly more in a smaller bandwidth. Compare the output SNR obtainable with three modulation methods: (a) AM, (b) SSB/SC, and (c) FM. Assume all filter bandwidths optimized for the modulation used.

3.2-9 A zero-mean Gaussian random process $m(t)$ is used to frequency-modulate a carrier with a frequency deviation constant of 10^5 Hz/V. $m(t)$ has a uniform spectral density of 10^{-4} V^2/Hz in the interval $|f| \leq 4$ kHz and zero elsewhere. Find the rms bandwidth of the modulated signal.

3.2-10 The signal in Problem 3.2-9 is received with a carrier amplitude of 1×10^{-6} volts and the receiver noise spectral density is 4×10^{-20} V^2/Hz. Find the output SNR in decibels when this signal is demodulated in a frequency discriminator.

3.2-11 A preemphasis/deemphasis system is to be designed to work with noise having a spectral density

$$S_n(f) = \exp(5 \times 10^{-4}|f|) \quad \text{V}^2/\text{Hz}$$

The deemphasis filter is designed to whiten the noise spectral density over the range of 0–15 kHz. Find the power transfer function of the preemphasis filter required to prevent signal distortion. What SNR improvement can be obtained with this system as compared with a system with no preemphasis/deemphasis?

3.2-12 Preemphasis/deemphasis filters are to be used in a PM system. The filter functions are

$$H_{pe}(f) = \frac{1}{H_{de}(f)} = 1 + j(f/f_0)$$

and extend over the frequency interval $|f| \leq W$. Find the improvement in output SNR obtained with this system as compared with one without preemphasis/deemphasis. Compute the improvement that would be obtained in stereo broadcasting; that is, let $W = 15$ kHz and $f_0 = 2.1$ kHz.

3.2-13 An equivalent baseband phase-tracking loop, such as shown in Fig. 3-19 in the text, has the following parameters.

$$K_m = K_2 = 2$$
$$F(\omega) = 1 \quad |\omega| \leq 20,000\pi$$
$$= 0 \quad |\omega| > 20,000\pi$$
$$K_1 = 200$$

If the modulated signal is

$$s(t) = A \cos[\omega_0 t + \theta(t)]$$

and

$$\theta(t) = 100\pi t + 20\pi \quad t \geq 0$$
$$= 0 \quad t < 0$$

what value of A is required to keep the final steady-state phase error, $\theta_e(\infty)$, less than $\pi/6$?

3.2-14 For the loop shown in Fig. 3-17 and the parameters of Problem 3.2-13 assume that the signal is frequency-modulated with a frequency deviation constant of 10^5 rad/s/V. If the message is band-limited to 10 kHz, what value of A is required to achieve an output SNR of 30 dB when $N_0 = 10^{-4}$ V^2/Hz?

3.2-15 Repeat Problem 3.2-13 assuming the phase-lock loop of Fig. 3-20 is used.

3.2-16 Repeat Problem 3.2-14 assuming the phase-lock loop of Fig. 3-20 is used.

3.2-17 It is observed that a first-order phase-lock loop ($K_m = K_1 = 1$) that has a 15-kHz bandwidth develops a steady-state error of $\pi/100$ rad when an unmodulated sinusoid having a frequency deviation of 15 kHz from the carrier is applied. When the same signal with no frequency offset is applied in the presence of receiver noise having a one-sided spectral density of $N_0 = 4 \times 10^{-13}$ V^2/Hz, the phase error has variance of $\pi/10$ $(rads)^2$. Find the carrier power.

3.2-18 A PLL is operating in the presence of white Gaussian noise having a one-sided spectral density of $N_0 = 10^{-6}$ V^2/Hz in the RF band. It is desired to operate the loop so that the probability of noise producing a phase error of 30 degrees is less than 0.001. Assuming that no signal-tracking error is present, what carrier amplitude would be required for a loop with a natural frequency of 6 kHz and a damping constant of 0.707?

3.2-19 A passive filter that is often used in phase-lock loops is of the form shown. Calculate the transfer function for this filter. Show that the PLL transfer function of Fig. 3-20 can be expressed as

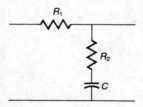

$$H(s) = \frac{\theta_v(s)}{\theta(s)} = \frac{AK(s\tau_2 + 1)/\tau_1}{s^2 + s(1 + AK\tau_2)/\tau_1 + AK/\tau_1}$$

where

$$\tau_1 = (R_1 + R_2)C$$
$$\tau_2 = R_2 C$$

3.2-20 Show that the transfer function in Problem 3.2-19 can be put into the form

$$H(s) = \frac{s(2\zeta\omega_n - \omega_n^2/AK) + \omega_n^2}{s^2 + 2\zeta\omega_n s + \omega_n^2}$$

Define the quantities ω_n and ζ. For what condition does this transfer function closely approximate that obtained with an active filter of the following form?

$$F_2(s) = \frac{s\tau_2 + 1}{s\tau_1}$$

3.2-21 The phase-lock loop shown has parameters $F(s) = 10^3$ and $G(s) = 10^3/s$. Assuming that the input phase noise has a uniform spectral density of 10^{-8} rad^2/Hz, find the noise power at the output of the low-pass filter.

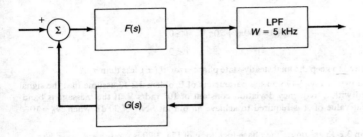

DETECTION OF BINARY SIGNALS AT BASEBAND

4-1 OPTIMUM DETECTION METHODS

Statistical decision theory provides the mathematical basis for designing optimum detection procedures for signals embedded in noise. A receiver for binary signals is required to make a decision on the basis of the available waveform (signal plus noise) as to which of two possible signals $s_1(t)$ or $s_0(t)$ was sent. The uncertainty is caused by the addition of noise during transmission through the channel. The actual received waveform is

$$r(t) = s_i(t) + n(t) \qquad i = 0, 1 \qquad (4\text{-}1)$$

To understand the underlying concepts, it is helpful to consider the most elementary case of binary hypothesis testing. Consider two hypotheses:

$$H_0: \quad \text{source produces } s_0 \qquad (4\text{-}2)$$

$$H_1: \quad \text{source produces } s_1 \qquad (4\text{-}3)$$

After the signal goes through the channel to the receiver, there is noise added and the received quantity is

$$r = s_i + n \qquad i = 0, 1 \qquad (4\text{-}4)$$

The receiver must now decide on the basis of r whether H_1 or H_0 is true; that is, whether s_1 was sent or s_0 was sent. Several different bases for making such a decision are possible. Many are special cases of Bayes' rule, which is based on the following two underlying assumptions.

1. The source outputs are governed by probability assignments P_1 and P_0 called the *a priori* probabilities.
2. A cost is associated with each decision that is made.

The decision rule must be chosen to make decisions in such a manner as to minimize the average cost; that is, the cost resulting from repeating the process many times. Let

$$C_{00} = \text{cost of deciding } H_0 \text{ when } H_0 \text{ is true}$$

$$C_{10} = \text{cost of deciding } H_1 \text{ when } H_0 \text{ is true}$$

$$C_{11} = \text{cost of deciding } H_1 \text{ when } H_1 \text{ is true} \quad (4\text{-}5)$$

$$C_{01} = \text{cost of deciding } H_0 \text{ when } H_1 \text{ is true}$$

The expected value of the cost is called the risk R and is

$$R = C_{00}P_0 P_r[H_0|H_0] + C_{10}P_0 P_r[H_1|H_0] + C_{11}P_1 P_r[H_1|H_1] + C_{01}P_1 P_r[H_0|H_1] \quad (4\text{-}6)$$

where $P_r[H_j|H_k]$ is the probability of choosing H_j when H_k is true. To make a decision, the observation space Z (i.e., the space occupied by all possible values r can take on) must be partitioned into two separate regions:

$$Z_0 \text{—choose } H_0 \quad (4\text{-}7)$$

$$Z_1 \text{—choose } H_1 \quad (4\text{-}8)$$

where

$$Z_0 + Z_1 = Z \quad (4\text{-}9)$$

The probability of choosing H_0 is the probability that the received quantity r falls in Z_0 and the probability of choosing H_1 is the probability that r falls in Z_1. These probabilities can be computed from the conditional probability density functions and incorporated into the risk as

$$R = C_{00}P_0 \int_{Z_0} p(r|H_0)\, dr + C_{10}P_0 \int_{Z_1} p(r|H_0)\, dr$$

$$+ C_{11}P_1 \int_{Z_1} p(r|H_1)\, dr + C_{01}P_1 \int_{Z_0} p(r|H_1)\, dr \quad (4\text{-}10)$$

In general $C_{10} > C_{00}$ and $C_{01} > C_{11}$. Making use of this relationship between costs of correct versus incorrect decisions and the relationship in Eq. (4-9), the risk can be rewritten as

$$R = P_0 C_{00} \int_{Z_0} p(r|H_0)\, dr + P_0 C_{10} \int_{Z-Z_0} p(r|H_0)\, dr$$

$$+ P_1 C_{01} \int_{Z_0} p(r|H_1)\, dr + P_1 C_{11} \int_{Z-Z_0} p(r|H_1)\, dr \quad (4\text{-}11)$$

Now

$$\int_Z p(r|H_0)\, dr = \int_Z p(r|H_1)\, dr = 1 \quad (4\text{-}12)$$

so the risk becomes

$$R = P_0 C_{10} + P_1 C_{11}$$

$$+ \int_{Z_0} \{[P_1(C_{01} - C_{11})p(r|H_1)] - [P_0(C_{10} - C_{00})p(r|H_0)]\} \, dr \qquad (4\text{-}13)$$

The first two terms of Eq. (4-13) are called the "fixed cost" and do not depend on the choice of Z_0 and Z_1. The integral represents costs controlled by the partitioning of Z into Z_0 and Z_1 and it is what must be minimized. Each of the bracketed terms is positive and, therefore, the desired minimization can be obtained by assigning all values of r for which the second term is larger than the first term to region Z_0. In equation form this becomes

$$P_1(C_{01} - C_{11})p(r|H_1) \underset{H_0}{\overset{H_1}{\gtrless}} P_0(C_{10} - C_{00})p(r|H_0) \qquad (4\text{-}14)$$

or

$$\frac{p(r|H_1)}{p(r|H_0)} \underset{H_0}{\overset{H_1}{\gtrless}} \frac{P_0(C_{10} - C_{00})}{P_1(C_{01} - C_{11})} \qquad (4\text{-}15)$$

The quantity on the left-hand side of Eq. (4-15) is called the *likelihood ratio* and the decision process involves determining whether or not the likelihood ratio exceeds the (positive) quantity on the right-hand side. This is called a threshold test and can be written as

$$\Lambda(r) = \frac{p(r|H_1)}{p(r|H_0)} \underset{H_0}{\overset{H_1}{\gtrless}} \eta \qquad (4\text{-}16)$$

where $\Lambda(r)$ is the likelihood ratio and η is the threshold. Since logarithms are monotonic functions of their arguments, an equivalent test is

$$\ln[\Lambda(r)] \underset{H_0}{\overset{H_1}{\gtrless}} \ln(\eta) \qquad (4\text{-}17)$$

The decision procedure is illustrated in Fig. 4-1.

As an illustration of the decision process consider the following example. A measurement of a waveform is made to determine whether it consists of a signal plus noise or noise only. The two hypotheses are

$$H_0: \quad r = n \qquad (4\text{-}18)$$

$$H_1: \quad r = a + n \qquad (4\text{-}19)$$

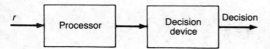

Figure 4-1 Optimum detection.

It is assumed that the additive noise n is a Gaussian random variable with zero mean and variance σ_n^2. The PDF of the noise is, therefore,

$$p(n) = \frac{1}{\sqrt{2\pi}\sigma_n} e^{-(n^2/2\sigma_n^2)} \tag{4-20}$$

The conditional PDFs required to compute the likelihood ratio are derived from Eq. (4-20) by recognizing that under H_0, r is the same as n; and under H_1, r is n plus the constant a. Thus

$$p(r|H_0) = p_n(r) = \frac{1}{\sqrt{2\pi}\sigma_n} e^{-r^2/2\sigma_n^2} \tag{4-21}$$

$$p(r|H_1) = p_n(r-a) = \frac{1}{\sqrt{2\pi}\sigma_n} e^{-[(r-a)^2/2\sigma_n^2]} \tag{4-22}$$

The likelihood ratio test then becomes

$$\Lambda(r) = \frac{\dfrac{1}{\sigma_n\sqrt{2\pi}} \exp[-(r-a)^2/2\sigma_n^2]}{\dfrac{1}{\sigma_n\sqrt{2\pi}} \exp[-r^2/2\sigma_n^2]} \underset{H_0}{\overset{H_1}{\gtrless}} \eta \tag{4-23}$$

$$\Lambda(r) = \exp\left[\frac{ar}{\sigma_n^2} - \frac{a^2}{2\sigma_n^2}\right] \underset{H_0}{\overset{H_1}{\gtrless}} \eta \tag{4-24}$$

$$\ln \Lambda(r) = \frac{ar}{\sigma_n^2} - \frac{a^2}{2\sigma_n^2} \underset{H_0}{\overset{H_1}{\gtrless}} \ln \eta \tag{4-25}$$

$$r \underset{H_0}{\overset{H_1}{\gtrless}} \frac{\sigma_n^2}{a} \ln \eta + \frac{a}{2} \tag{4-26}$$

where, as before,

$$\eta = \frac{P_0(C_{10} - C_{00})}{P_1(C_{01} - C_{11})} \tag{4-27}$$

For most problems it is not possible to assign numerical values to the costs. A frequent assumption is that there is no cost associated with a correct decision and that incorrect decisions have equal costs; that is,

$$C_{00} = C_{11} = 0$$

$$C_{01} = C_{10} = 1$$

For this case,

$$\eta = \frac{P_0}{P_1}$$

Substituting this set of costs into the expression for the risk, Eq. (4-10), gives

$$R = P_0 \int_{Z_1} p(r|H_0) \, dr + P_1 \int_{Z_0} p(r|H_1) \, dr \tag{4-28}$$

The first term is the probability of choosing H_1 when H_0 is true and the second term is the probability of choosing H_0 when H_1 is true. Therefore, the risk R is equal to the total probability of making an error. It is this error probability that the *likelihood ratio test* (LRT) minimizes for this set of costs. The decision rule is

$$\Lambda(r) = \frac{p(r|H_1)}{p(r|H_0)} \mathop{\gtrless}_{H_0}^{H_1} \frac{P_0}{P_1} \tag{4-29}$$

This is called the *ideal observer* rule and minimizes the total error probability.

When the two *a priori* probabilities P_0 and P_1 are equal, as is generally true in digital communications, the threshold is unity and

$$\Lambda(r) \mathop{\gtrless}_{H_0}^{H_1} 1$$

$$\ln \Lambda(r) \mathop{\gtrless}_{H_0}^{H_1} 0$$

A receiver operating under these assumptions is called a *minimum probability of error receiver*.

Another approach to the decision process is to choose the hypothesis that has the highest probability of being true based on the information contained in the received signal. This is called a *maximum a posteriori probability (MAP) receiver*. For the binary case, the decision rule is

$$Pr[H_1|r] \mathop{\gtrless}_{H_0}^{H_1} Pr[H_0|r] \tag{4-30}$$

But

$$Pr[H_i|r] = \frac{Pr[H_i]p(r|H_i)}{p(r)} \tag{4-31}$$

so the rule becomes

$$\frac{Pr[H_1]p(r|H_1)}{Pr[H_0]p(r|H_0)} = \frac{P_1 p(r|H_1)}{P_0 p(r|H_0)} \mathop{\gtrless}_{H_0}^{H_1} 1 \tag{4-32}$$

$$\frac{p(r|H_1)}{p(r|H_0)} \mathop{\gtrless}_{H_0}^{H_1} \frac{P_0}{P_1} \tag{4-33}$$

It is seen that Eq. (4-33) is the same decision rule as that of the minimum probability of error receiver. If the *a priori* probabilities are equal, this is called a *maximum likelihood (ML) receiver*.

4-2 BINARY SIGNAL DECODING AT BASEBAND

The binary encoded baseband signal consists of a sequence of pulses (or other waveforms) representing the binary digit 1 or 0.

$$1 \rightarrow s_1(t) \tag{4-34}$$

$$0 \rightarrow s_0(t) \tag{4-35}$$

It is assumed that each of these pulses is T seconds long and that the source bits occur independently. Further, it is assumed that the energies of $s_1(t)$ and $s_0(t)$ are equal and are given by

$$E_s = \int_0^T s_i^2(t)\, dt = P_m T \tag{4-36}$$

where P_m is the average signal power transmitted at baseband.

The channel is modeled as that of additive Gaussian noise and results in demodulated baseband noise having a spectral density of one of the following forms, depending on the modulation employed.

$$S_{nb}(\omega) = N_0 |\tilde{H}_{IF}(\omega)|^2 |H(\omega)|^2 \qquad \text{AM} \tag{4-37}$$

$$= \frac{N_0}{A^2} \omega^2 |\tilde{H}_{IF}(\omega)|^2 |H(\omega)|^2 \qquad \text{FM} \tag{4-38}$$

$$= \frac{N_0}{A^2} |\tilde{H}_{IF}(\omega)|^2 |H(\omega)|^2 \qquad \text{PM} \tag{4-39}$$

$$= \frac{N_0}{A^2} |H(\omega)|^2 \qquad \text{Phase tracking} \tag{4-40}$$

The random sequence of binary encoded signals can be written as

$$a(t) = \sum_{k=-\infty}^{\infty} a_k s_1(t - kT + t_0) + (1 - a_k)s_0(t - kT + t_0) \tag{4-41}$$

where

$$t_0 = \text{random time shift over } (0, T)$$

$$\{a_k\} = \text{independent sequence of binary random variables } 0, 1.$$

If the probability of a 1 is p, then the probability of a 0 is $(1 - p)$ and the statistics of a_k are

$$E\{a_k\} = p \tag{4-42}$$

$$E\{a_k^2\} = p \tag{4-43}$$

$$E\{a_k a_j\} = p^2 \qquad k \neq j \tag{4-44}$$

The autocorrelation function of $a(t)$ can then be written as

$$R_a(\tau) = E\{a(t)a(t+\tau)\}$$

$$= \sum_j \sum_k E\left\{\frac{1}{T} \int_0^T [a_k s_1(t - kT + t_0) + (1 - a_k)s_0(t - kT + t_0)]\right.$$

$$\left. \cdot [a_j s_1(t + \tau - jT + t_0) + (1 - a_j)s_0(t + \tau - jT + t_0)] \, dt_0\right\} \quad (4\text{-}45)$$

where the expectation is over a_k and the PDF of t_0 is $1/T$. Changing the variable of integration to

$$u = t_0 - kT + t \quad (4\text{-}46)$$

and using the statistics of a_k yields

$$R_a(\tau) = \sum_{k=-\infty}^{\infty} \left[\frac{p}{T} \int_{t-kT}^{t-(k-1)T} s_1(u)s_1(u+\tau) \, du\right.$$

$$\left. + \frac{1-p}{T} \int_{t-kT}^{t-(k-1)T} s_0(u)s_0(u+\tau) \, du\right]$$

$$+ \frac{1}{T} \sum_{\substack{k=-\infty \\ k \neq j}}^{\infty} \sum_{j=-\infty}^{\infty} \int_{t-kT}^{t-(k-1)T} s(u)s[u + \tau + (k-j)T] \, du \quad (4\text{-}47)$$

where

$$s(t) = ps_1(t) + (1-p)s_0(t) \quad (4\text{-}48)$$

The integrals extend over an interval T seconds long. The summation adds together all of the intervals and is equivalent to extending the limits of the integrals to $\pm \infty$. Rearranging indices, $R_a(\tau)$ can be written as

$$R_a(\tau) = \frac{p}{T} r_{11}(\tau) + \frac{1-p}{T} r_{00}(\tau) + \frac{1}{T} \sum_{\substack{j=-\infty \\ j \neq 0}}^{\infty} r_s(\tau + jT) \quad (4\text{-}49)$$

$$= \frac{p}{T} r_{11}(\tau) + \frac{1-p}{T} r_{00}(\tau) - \frac{1}{T} r_s(\tau) + \frac{1}{T} \sum_{j=-\infty}^{\infty} r_s(\tau + jT) \quad (4\text{-}50)$$

where

$$r_{11}(\tau) = \int_{-\infty}^{\infty} s_1(t)s_1(t+\tau) \, dt \quad (4\text{-}51)$$

$$r_{00}(\tau) = \int_{-\infty}^{\infty} s_0(t)s_0(t+\tau) \, dt \quad (4\text{-}52)$$

$$r_s(\tau) = \int_{-\infty}^{\infty} s(t)s(t+\tau) \, dt \quad (4\text{-}53)$$

Thus the autocorrelation function of the binary encoded baseband signal is determined by the time-correlation functions of the individual waveforms.

The spectral density of the binary modulated waveform can be obtained by taking the Fourier transform of Eq. (4-50), using the property that the Fourier transform of a time autocorrelation function is the magnitude squared of the Fourier transform of the time function.

$$S_a(\omega) = \frac{p(1-p)}{T} \left[|S_1(\omega) - S_0(\omega)|^2 \right]$$

$$+ \frac{2\pi}{T^2} \sum_{k=-\infty}^{\infty} \left| pS_1\left(\frac{2\pi k}{T}\right) + (1-p)S_0\left(\frac{2\pi k}{T}\right) \right|^2 \delta\left(\omega - \frac{2\pi k}{T}\right) \qquad (4\text{-}54)$$

where

$$S_1(\omega) = \mathscr{F}\{s_1(t)\}$$
$$S_0(\omega) = \mathscr{F}\{s_0(t)\}$$

The first term in Eq. (4-54) is a continuous spectrum determined by the spectra of the individual bit waveforms. The second term is a line spectrum resulting from the periodicity of the sequence.

The data bits are recovered from the encoded baseband signals by a filtering, sampling, and thresholding operation as shown in Fig. 4-2. It is assumed here that the received signal is passed through a linear filter before the decision is made. It can be shown that linear processing is optimum when the additive noise is Gaussian.[1] For other types of noise, linear processing may not be optimum but often is used because of its ease of implementation. The recovered baseband signal is processed during the bit time and a decision made as to which bit was transmitted based on the filter output voltage at the end of the processing time ($t = T$). This process is repeated periodically in synchronism with the bit sequence.

Assuming that the decoder is in perfect synchronism with the bit sequence, the signal to be decoded is

$$x(t) = s_i(t) + n_b(t) \qquad (4\text{-}55)$$

After filtering,

$$y(t) = \int_0^\infty h(\lambda)x(t-\lambda)\, d\lambda$$

$$= \int_0^\infty h(\lambda)s_i(t-\lambda)\, d\lambda + \int_0^\infty h(\lambda)n_b(t-\lambda)\, d\lambda \qquad (4\text{-}56)$$

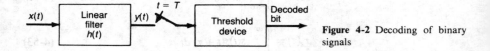

Figure 4-2 Decoding of binary signals

The sampled value at $t = T$ is

$$y(T) = \int_0^\infty h(\lambda) s_i(T - \lambda) \, d\lambda + \int_0^\infty h(\lambda) n_b(T - \lambda) \, d\lambda \qquad (4\text{-}57)$$

The first term in Eq. (4-57) is a constant depending on which signal, $s_0(t)$ or $s_1(t)$, is present. The second term is a Gaussian random variable since $n_b(t)$ is a Gaussian process. The mean of $y(t)$ depends on whether $s_0(t)$ or $s_1(t)$ was received. Calling $y_1(T)$ the value when $s_1(t)$ is received and $y_0(T)$ the value when $s_0(t)$ is received gives

$$m_1 = \text{mean of } y_1(T) = \int_0^\infty h(\lambda) s_1(T - \lambda) \, d\lambda \qquad (4\text{-}58)$$

$$m_0 = \text{mean of } y_0(T) = \int_0^\infty h(\lambda) s_0(T - \lambda) \, d\lambda \qquad (4\text{-}59)$$

The variance is the same for either bit and is

$$\sigma_y^2 = \frac{1}{2\pi} \int_{-\infty}^\infty |H(\omega)|^2 S_{nb}(\omega) \, d\omega \qquad (4\text{-}60)$$

The receiver must now make a decision on the basis of $y(T)$ as to which bit was sent. This is done using the hypothesis testing procedures described earlier. For the case of most importance in binary communication, the *a priori* probabilities are equal ($P_0 = P_1 = 1/2$) and the optimum receiver makes an LRT with unity threshold; that is,

$$\Lambda[y(T)] = \frac{p[y(T)|s = s_1]}{p[y(T)|s = s_0]}$$

$$= \frac{\dfrac{1}{\sigma_y\sqrt{2\pi}} \exp[-(y(T) - m_1)^2 / 2\sigma_y^2]}{\dfrac{1}{\sigma_y\sqrt{2\pi}} \exp[-(y(T) - m_0)^2 / 2\sigma_y^2]} \underset{s_0}{\overset{s_1}{\gtrless}} 1$$

After canceling common factors and taking logarithms, this becomes

$$(y(T) - m_0)^2 - (y(T) - m_1)^2 \underset{s_0}{\overset{s_1}{\gtrless}} 0 \qquad (4\text{-}61)$$

$$y(T) \underset{s_0}{\overset{s_1}{\gtrless}} \frac{m_0 + m_1}{2} \qquad \text{(assuming } m_1 > m_0) \qquad (4\text{-}62)$$

Whenever $y(T)$ exceeds the average of m_0 and m_1, the receiver decides $s_1(t)$ was received; otherwise it decides $s_0(t)$ was received.

An error is made whenever $s_1(t)$ was received but the threshold was not exceeded due to noise, or whenever $s_0(t)$ was received but the threshold was exceeded due to noise. The total probability of error is given by

$$P_e = P_0 Pr\left[y(T) > \frac{m_0 + m_1}{2} \,\bigg|\, s = s_0\right] + P_1 P_r\left[y(T) < \frac{m_0 + m_1}{2} \,\bigg|\, s = s_1\right] \qquad (4\text{-}63)$$

$$= \frac{1}{2} Q\left[\frac{\dfrac{m_0 + m_1}{2} - m_0}{\sigma_y}\right] + \frac{1}{2}\left\{1 - Q\left[\frac{\dfrac{m_0 + m_1}{2} - m_1}{\sigma_y}\right]\right\}$$

$$= Q\left[\frac{|m_1 - m_0|}{2\sigma_y}\right] \qquad (4\text{-}64)$$

The probability of error is thus determined by the distance between the means under the two hypotheses normalized by the standard deviation of the noise. To obtain the best performance, the filter $h(t)$ must be designed to maximize

$$d = \frac{|m_1 - m_0|}{2\sigma_y} \qquad (4\text{-}65)$$

For the case of the baseband noise being white with spectral density $N_{b0}/2$, it follows that

$$\sigma_y^2 = \frac{N_{b0}}{2}\left[\frac{1}{2\pi}\int_{-\infty}^{\infty}|H(\omega)|^2\,d\omega\right]$$

$$= \frac{N_{b0}}{2}\int_0^{\infty} h^2(t)\,dt \qquad (4\text{-}66)$$

$$d^2 = \frac{\left|\displaystyle\int_0^{\infty} h(\lambda)[s_1(T-\lambda) - s_0(T-\lambda)]\,d\lambda\right|^2}{2N_{b0}\displaystyle\int_0^{\infty} h^2(t)\,dt} \qquad (4\text{-}67)$$

Choosing $h(t)$ to maximize Eq. (4-67) is the same problem that was previously solved in connection with the matched filter where now the signal is $s_1(T-t) - s_0(T-t)$ and the noise spectral density is $2N_{b0}$. From the previous results, it follows that the optimum filter is

$$h(t) = s_1(T-t) - s_0(T-t) \qquad 0 < t < T \qquad (4\text{-}68)$$

$$d_{\max}^2 = \frac{2E}{N_0} = \frac{\displaystyle\int_0^T [s_1(T-t) - s_0(T-t)]^2\,dt}{2N_{b0}} \qquad (4\text{-}69)$$

Expanding the numerator of Eq. (4-69) gives

$$\int_0^T [s_1^2(T-t) - 2s_1(T-t)s_0(T-t) + s_0^2(T-t)]\,dt = E_{s_1} + E_{s_0} - 2\rho\sqrt{E_{s_1}E_{s_0}} \qquad (4\text{-}70)$$

where

$$\rho \triangleq \frac{\int_0^T s_1(t)s_0(t)\, dt}{\sqrt{E_{s_1}E_{s_0}}} \tag{4-71}$$

is the *correlation coefficient* and takes on values from $(-1, 1)$, depending on the similarity between $s_1(t)$ and $s_2(t)$. The error probability can then be found as:

$$d_{max} = \left[\frac{E_{s_1} + E_{s_0} - 2\rho\sqrt{E_{s_1}E_{s_0}}}{2N_{b0}}\right]^{1/2} \tag{4-72}$$

$$P_e = Q\left\{\left[\frac{E_{s_1} + E_{s_0} - 2\rho\sqrt{E_{s_1}E_{s_0}}}{2N_{b0}}\right]^{1/2}\right\} \tag{4-73}$$

For the case of equal signal energies $E_{s_1} = E_{s_0} = E_s$ this becomes

$$P_e = Q\left[\sqrt{\frac{E_s}{N_{b0}}(1-\rho)}\right] \tag{4-74}$$

It is evident from Eq. (4-74) that the ratio of the signal energy to the noise spectral density and the correlation between $s_1(t)$ and $s_0(t)$ determine the detection error performance. The case of equal signal energies given by Eq. (4-74) is of most importance since this gives the lowest error probability for a given average transmitted power.

The ratio E_s/N_{b0} depends on the carrier modulation system employed and is summarized in Table 4-1.

The optimum decoder has an impulse response of the form

$$h(t) = h_1(t) - h_0(t)$$

where

$$h_1(t) \triangleq s_1(T-t) \qquad 0 < t < T$$

$$h_0(t) \triangleq s_0(T-t) \qquad 0 < t < T$$

This is illustrated in Fig. 4-3.

Table 4-1 E_s/N_{b0} **for different carrier modulation systems**

Modulation system	E_s	N_{b0}	E_s/N_{b0}
AM	$\dfrac{A^2 P_m T}{4}$	$\dfrac{N_0}{2}$	$\dfrac{A^2 P_m T}{2N_0}$
PM	$k_p^2 P_m T$	$\dfrac{N_0}{A^2}$	$\dfrac{k_p^2 A^2 P_m T}{N_0}$
$N_0 = kT_0F$	$P_m = $ modulating waveform power		

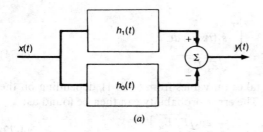

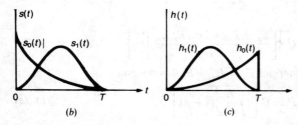

Figure 4-3 Binary signal matched filter. (*a*) Block diagram; (*b*) binary signals; (*c*) impulse responses.

The matched filtering operation can also be carried out with an active correlator as shown in Fig. 4-4. It should be emphasized, however, that the matched filter and the active correlator produce equivalent results only at $t = T$. At other times their output is not the same.

If a filter impulse response $\hat{h}(t)$ different from the optimum is used, the normalized distance in the probability of error expression given by Eq. (4-67) becomes

$$d = \frac{\left| \int_0^T \hat{h}(\lambda)[s_1(T - \lambda) - s_0(T - \lambda)] \, d\lambda \right|}{\left[2N_{b0} \int_0^T \hat{h}^2(t) \, dt \right]^{1/2}} \tag{4-75}$$

This can be put into the form

$$d = d_{\max} \left\{ \left| \int_0^T \frac{\hat{h}(\lambda)}{(E_{\hat{h}})^{1/2}} \frac{h(\lambda)}{(E_h)^{1/2}} \, d\lambda \right| \right\} \tag{4-76}$$

$$= d_{\max} \left\{ \left| \int_0^T \frac{s(t)}{(E_s)^{1/2}} \frac{\hat{s}(t)}{(E_{\hat{s}})^{1/2}} \, dt \right| \right\} \tag{4-77}$$

where $s(t)$ and $\hat{s}(t)$ are the signal waveforms to which $h(t)$ and $\hat{h}(t)$ are matched and E_s and E_s are their respective energies. This expression allows determination of performance when a suboptimum processor is used in the receiver.

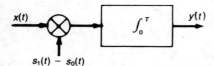

Figure 4-4 Correlator matched filter for binary signals.

4-3 SIGNAL SELECTION FOR BINARY SYSTEMS

The only limitations imposed on the binary signals considered so far are that they have equal energy and have durations of T seconds. The actual waveshapes entered only through the expression used to determine the correlation coefficient ρ. From the error expression Eq. (4-74),

$$P_e = Q\left\{\sqrt{\frac{E}{N_{b0}}(1-\rho)}\right\}$$

it is evident that the best performance is obtained for $\rho = -1$. The expression for ρ is

$$\rho = \frac{\int_0^T s_1(t)s_0(t)\,dt}{E}$$

and it is seen that this will be equal to -1 for

$$s_0(t) = -s_1(t) \tag{4-78}$$

Signal pairs having this property are called *antipodal signals* and for them

$$P_e = Q\left\{\sqrt{\frac{2E}{N_{b0}}}\right\} \tag{4-79}$$

For such signals the PE depends only on the signal energy and not the waveshape.

Another important class of signals is those having the property of being uncorrelated; that is,

$$\rho = \int_0^T s_1(t)s_0(t)\,dt = 0 \tag{4-80}$$

Such signals are called *orthogonal* and for them

$$P_e = \left\{Q\sqrt{\frac{E}{N_{b0}}}\right\} \tag{4-81}$$

By comparison of Eqs. (4-79) and (4-81) it is seen that twice the signal energy is required for orthogonal signals to achieve the same error performance as antipodal signals. The error performance for both these signals is shown in Fig. 4-5.

For antipodal signals $m_0 = -m_1$ so the threshold for the LRT is zero. Also, since the outputs of the two filters $h_1(t)$ and $h_0(t)$ are the negatives of each other, only one output needs to be computed and the processor would be as shown in Fig. 4-6. The bandwidth required for this type of signaling can be determined from the general expression derived earlier for the spectral density of a sequence of

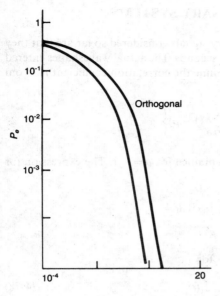

Figure 4-5 Probability of error for antipodal and orthogonal binary signals.

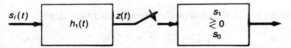

Figure 4-6 Simplified processor for antipodal signals.

binary waveforms. From Eq. (4-54) with $p = 1/2$:

$$S_a(\omega) = \frac{1}{4T} \left[|S_1(\omega) - S_0(\omega)|^2 \right]$$

$$+ \frac{2\pi}{T^2} \sum_{k=-\infty}^{\infty} \left| \frac{1}{2} S_1\left(\frac{2\pi k}{T}\right) + \frac{1}{2} S_0\left(\frac{2\pi k}{T}\right) \right|^2 \delta\left(\omega - \frac{2\pi k}{T}\right)$$

In the case of antipodal signals, $S_1(\omega) = -S_0(\omega)$, so the discrete spectrum is not present and

$$S_a(\omega) = \frac{1}{T} |S_1(\omega)|^2 \qquad (4\text{-}82)$$

Therefore, the only bandwidth requirement is that it be great enough to pass the signal waveforms without distortion. This minimizes the power requirements but makes synchronization more difficult. Components of the line spectrum can be introduced by forcing $p \neq 1/2$ through addition of extra 1's at prescribed intervals. This provides a synchronization component.

Some typical antipodal signals are shown in Fig. 4-7 along with the optimum filter impulse responses and correlation processor configurations. These signals can be put into more quantitative form as follows.

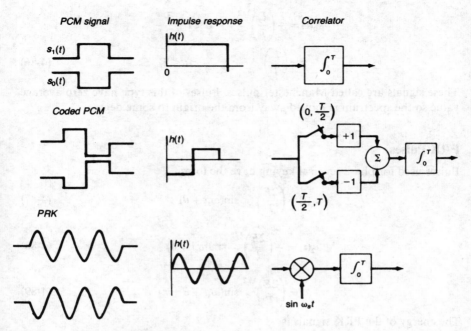

Figure 4-7 Typical antipodal signals.

PCM Pulses

Let the two signals be

$$s_1(t) = \left(\frac{E}{T}\right)^{1/2} \qquad 0 \le t \le T \qquad (4\text{-}83)$$

$$s_0(t) = -\left(\frac{E}{T}\right)^{1/2} \qquad 0 \le t \le T \qquad (4\text{-}84)$$

Signals of this type are referred to as *bipolar* or nonreturn-to-zero (NRZ) pulses. The decoder for pulses of this type is called an *integrate-and-dump* decoder.

Coded PCM Pulses

A typical set of these pulses is of the form

$$
\begin{aligned}
s_1(t) &= \left(\frac{E}{T}\right)^{1/2} & 0 < t \le \frac{T}{2} \\[2mm]
&= -\left(\frac{E}{T}\right)^{1/2} & \frac{T}{2} < t \le T
\end{aligned}
\qquad (4\text{-}85)
$$

$$s_0(t) = -\left(\frac{E}{T}\right)^{1/2} \qquad 0 < t \leq \frac{T}{2}$$

$$= \left(\frac{E}{T}\right)^{1/2} \qquad \frac{T}{2} < t \leq T \tag{4-86}$$

These signals are called Manchester pulses. Pulses of this type have zero average value so the spectrum is moved away from the origin to some degree.

PRK Pulses

Pulses used for phase reversal keying have the form

$$s_1(t) = \left(\frac{2E}{T}\right)^{1/2} \sin(\omega_s t + \theta) \tag{4-87}$$

$$s_2(t) = -\left(\frac{2E}{T}\right)^{1/2} \sin(\omega_s t + \theta) \tag{4-88}$$

$$= \left(\frac{2E}{T}\right)^{1/2} \sin(\omega_s t + \theta \pm \pi) \tag{4-89}$$

The energy of the PRK signals is

$$\frac{2E}{T} \int_0^T \sin^2(\omega_s t + \theta)\, dt = \frac{2E}{T}\left[\frac{T}{2} - \frac{1}{2}\int_0^T \cos(2\omega_s t + 2\theta)\, dt\right]$$

$$= E \qquad \text{if } \omega_s \text{ is harmonically related to } 2\pi/T$$

$$\simeq E \qquad \text{if } T \gg 2\pi/\omega_s \tag{4-90}$$

A major problem is establishing the correct phase for use in the processor. Assume that a phase angle $\hat{\theta}$ different from the true phase θ is employed by the processor. This corresponds to using a mismatched filter and from Eq. (4-77) the performance parameter d is given by

$$d = d_{\max}\left[\left|\int_0^T \frac{\sin(\omega_s t + \theta)}{(T/2)^{1/2}} \cdot \frac{\sin(\omega_s t + \hat{\theta})}{(T/2)^{1/2}}\, dt\right|\right] \tag{4-91}$$

$$= \left(\frac{2E}{N_{b0}}\right)^{1/2}\left[\frac{2}{T}\int_0^T \frac{\cos(\theta - \hat{\theta})}{2}\, dt\right] \tag{4-92}$$

$$= \left(\frac{2E}{N_{b0}}\right)^{1/2} \cos\theta_e \tag{4-93}$$

$$P_e = Q\left[\left(\frac{2E}{N_{b0}}\right)^{1/2} \cos\theta_e\right] \tag{4-94}$$

where

$$\theta_e = \theta - \hat{\theta} \tag{4-95}$$

4-4 EFFECTS OF BIT TIMING ERRORS FOR ANTIPODAL SIGNALS

Assume that the actual timing of the pulse sequence and that assumed by the decoder are as shown in Fig. 4-8. The decoder operates on the assumption that the timing of each pulse arrival is correct. Since the waveshapes in adjacent intervals differ only in polarity, the output of the matched filter at decoder time T for signal only can be expressed as

$$z_1(T, \varepsilon) = \pm \int_0^{T-\varepsilon} s(t+\varepsilon)s(t)\, dt \pm \int_{T-\varepsilon}^{T} s(t+\varepsilon - T)s(t)\, dt \tag{4-96}$$

The first $\pm$ depends on the bit in the interval $(0, T)$ and the second $\pm$ depends on the bit in the interval $(T, 2T)$. Let

$$\rho(\tau) = \frac{1}{E}\int_0^T s(t+\tau)s(t)\, dt$$

be the normalized autocorrelation function for the periodic signal $s(t)$ consisting of replications of the bit waveforms. Then

$$z_1(T, \varepsilon) = \pm \int_0^T s(t+\varepsilon)s(t)\, dt = \pm E\rho(\varepsilon) \qquad \text{bits same}$$

$$= \pm \left\{ E\rho(\varepsilon) - 2\left|\int_{T-\varepsilon}^{T} s(t+\varepsilon - T)s(t)\, dt\right| \right\} \quad \text{bits different} \tag{4-97}$$

It is seen from this equation that bit timing errors cause a reduction in the matched filter output that is equivalent to a reduction in effective signal energy. The actual magnitude of this reduction depends on the sign of the adjacent bit. By averaging

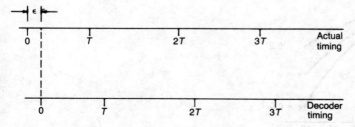

Figure 4-8 Decoder timing errors.

over the PDF of adjacent bits, an average PE can be obtained and for the equal-probability case this becomes

$$P_e = \frac{1}{2} Q \left\{ \left[\frac{2E}{N_{bo}} \right]^{1/2} \rho(\varepsilon) \right\}$$

$$+ \frac{1}{2} Q \left\{ \left(\frac{2E}{N_{bo}} \right)^{1/2} \left[\rho(\varepsilon) - \frac{2}{E} \left| \int_{T-\varepsilon}^{T} s(t + \varepsilon - T)s(t)\, dt \right| \right] \right\} \qquad (4\text{-}98)$$

Evaluation of Eq. (4-98) requires specification of the particular antipodal waveforms used.

Timing Errors in PCM

For a PCM signal consisting of rectangular pulses extending over the entire signal interval:

$$s(t) = \pm \left(\frac{E}{T} \right)^{1/2} \qquad 0 < t \le T$$

For adjacent bits the same:

$$\rho(\tau) = 1$$

For adjacent bits different:

$$\frac{2}{E} \left| \int_{T-\varepsilon}^{T} s(t + \varepsilon - T)s(t)\, dt \right| = \frac{2|\varepsilon|}{T}$$

$$z_1(T, \varepsilon) = \begin{cases} E & \text{bits same} \\ E\left(1 - \dfrac{2|\varepsilon|}{T} \right) & \text{bits different} \end{cases}$$

$$P_e = \frac{1}{2} Q \left[\left(\frac{2E}{N_{bo}} \right)^{1/2} \right] + \frac{1}{2} Q \left\{ \left(\frac{2E}{N_{bo}} \right)^{1/2} \left(1 - \frac{2|\varepsilon|}{T} \right) \right\} \qquad (4\text{-}99)$$

The error performance degradation will be small if $|\varepsilon| \ll T$. Representative values are shown in Fig. 4-9.

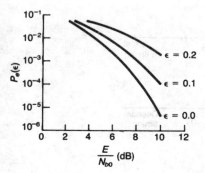

Figure 4-9 PCM error probability with timing offset (ε).

4-5 DETECTION IN NONWHITE NOISE

Matched Filter in Nonwhite Noise

In previous chapters it was assumed that the binary signals were to be detected in the presence of white noise and under conditions in which no symbol interfered with any other symbol. There are many circumstances under which this is not true. Therefore, it is necessary to consider what happens when the signals are to be detected under less than ideal conditions.

Assume that the noise is not white, but has a two-sided spectral density $S_n(\omega)$. Since the procedure for the design of a matched filter for white noise has already been established, a reasonable thing to do is to incorporate a whitening filter in the system such that the output is white when the input is $S_n(\omega)$. Such a filter is shown in Fig. 4-10. The power transfer function of the whitening filter should be

$$|H_1(\omega)|^2 = \frac{N_{b0}}{2S_n(\omega)} \tag{4-100}$$

It is convenient to let $N_{b0} = 2$, and since this is equivalent to changing the system gain, it does not affect the error performance or SNR.

The matched filter can now be separated into two parts—the first part is the whitening filter, and the second part a filter that is matched to the signal coming out of the whitening filter. This is illustrated in Fig. 4-11. The output of the whitening filter may be represented as

$$\hat{x}(t) = \hat{s}_i(t) + n(t) \tag{4-101}$$

where

$$\hat{m}(t) = \begin{array}{l} \hat{s}_0(t) = \text{response of } h_1(t) \text{ to } s_0(t) \\[2mm] \hat{s}_1(t) = \text{response of } h_1(t) \text{ to } s_1(t) \end{array}$$

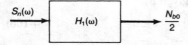

Figure 4-10 Whitening filter for nonwhite noise.

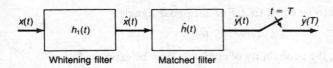

Figure 4-11 Matched filter for nonwhite noise.

It may be noted that even though the input signal is time limited, the signal out of the whitening filter may not be. To proceed it is necessary to determine the energies in the signals out of the whitening filter, and these are given by

$$\hat{E}_0 = \int_0^\infty \hat{s}_0^2(t)\, dt = \frac{1}{2\pi} \int_{-\infty}^\infty |S_0(\omega)|^2 |H_1(\omega)|^2 \, d\omega \tag{4-102}$$

$$\hat{E}_1 = \int_0^\infty \hat{s}_1^2(t)\, dt = \frac{1}{2\pi} \int_{-\infty}^\infty |S_1(\omega)|^2 |H_1(\omega)|^2 \, d\omega \tag{4-103}$$

where

$$S_0(\omega) = \mathscr{F}\{s_0(t)\}$$
$$S_1(\omega) = \mathscr{F}\{s_1(t)\}$$

are the Fourier transforms of the signals. For purposes of illustration, let us assume that $\hat{E}_0 = \hat{E}_1 = \hat{E}$, although this may not be true under all conditions. Hence the correlation coefficient of the signals out of the filter becomes

$$\hat{\rho} = \frac{1}{\hat{E}} \int_0^\infty \hat{s}_0(t)\hat{s}_1(t)\, dt$$
$$= \frac{1}{2\pi\hat{E}} \int_{-\infty}^\infty \hat{S}_0^*(\omega)\hat{S}_1(\omega)\, d(\omega) \tag{4-104}$$

where the * implies the complex conjugate. By using Eq. (4-100), this correlation coefficient can be written as

$$\hat{\rho} = \frac{1}{2\pi\hat{E}} \int_{-\infty}^\infty S_0^*(\omega)S_1(\omega)|H_1(\omega)|^2 \, d\omega$$
$$= \frac{1}{2\pi\hat{E}} \int_{-\infty}^\infty \frac{S_0^*(\omega)S_1(\omega)}{S_n(\omega)} \, d\omega \tag{4-105}$$

Since the filter must now be matched to the signal coming out of the whitening filters, the matched part of the matched filter is given by

$$\hat{h}(t) = \hat{s}_1(T-t) - \hat{s}_0(T-t) \qquad t \geq 0 \tag{4-106}$$

and the probability of error becomes

$$P_e = Q\left[\sqrt{\frac{\hat{E}(1-\hat{\rho})}{2}}\right] \tag{4-107}$$

since N_{b0} was assumed to be equal to 2. For antipodal signals

$$S_0(\omega) = -S_1(\omega)$$

and, since $\hat{\rho}$ equals -1, the probability of error can now be expressed as

$$P_e = Q\left[\sqrt{\frac{1}{2\pi} \int_{-\infty}^\infty \frac{|S(\omega)|^2}{S_n(\omega)} \, d\omega}\right] \tag{4-108}$$

It should also be noted that even though the original signals might be orthogonal, the signals coming out of the whitening filter are not necessarily orthogonal and, in fact, in general are not.

Example of Detection in Nonwhite Noise

For purposes of illustration, assume that we have two antipodal signals that are of raised-cosine form; that is, they are described by

$$s_i(t) = (-1)^i \left[1 - \cos \frac{2\pi t}{T} \right] \qquad 0 \le t \le T, \quad i = 0, 1 \qquad (4\text{-}109)$$

and are illustrated in Fig. 4-12.

If the bit rate of the signal is 30,000 bits per second (b/s), then

$$T = \frac{1}{30,000}$$

and the signal energy becomes

$$E = \int_0^T \left(1 - \cos \frac{2\pi t}{T} \right)^2 dt = \frac{3T}{2} = \frac{3}{2 \times 30,000} = 5 \times 10^{-5} \text{ V}^2 s \qquad (4\text{-}110)$$

Consider the white noise case first. Assume that the one-sided spectral density of the white noise is $N_{bo} = 4 \times 10^{-6}$ V^2/Hz. In this case the probability of error is given by

$$P_e = Q\left(\sqrt{\frac{2E}{N_0}} \right) \qquad \text{since } \rho = -1$$

$$= Q\left(\sqrt{\frac{2 \times 5 \times 10^{-5}}{4 \times 10^{-6}}} \right) = Q(5) = 2.87 \times 10^{-7} \qquad (4\text{-}111)$$

Now assume that the noise spectral density is

$$S_n(\omega) = \frac{8 \times 10^4}{\omega^2 + 4 \times 10^{10}} \qquad (4\text{-}112)$$

This corresponds to a first-order Butterworth spectrum having a half-power bandwidth of 31.8 kHz. Note that at zero frequency the value of the spectral density is exactly the same as it was in the white noise case considered above. The required

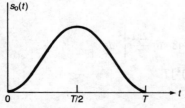

Figure 4-12 The raised-cosine signal.

power transfer function of the whitening filter is given by

$$|H_1(\omega)|^2 = \frac{N_{b0}}{2S_n(\omega)} = \frac{2 \times 10^{-6}(\omega^2 + 4 \times 10^{10})}{8 \times 10^4}$$

$$= 25 \times 10^{-12}(\omega^2 + 4 \times 10^{10})$$

and, therefore, the voltage transfer function is

$$H_1(\omega) = 1 + j5 \times 10^{-6}\omega \tag{4-113}$$

Because of the particular nature of this transfer function, it is easier to find the signals out of the whitening filter from the impulse response. For $H_1(\omega)$ this impulse response is

$$h_1(t) = \delta(t) + 5 \times 10^{-6}\dot{\delta}(t) \tag{4-114}$$

from which

$$\hat{s}_0(t) = s_0(t) * h_1(t) = s_0(t) + 5 \times 10^{-6}\dot{s}_0(t) \tag{4-115}$$

where the asterisk implies convolution and

$$\dot{s}_0(t) = \frac{ds_0(t)}{dt} \tag{4-116}$$

The signal out of the whitening filter may now be written as

$$\hat{s}_0(t) = \left(1 - \cos\frac{2\pi t}{T}\right) + \frac{5 \times 10^{-6} \times 2\pi}{T}\sin\frac{2\pi t}{T}$$

$$= 1 - \cos\frac{2\pi t}{T} + 0.94248\sin\frac{2\pi t}{T} = -\hat{s}_1(t) \tag{4-117}$$

Since the signals are still antipodal, the matched filter can be written as

$$\hat{h}(t) = \hat{s}_0(T - t) - \hat{s}_1(T - t) = 2\hat{s}_0(T - t) \tag{4-118}$$

$$\hat{h}(t) = 2 - 2\cos\frac{2\pi t}{T} + 1.88496\sin\frac{2\pi t}{T} \quad \begin{array}{l} 0 \leq t \leq T \\[4pt] T = 1/30,000 \end{array} \tag{4-119}$$

To find the probability of error, it is necessary to evaluate the energy of these signals. This energy is

$$\hat{E} = \int_0^T \hat{s}_0^2(t)\, dt$$

$$= \int_0^T \left[1 - \cos\frac{2\pi t}{T} + 0.94248\sin\frac{2\pi t}{T}\right]^2 dt$$

$$= [1 + 0.5 + 0.5 \times (0.94248)^2]T$$

$$= 6.480 \times 10^{-5}\ \text{V}^2s$$

and the probability of error is

$$P_e = Q\left[\sqrt{\frac{2\hat{E}}{N_{b0}}}\right]$$

$$= Q\left[\sqrt{\frac{2 \times 6.48 \times 10^{-5}}{4 \times 10^{-6}}}\right] = Q(5.692)$$

$$= 6.285 \times 10^{-9}$$

Note that the probability of error is smaller with nonwhite noise, even though it has the same spectral level at low frequencies as the white noise. This is so because the nonwhite noise has a lower spectral level at some of the signal frequencies.

Singular Detection in Nonwhite Noise

It may be noted from Eq. (4-108) that P_e is reduced by making $|S(\omega)|^2$ large where $S_n(\omega)$ is small. If there is any range of ω in which $S_n(\omega)$ is zero, but $|S(\omega)|$ is not zero, then the argument of the Q function becomes infinite and the probability of error becomes zero. This is known as singular detection.

Singular detection can also occur when $S_n(\omega)$ is not zero anywhere. This happens if the signal energy at the output of the whitening filter is infinite. As an example, let the signal be a rectangular pulse

$$s_i(t) = (-1)^i \qquad 0 \le t \le T, \quad i = 0, 1 \tag{4-120}$$

in the previous nonwhite noise situation. The signal out of the whitening filter now becomes

$$\hat{s}_i(t) = s_i(t) + 5 \times 10^{-6}\dot{s}_i(t)$$

$$= (-1)^i + 5 \times 10^{-6}(-1)^i[\delta(t) - \delta(t-T)] \tag{4-121}$$

since the derivative of a rectangular pulse is a pair of δ functions. The energy of either signal, say $\hat{s}_0(t)$, is

$$\hat{E}_0 = \int_0^T \hat{s}_0^2(t)\, dt$$

$$= \int_0^T [1 + 5 \times 10^{-6}\delta(t) - 5 \times 10^{-6}\delta(t-T)]^2\, dt$$

$$= \infty \tag{4-122}$$

This is infinite because the area of the square of any δ function is infinite. Since the energy is infinite, we again have singular detection.

More generally it can be shown that singular detection occurs whenever

$$\lim_{\omega \to \infty} \frac{|S(\omega)|^2}{S_n(\omega)} \neq 0 \tag{4-123}$$

This condition cannot occur with a finite energy signal if $S_n(\omega)$ has a white noise component. It is common, therefore, to model nonwhite noise as

$$S_n(\omega) = S_c(\omega) + N_{b0}/2 \qquad (4\text{-}124)$$

where N_{b0} is a white component. If singular detection does not exist, N_{b0} can be set to zero in the final solution. If N_{b0} cannot be set to zero in the final solution, then singular detection occurs. This is a realistic approach because there is always a thermal noise component present in any practical system.

4-6 BAND-LIMITING EFFECTS

All communication systems have a finite bandwidth, due either to filters at the transmitter or at the receiver, or to the characteristics of the transmission medium itself. To determine the effect of this finite bandwidth on the system performance, it is convenient to model the system in its equivalent baseband form. That is, the carrier frequency is set to zero and the transfer functions are those that would apply to a low-frequency signal. Such an equivalent baseband system model is shown in Fig. 4-13. Let the equivalent baseband transfer function of the modulator and demodulator be denoted as $H_b(\omega)$. When the input to the system $w(t)$ is time limited, the outputs $s(t)$ or $y(t)$ may not be. This is illustrated in Fig. 4-14. As a result of the system transfer function, the received signal is distorted in shape and stretched out in time. Since signals follow one another every T seconds, the

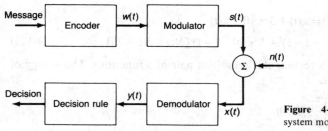

Figure 4-13 Equivalent baseband system model.

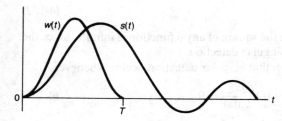

Figure 4-14 Signal stretching due to channel transfer function.

stretched signal will interfere with the following signals. This is known as inter-symbol interference. Intersymbol interference increases the probability of error for at least two different reasons. In the first place, the signal energy in each interval of T seconds is reduced because it has been spread out over a longer time interval, and second, at the instant of decision, the added interfering signals increase the effective noise level.

Probability of Error with Intersymbol Interference

The analysis of the effects of intersymbol interference on the probability of error in a practical communication system can be very difficult indeed. To carry out an analysis that illustrates some of the problems, it is necessary to introduce some simplifications. Therefore, consider an antipodal signal pair

$$\pm \sqrt{\frac{E}{T}} \, w(t)$$

where

$$w(t) = 1 \qquad 0 \le t \le T$$

$$= 0 \qquad \text{elsewhere} \tag{4-125}$$

Assume that the response of the channel transfer function, $H_b(\omega)$, to the input signal $w(t)$ is $s(t)$. At any time t, the total output can be represented as

$$x(t) = \sqrt{\frac{E}{T}} \sum_{j=0}^{\infty} d_j s(t + jT) + n(t) \tag{4-126}$$

where $d_j = \pm 1$. Note that d_0 represents the bit in the interval $(0, T)$ and d_j represents the bit in the interval j periods in the past. The Fourier transform of the signal $s(t)$ is given by

$$S(\omega) = W(\omega)H_b(\omega)$$

Next assume that the matched filter in the receiver is an integrate-and-dump circuit that produces at time $t = T$

$$y(T) = \int_0^T x(t)\,dt = \int_0^T \sqrt{\frac{E}{T}} \sum_{j=0}^{\infty} d_j s(t + jT)\,dt + \int_0^T n(t)\,dt \tag{4-127}$$

The signal response can be separated into two parts. One of these is the response to the present bit, which is

$$m_0 = \sqrt{\frac{E}{T}} \int_0^T d_0 s(t)\,dt \tag{4-128}$$

and the rest is the response to all previous bits, which is

$$m_I = \sqrt{\frac{E}{T}} \sum_{j=1}^{\infty} d_j \int_0^T s(t + jT)\,dt \tag{4-129}$$

The noise at the output of the integrate-and-dump circuit is given by

$$n_0(T) = \int_0^T n(t)\,dt \qquad (4\text{-}130)$$

and the variance of this noise is given by

$$
\begin{aligned}
\sigma_n^2 &= E[n_0^2(T)] \\
&= E\left[\int_0^T \int_0^T n(t)n(u)\,dt\,du\right] \\
&= \int_0^T \int_0^T R_n(t-u)\,dt\,du \\
&= \frac{N_{b0}}{2}\int_0^T \int_0^T \delta(t-u)\,dt\,du = \frac{N_{b0}T}{2} \qquad (4\text{-}131)
\end{aligned}
$$

if the noise is assumed to be white and stationary.

The conditional mean value of the receiver output $y(T)$ is

$$E[y(T)] = [m_0 + m_I] \qquad (4\text{-}132)$$

and for a particular bit history

$$\mathbf{d} = (\ldots d_2, d_1, d_0)$$

has a specified value that depends upon whether d_0 is $+1$ or -1. Since the noise is Gaussian, the conditional density functions of the output are as shown in Fig. 4-15, where a threshold is indicated at zero for the case where the signals are equally probable. It is clear, therefore, that the probability of error, given a particular bit history, is

$$P_e(\mathbf{d}) = \frac{1}{2}Q\left[\frac{(m_0+m_I)}{\sigma_n}\right] + \frac{1}{2}Q\left[\frac{(m_0-m_I)}{\sigma_n}\right] \qquad (4\text{-}133)$$

assuming $+1$ and -1 are equally probable.

To evaluate the significance of the result, it is desirable to look at a "worst-case" bit history. This is defined by

$$d_i = d_0 \quad \text{if} \int_0^T s(t+iT)\,dt \int_0^T s(t)\,dt < 0 \qquad (4\text{-}134)$$

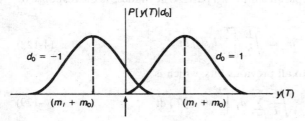

$P[y(T)|d_0]$

$d_0 = -1$ $d_0 = 1$

$(m_I + m_0)$ $(m_I + m_0)$ $y(T)$

Figure 4-15 Conditional probability density functions.

$$d_i = -d_0 \qquad \text{if} \int_0^T s(t+iT)\,dt \int_0^T s(t)\,d(t) > 0 \qquad (4\text{-}135)$$

Since this particular choice of prior bits makes all of the interference tails have a sign opposite to that of the present bit, it produces the greatest cumulative interference effect. It follows, therefore, that in this worst-case situation

$$\sqrt{\frac{T}{E}}\,(m_0 \pm m_I) = \int_0^T s(t)\,dt - \int_T^\infty |s(t)|\,dt \qquad (4\text{-}136)$$

and the probability of error is bounded by

$$P_e \leq Q\left[\frac{(m_0 + m_I)}{\sigma_n}\right] \qquad (4\text{-}137)$$

By using Eq. (4-131), the probability of error can now be expressed as

$$P_e \leq Q\left[\sqrt{\frac{2E}{N_{bo}}}\left[\frac{1}{T}\int_0^T s(t)\,dt - \frac{1}{T}\int_T^\infty |s(t)|\,dt\right]\right] \qquad (4\text{-}138)$$

If there were no intersymbol interference, the probability of error would be

$$P_e = Q\left[\sqrt{\frac{2E}{N_{bo}}}\right]$$

Comparing this with Eq. (4-138), it is clear that the probability of error is always increased (in the worst case) by intersymbol interference.

Example of Intersymbol Interference

To make some numerical calculations of the effect of intersymbol interference, it is assumed that the equivalent baseband system is a simple RC circuit as illustrated in Fig. 4-16. If $w(t)$ is a rectangular pulse as shown, then the output $s(t)$ is given by

$$\begin{aligned}
s(t) &= 1 - e^{-\alpha t} & 0 \leq t \leq T \\
&= (1 - e^{-\alpha T})e^{-\alpha(t-T)} & t > T
\end{aligned} \qquad (4\text{-}139)$$

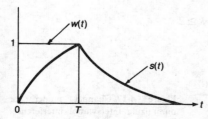

Figure 4-16 The input and output channel signals.

The integrals needed to evaluate the error probability are

$$\frac{1}{T}\int_0^T s(t)\,dt = \frac{1}{T}\int_0^T (1 - e^{-\alpha t})\,dt = 1 - \frac{1 - e^{-\alpha T}}{\alpha T}$$

$$\frac{1}{T}\int_T^\infty |s(t)|\,dt = \frac{1}{T}\int_T^\infty (1 - e^{-\alpha T})e^{-\alpha(t-T)}\,dt = \frac{1 - e^{-\alpha T}}{\alpha T}$$

Hence from Eq. (4-138), the probability of error becomes

$$P_e \leq Q\left\{\sqrt{\frac{2E}{N_{b0}}}\left[1 - \frac{2(1 - e^{-\alpha T})}{\alpha T}\right]\right\} \tag{4-140}$$

To illustrate this numerically, let $2E/N_{b0} = 25$. Without intersymbol interference, the probability of error would be

$$P_e = Q\left[\sqrt{\frac{2E}{N_{b0}}}\right] = Q(5) = 2.87 \times 10^{-7}$$

With intersymbol interference, and assuming $\alpha T = 5$ (that is, one bit interval is five time constants long), the probability of error becomes

$$P_e \leq Q\left\{5\left[1 - \frac{2(1 - e^{-5})}{5}\right]\right\} = Q(3.013) = 1.29 \times 10^{-3}$$

Minimizing Intersymbol Interference

It should be noted that if the signal is not band-limited, intersymbol interference can be made arbitrarily small because the tails can be made as short as necessary. However, since all channels are actually band-limited, the problem is to find a band-limited signal that will produce minimum intersymbol interference. To do this let the channel be modeled as shown in Fig. 4-17, where $h_b(t)$ is the baseband impulse response of the channel and $h_s(t)$ is an added filter that is used to shape the signal. The output at time T is again represented as

$$y(T) = \sqrt{\frac{E}{T}}\left[d_0 s(T) + \sum_{j=1}^\infty d_j s[(j+1)T]\right] + n_y(T) \tag{4-141}$$

Upon examining this result, it is clear that intersymbol interference is eliminated if

$$s(T) \neq 0$$

$$s(kT) = 0 \qquad k \geq 2$$

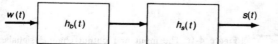

$w(t) \longrightarrow \boxed{h_b(t)} \longrightarrow \boxed{h_s(t)} \xrightarrow{\ s(t)\ }$

Figure 4-17 Channel model for minimizing intersymbol interference.

A general form of $s(t)$ that would accomplish this is

$$s(t) = a(t)\text{sinc}\left(\frac{t}{T} - 1\right) \tag{4-142}$$

where $a(t)$ is an arbitrary time function with $a(T) \neq 0$. Hence the Fourier transform of $s(t)$ is given by

$$S(f) = A(f) * \frac{1}{T} \text{rect}(Tf)e^{-j2\pi fT} \tag{4-142a}$$

Thus $s(t)$ has a spectrum that is given by the convolution of the spectrum $A(f)$ of an arbitrary function with a rectangular spectrum extending over the interval $|f| \leq 1/2T$. It should also be noted that the Fourier transform of $s(t)$ is given by

$$S(\omega) = W(\omega)H_b(\omega)H_s(\omega)$$

and the channel bandwidth B_m is determined by the bandwidth of $H_b(\omega)H_s(\omega)$. The minimum possible value for the bandwidth B_m is $1/2T$, which corresponds to signaling at the Nyquist rate. To accomplish signaling at this rate, it is necessary that $A(\omega)$ be a δ function, which implies that $a(t)$ is a constant. The result is as illustrated in Fig. 4-18, in which it is seen that every signal waveform is zero at all of the sampling instants other than the one at which it is transmitted.

Although signaling at the Nyquist rate theoretically is possible, it does not represent a practical system design because of the sharp transitions in the spectrum of $s(t)$ that must be achieved by the filter $H_s(\omega)$. Also small timing errors greatly increase the error rate. It is preferable, therefore, to let the bandwidth become somewhat greater than $1/2T$ but still preserve the condition of there being no intersymbol interference. It turns out that there are an infinite number of possible solutions in this case, but a convenient one is the cosine rolloff frequency function

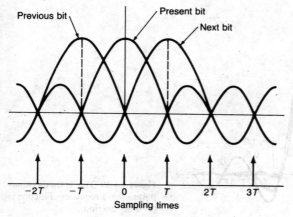

Figure 4-18 Signaling at the Nyquist rate.

given by

$$S(f) = e^{-j2\pi fT} \qquad |f| \le \frac{1}{2T}(1-\alpha)$$

$$= \frac{1}{2}\left\{1 + \cos\left[\frac{\pi(2fT - 1 + \alpha)}{2\alpha}\right]\right\} e^{-j2\pi fT} \qquad \frac{1-\alpha}{2T} \le |f| \le \frac{1+\alpha}{2T}$$

$$= 0 \qquad \text{elsewhere} \qquad (4\text{-}143)$$

This function has a spectrum that extends out to

$$W_m = \frac{1}{2T}(1+\alpha) \qquad (4\text{-}144)$$

where α is a measure of the excess bandwidth required. $\alpha = 0$ corresponds to signaling at the Nyquist rate, while $\alpha = 1$ corresponds to signaling at one half the Nyquist rate. Spectral shaping of this type is often referred to in terms of a rolloff factor that is the fraction of the spectrum involved in the transition from full amplitude to half amplitude. Thus α is the rolloff factor. The time function corresponding to

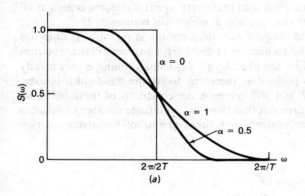

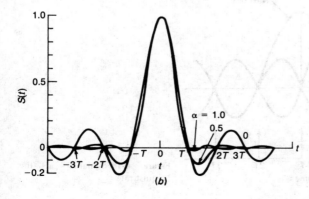

Figure 4-19 The raised-cosine pulse. (a) Frequency function; (b) time function.

Eq. (4-143) is

$$s(t) = \frac{1}{T} \frac{\cos[\pi\alpha(t-T)/T]}{1-[2\alpha(t-T)/T]^2} \operatorname{sinc}\left(\frac{t-T}{T}\right) \tag{4-145}$$

The frequency functions and time functions for this particular signal are illustrated in Fig. 4-19.

Partial Response (Duobinary) Signaling

An alternate approach to eliminating the effects of intersymbol interference is to control the amount of interference rather than to suppress it entirely. To illustrate this let

$$s(kT) = s_0 \qquad k = 0, 1$$
$$= 0 \qquad k \geq 2 \tag{4-146}$$

Note that two sequential samples of each signal are nonzero, and thus at any sampling instant, the value from the present bit is added to the value from the previous bit. If the previous bit is the same as the present one, the sample will be $\pm 2s_0$. If the previous bit is different from the present one, the sample will be zero. Since we have already decided the state of the previous bit, the present one can be determined uniquely. Waveforms for partial response signaling are illustrated in Fig. 4-20.

A convenient frequency function for partial response signaling is

$$S(\omega) = 2T \cos\left(\frac{\omega T}{2}\right) \qquad |\omega| \leq \frac{\pi}{T}$$
$$= 0 \qquad \text{elsewhere} \tag{4-147}$$

For this function the spectrum is contained within bandwidth

$$W_m = 1/2T$$

Thus there is no excess bandwidth and signaling occurs at the Nyquist rate. Frequency and time functions for this particular form of partial response signaling are illustrated in Fig. 4-21. It may be noted from this figure that at the sampling instants, the time function has an amplitude of unity that is down by a factor of $\pi/4$ from its peak value. Thus there is an effective loss of 2 dB in detection energy.

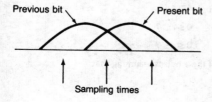

Figure 4-20 Partial response signaling.

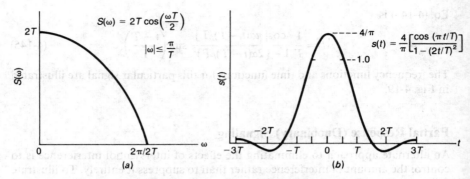

Figure 4-21 The cosine partial response pulse. (a) Frequency function; (b) time function.

4-7 REFERENCES

1. Davenport, W. B., and W. L. Root: In *Introduction to Random Signals and Noise*, McGraw-Hill Book Company, New York, 1958.
2. Van Trees, H. L.: *Detection, Estimation and Modulation Theory, Part I*, John Wiley & Sons, Inc., New York, 1968.
3. Wozencraft, J. M., and I. M. Jacobs: *Principles of Communication Engineering*, John Wiley & Sons, Inc., New York, 1965.
4. Gagliardi, R. M.: *Introduction to Communications Engineering*, John Wiley & Sons, Inc., New York, 1978.
5. Bennett, W. R., and J. R. Davey: *Data Transmission*, McGraw-Hill Book Company, New York, 1965.
6. Lucky, R. W., J. Salz, and E. J. Weldon, Jr.: *Principles of Data Communication*, McGraw-Hill Book Company, New York, 1968.

4-8 PROBLEMS

4.1-1 A simple alarm system is designed to produce a signal of $+1$ volt during quiescent conditions and a signal of -1 volt when an alarm is given. The signals are transmitted through a channel that adds Gaussian noise having zero mean and a variance of $1\ V^2$.

(a) Find the optimum threshold setting for a detector that measures the received voltage and decides that either there is an alarm or there is no alarm. Assume that the cost of missing an alarm is twice the cost of a false alarm and the costs of correct decisions are zero. Also assume that the *a priori* probability of an alarm is 0.01.

(b) Find the false alarm probability for the optimum threshold setting.

4.1-2 For the alarm system of Problem 4.1-1, determine the optimum threshold for a maximum likelihood detector and find the false alarm probability.

4.2-1 A binary sequence is encoded into pulses of the form

$$s_1(t) = 1 - \cos(2\pi t/T) \qquad 0 \le t \le T$$
$$s_0(t) = 1 + \cos(2\pi t/T) \qquad 0 \le t \le T$$

Find the spectral density $S_s(f)$ of a random sequence of these pulses assuming that:

(a) $s_0(t)$ and $s_1(t)$ occur with equal probability.

(b) $s_0(t)$ occurs twice as often as $s_1(t)$.

4.2-2 The output of a matched filter is sampled to determine which of two equally likely signals is present. The mean values for the two signals at the filter output are $+10$ volts and -5 volts respectively. Find the decision threshold that minimizes the error probability under the assumption that the noise present during the observation has the probability density function given below and determine the minimum probability of error.

$$p(n) = 0.1(1 - 0.1|n|) \qquad |n| \leq 10$$
$$= 0 \qquad |n| > 10$$

4.2-3 A random binary sequence is encoded into pulses of the form

$$s_1(t) = \text{rect}(5 \times 10^6 t)$$
$$s_0(t) = -\text{rect}(5 \times 10^6 t)$$

This pulse sequence occurs in noise having a one-sided spectral density $N_0 = 10^{-7}$ V²/Hz. The signal-plus-noise sequence is processed by a filter having impulse response

$$h(t) = \text{rect}(5 \times 10^6 t)$$

and sampled at the end of each bit period. Find the bit error probability assuming maximum-likelihood decoding is employed.

4.2-4 A binary signal set is of the form shown.

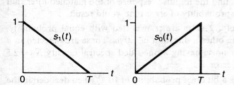

A sequence of these signals is received and decoded in the presence of white noise. If the noise has a one-sided spectral density $N_0 \simeq 10^{-7}$ V²/Hz and the bit duration is $T = 10^{-7}$ second, find the minimum probability of error with which they can be decoded. Sketch the matched filter impulse response for this signal set.

4.2-5 Repeat Problem 4.2-4 for signal $s_1(t)$ the same but signal $s_0(t)$ changed to that shown in the accompanying figure.

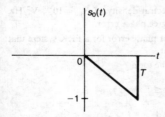

4.2-6 A binary sequence is encoded into a rectangular signal set of the form

$$s_1(t) = \text{rect } 10^6 t$$
$$s_0(t) = -\text{rect } 10^6 t$$

The binary sequence is received in the presence of white Gaussian noise with one-sided spectral density $N_0 = 10^{-7}$ V²/Hz. The decoding is done with an RC low-pass filter that has the capacitor discharged after each decision is made. Assuming a time constant of $RC = 10^{-6}$, how would the error performance of this system compare with use of a matched filter detector?

4.2-7 In the binary decoder shown, the noise is white with a one-sided spectral density of 0.04 V^2/Hz. Assuming the two signals are equally probable, find the threshold with which $y(T)$ should be compared and find the resulting probability of error.

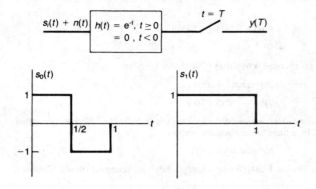

4.2-8 For the signals and noise of Problem 4.2-7, find the impulse response of the matched filter that should be used for decoding these signals and the probability of error that would result.

4.2-9 Two rectangular pulses with amplitudes of $\sqrt{2}$ volts are transmitted with equal probability. One of the pulses has a duration of 0.5 μs and the other a duration of 1.0 μs. These are detected in a matched filter receiver in the presence of receiver noise having a one-sided spectral density $N_0 = 2.5 \times 10^{-8}$ V^2/Hz. Find the minimum probability of error.

4.3-1 A binary sequence of antipodal signals gives a bit error probability of 10^{-3} when decoded in the presence of noise having a two-sided spectral density $S_n(f) = 10^{-4}$ V^2/Hz. What would be the signal energy, E_s, of a set of equal energy orthogonal signals that give an error probability of 10^{-4} in the same noise environment?

4.3-2 A PSK sequence consists of single cycles of a carrier and has signals of the form

$$s_1(t) = \cos 2\pi \times 10^7 t \qquad 0 \le t \le 10^{-7}$$
$$s_0(t) = -\cos 2\pi \times 10^7 t \qquad 0 \le t \le 10^{-7}$$

These signals are received in white noise having a one-sided spectral density of $N_{bo} = 10^{-8}$ V^2/Hz. Find the error probability when a correlator is used with a 30-degree phase error.

4.3-3 Make a graph of error probability versus reference oscillator phase error for a PRK system that operates with an error rate of 10^{-6} when no phase error is present.

4.4-1 A binary communication system uses antipodal signals of the form

$$s_i(t) = (-1)^i \left[1 - \cos \frac{2\pi t}{T} \right] \qquad 0 \le t \le T$$

$$i = 0, 1$$

and the two signals are equally probable. These signals are received in the presence of white noise having a one-sided spectral density of 4×10^{-6} V^2/Hz. The bit rate is 30,000 b/s.

    (a) The signals are detected at the output of a matched filter that is sampled at the correct time. Find the probability of error.

    (b) Repeat (a) assuming the sampling times are in error by 5 μs.

4.4-2 A PCM system with signals $\pm(E/T)^{1/2}$, $0 \le t \le T$, is designed to operate at a bit error rate of 10^{-6} when no bit timing errors are present. How much of an increase in bit energy would be required to maintain the same error probability when there is a 10 percent error in bit timing?

4.2-6 A binary sequence is received in the presence of white noise. The encoding waveforms are

$$s_1(t) = \text{rect } 10^6 t$$
$$s_0(t) = -\text{rect } 10^6 t$$

Assuming the noise has a one-sided spectral density of $N_0 = 10^{-7}$ V²/Hz and the decoding is done with a matched filter, find the probability of error. If there is a timing error of 20 percent in the demodulator, what would the error probability be?

4.5-1 Binary signals of the form

$$s_i(t) = (-1)^i(1 - |t|) \qquad |t| \le 1, \quad i = 0, 1$$

are to be detected in the presence of nonwhite noise having a two-sided spectral density of

$$S_{nb}(\omega) = \frac{1}{\omega^4 + 1}$$

Find the resulting probability of error.

4.5-2 A binary communication system uses antipodal signals of the form

$$s_i(t) = (-1)^i \left[1 - \cos \frac{2\pi t}{T} \right] \qquad 0 \le t \le T$$
$$i = 0, 1$$

and the two signals are equally probable. These signals are received in the presence of nonwhite noise having a two-sided spectral density of

$$S_{nb}(\omega) = \frac{8 \times 10^4}{\omega^2 + 4 \times 10^{10}}$$

The bit rate is 30,000 b/s.

(a) Find the transfer function of the whitening filter that is needed.

(b) Find the impulse response of the matched filter that follows the whitening filter.

4.5-3 Find the probability of error for the signals and noise of Problem 4.5-2. (This problem is more easily worked in the time domain by looking at the *impulse response* of the whitening filter.)

4.5-4 Repeat Problem 4.5-3 assuming the noise is the same as in Problem 4.5-2 but signals are

$$s_i(t) = (-1)^i \qquad 0 \le t \le T$$
$$i = 0, 1$$

Explain the result.

4.6-1 A binary communication system uses an infinite sequence of antipodal waveforms proportional to

$$w(t) = 1 \qquad 0 \le t \le T$$
$$= 0 \qquad \text{elsewhere}$$

These signals pass through a channel whose equivalent baseband impulse response is

$$h_{bc}(t) = \frac{\pi}{T_1} \sin \frac{2\pi t}{T_1} \qquad 0 \le t \le T_1/2$$
$$= 0 \qquad \text{elsewhere}$$

and are demodulated in an integrate-and-dump circuit. The noise into the demodulator is white. When $T_1 = 0$, it is found that the probability of error is 10^{-6}. Find the probability of error assuming $T_1/T = 0.1$.

4.6-2 Repeat Problem 4.6-1 assuming

$$h_{bc}(t) = \delta(t) + \delta(t - T_1)$$

(This might correspond to a multipath channel.)

4.6-3 In duobinary signaling a decision is made as to the present bit based on the previous bit. Thus once an error is made, it tends to propagate through the bit stream. One way to overcome this is to encode the message bits before duobinary encoding using the relationship $b_c = x_k \oplus b_{k-1}$ where b_k is the kth bit transmitted, x_k is the kth message bit, and $\oplus$ indicates addition modulo 2. The transmitted (or received) signal is then $y_k = b_k + b_{k-1}$. Show that the decoding algorithm is $\hat{x} = y_k$ mod 2. Set up a table showing the results of decoding when an error is made in a previous bit.

DETECTION OF BINARY SIGNALS AT RADIO FREQUENCY

5-1 COHERENT DETECTION

When the phase of the RF carrier is incorporated into the demodulation process, it is referred to as coherent detection. This type of detection generally leads to the optimum performance achievable for a given modulation scheme, but the performance does differ for different modulations as discussed in the following subsections.

On–Off Keying

The simplest, though far from the most efficient, modulation system for binary signals employs a signal set of the form

$$s_1(t) = \sqrt{\frac{2E}{T}} \cos(\omega_c t + \theta) \qquad 0 \leq t \leq T \tag{5-1}$$

$$s_0(t) = 0 \qquad 0 \leq t \leq T \tag{5-2}$$

The normalized distance between the means of these signals is, from Eq. (4-69),

$$d_{max}^2 = \frac{\int_0^T \frac{2E}{T} \cos^2[\omega_c(T-t) + \theta]\, dt}{2N_0}$$

$$= \frac{E}{2N_0} \tag{5-3}$$

From Eq. (4-64) the probability of error is

$$P_e = Q\left(\sqrt{\frac{E}{2N_0}}\right) \tag{5-4}$$

The error performance of this system is the same as that of an antipodal system with signals

$$s_1(t) = \sqrt{\frac{E}{2T}} \cos(\omega_c t + \theta) \tag{5-5}$$

$$s_0(t) = -\sqrt{\frac{E}{2T}} \cos(\omega_c t + \theta) \tag{5-6}$$

It follows, therefore, that the on–off keying system requires four times the peak power of the equivalent antipodal system and an average power twice as great for the same error performance.

Phase Shift Keying

The signals used for phase shift keying are of the form

$$s_1(t) = A \sin(\omega_c t + \cos^{-1} \alpha) \qquad 0 \le t \le T \tag{5-7}$$

$$s_0(t) = A \sin(\omega_c t - \cos^{-1} \alpha) \qquad 0 \le t \le T \tag{5-8}$$

where $\cos^{-1} \alpha$ is the modulation angle. Assume that $\omega_c = 2\pi n/T$ where n is an integer and T is the signal duration. This assumes that there will be an integral number of cycles in each waveform. Using appropriate trigonometric identities, the signals can be put into the form

$$s_1(t) = A\alpha \sin \omega_c t + A\sqrt{1 - \alpha^2} \cos \omega_c t \tag{5-9}$$

$$s_0(t) = A\alpha \sin \omega_c t - A\sqrt{1 - \alpha^2} \cos \omega_c t \tag{5-10}$$

The first term is a carrier component and is used in some systems for synchronization of the local reference oscillator. The power in the carrier component is $(A\alpha)^2/2$ and that in the modulation component is $A^2(1 - \alpha^2)/2$. Thus α^2 is the fraction of the total power that is contained in the carrier. The detector for waveforms of this type is shown in Fig. 5-1.

The signals $s_1(t)$ and $s_0(t)$ have a correlation coefficient given by

$$\rho = \frac{\displaystyle\int_0^T s_1(t)s_0(t)\, dt}{E}$$

$$= \frac{\displaystyle\int_0^T [A\alpha \sin \omega_c t + A\sqrt{1 - \alpha^2} \cos \omega_c t][A\alpha \sin \omega_c t - A\sqrt{1 - \alpha^2} \cos \omega_c t]\, dt}{\left[\dfrac{A^2\alpha^2}{2} + \dfrac{A^2(1 - \alpha^2)}{2}\right] T}$$

$$= 2\alpha^2 - 1 \tag{5-11}$$

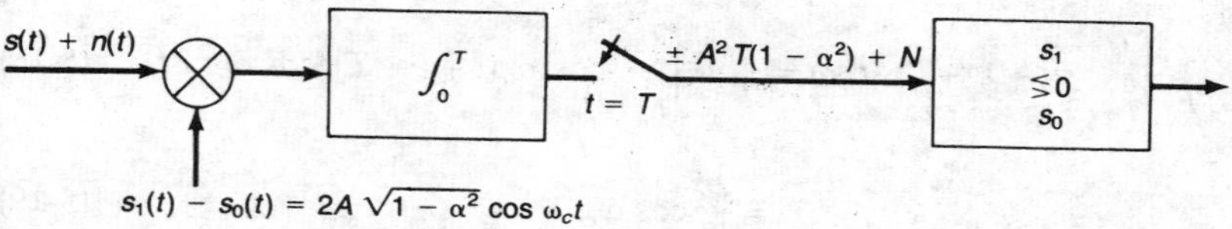

Figure 5.1 Correlation receiver for phase shift keying modulation.

Accordingly the error probability is given by

$$P_e = Q \left\{ \left[\frac{E}{N_0} (1 - \rho) \right]^{1/2} \right\} \tag{5-12}$$

$$= Q \left\{ \left[\frac{2E}{N_0} (1 - \alpha^2) \right]^{1/2} \right\} \tag{5-13}$$

It is seen that the effect of allocating α^2 of the total transmitted power to the carrier component is to degrade the performance an amount equivalent to a power reduction of $10 \log(1 - \alpha^2)$ dB.

Effect of Bit Timing Errors on PSK

To simplify the analysis, let $\alpha = 0$ so that there is no carrier component. The signals can then be written as

$$s_1(t) = \sqrt{\frac{2E}{T}} \cos \omega_c t \qquad 0 \le t \le T \tag{5-14}$$

$$s_0(t) = -\sqrt{\frac{2E}{T}} \cos \omega_c t \qquad 0 \le t \le T \tag{5-15}$$

It is also assumed that a matched filter receiver is being used to detect the signals. The results for a correlator receiver are the same as those obtained at baseband in Sec. 4.4. Following the development given in connection with Eq. (4-98), the time correlation function is

$$\rho(\tau) = \frac{1}{E} \int_0^T s(t + \tau)s(t) \, dt \tag{5-16}$$

$$= \frac{2}{T} \int_0^T \cos(\omega_c t + \omega_c \tau)\cos \omega_c t \, dt$$

$$= \frac{1}{T} \int_0^T \left[\cos \omega_c \tau + \cos(2\omega_c t + \omega_c \tau) \right] dt$$

$$= \cos \omega_c \tau \tag{5-17}$$

when a matched filter is used.

$$\frac{2}{E} \int_{T-\varepsilon}^{T} s(t+\varepsilon-T)s(t)\,dt = \frac{4}{T} \int_{T-\varepsilon}^{T} \cos[\omega_c(t+\varepsilon-T)]\cos \omega_c t\,dt \qquad (5\text{-}18)$$

$$= \frac{2\varepsilon}{T} \cos \omega_c \varepsilon \qquad (5\text{-}19)$$

The filter output for the bits is then

$$z_1(T, \varepsilon) = E \cos \omega_c \varepsilon \qquad \text{adjacent bit same} \qquad (5\text{-}20)$$

$$= E\left(\cos \omega_c \varepsilon - \left|\frac{2\varepsilon}{T} \cos \omega_c \varepsilon\right| \right) \qquad \text{adjacent bit different} \qquad (5\text{-}21)$$

The error probability is, therefore,

$$P_e = \frac{1}{2} Q\left\{ \sqrt{\frac{2E}{N_0}} \cos \omega_c \varepsilon \right\} + \frac{1}{2} Q\left\{ \sqrt{\frac{2E}{N_0}} \left[\cos \omega_c \varepsilon - \left|\frac{2\varepsilon}{T} \cos \omega_c \varepsilon\right| \right] \right\} \qquad (5\text{-}22)$$

The energy degrades as ε gets larger so ε must be kept small compared with the period of ω_c. The error probability of Eq. (5-22) is shown in Fig. 5-2. It should be noted that waveforms different than Eqs. (5-14) and (5-15) give different results.

Frequency Shift Keying

The signals employed in FSK can be represented as

$$s_1(t) = \left(\frac{2E}{T}\right)^{1/2} \sin(\omega_1 t + \theta) \qquad 0 \le t \le T \qquad (5\text{-}23)$$

$$s_0(t) = \left(\frac{2E}{T}\right)^{1/2} \sin(\omega_0 t + \theta) \qquad 0 \le t \le T \qquad (5\text{-}24)$$

Such signals can be readily generated by switching an oscillator or by switching between two triggered oscillators. The signals are not antipodal and have a

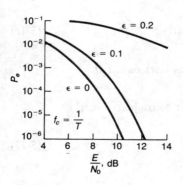

Figure 5-2 Error probability of PSK with timing error present.

correlation coefficient given by

$$\rho = \frac{1}{E} \int_0^T \frac{2E}{T} \sin(\omega_1 t + \theta)\sin(\omega_0 t + \theta) \, dt \tag{5-25}$$

$$= \frac{1}{T} \int_0^T \left[\cos(\omega_1 - \omega_0)t - \cos(\omega_1 + \omega_0)t \right] dt \tag{5-26}$$

Assuming $(\omega_1 + \omega_0) \gg 2\pi/T$, the second term will be essentially zero so that

$$\rho = \frac{\sin(\omega_1 - \omega_0)t}{(\omega_1 - \omega_0)T} \Big|_0^T = \frac{\sin(\omega_1 - \omega_0)T}{(\omega_1 - \omega_0)T} \tag{5-27}$$

This is sketched in Fig. 5-3.

The correlation coefficient is strongly dependent on the quantity $(\omega_1 - \omega_0)T$. It is evident from the figure and from Eq. (5-27) that ρ never reaches -1. In fact the most negative correlation occurs when

$$(\omega_1 - \omega_0)T \doteq 1.43\pi \tag{5-28}$$

$$(f_1 - f_0) \doteq 0.71\left(\frac{1}{T}\right) \tag{5-29}$$

For this frequency difference, $\rho = -0.22$ so that E/N_0 is multiplied by $(1 - \rho) = 1.22$. This frequency separation gives a 0.86-dB improvement over orthogonal signaling. For $(f_1 - f_0) = n/2T$, the signals are orthogonal and this is also essentially the case whenever $f_1 - f_0$ is large.

Detection of FSK signals requires a pair of coherent correlators that are phase synchronized to the carrier phase angle θ. This is called *coherent FSK*. When $(f_1 - f_0) = 1/2T$, the FSK system is operating at the first zero crossing of the correlation coefficient. This is the minimum frequency difference that will provide orthogonal signals and is called *minimum shift keying* (MSK). Under this condition the frequency difference between the 1-bit signal and the 0-bit signal is $1/2T$, or one half the bit rate frequency. By making $(f_1 - f_0) = 1/T$, there will be exactly one cycle more in $s_1(t)$ than in $s_0(t)$. This allows phase continuity to be maintained for adjacent waveforms and gives a minimum bandwidth signal. Theoretical analysis of such signals is quite complicated.[1]

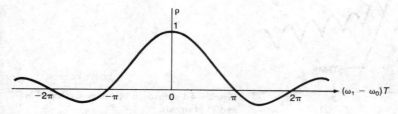

Figure 5-3 Correlation coefficient of FSK signals.

One of the problems that is addressed by MSK signaling is bandwidth reduction while retaining the performance of orthogonal signals. The increase in bandwidth over and above $(f_1 - f_0)$ is brought about because of abrupt phase changes that occur in going from $s_1(t)$ to $s_0(t)$ and vice versa. One approach to solving this problem is to make the phase of the waveform dependent on the sequence following the given bit. If the data set is defined by

$$a(t) = \sum_{n=-\infty}^{\infty} a_n \, \delta(t - nT) \tag{5-30}$$

where $a_n = -1$ or 1, then for MSK

$$\phi(mT - T) - \phi(mT) = \frac{\pi}{2} a_m \tag{5-31}$$

If, instead, the phase is obtained from the following rule,

$$\phi(mT + T) - \phi(mT) = \frac{\pi}{2}\left(\frac{a_{m-1}}{4} + \frac{a_m}{2} + \frac{a_{m+1}}{4}\right) \tag{5-32}$$

with

$$\phi(0) = 0 \qquad \text{if } a_0 a_1 = 1 \tag{5-33}$$

$$\phi(0) = \pi/4 \qquad \text{if } a_0 a_1 = -1 \tag{5-34}$$

then a reduction in phase discontinuities can be obtained. From the rule a change of $\pi/2$ occurs if three succeeding bits have the same polarity and the phase remains the same if the three succeeding bits have alternating polarities. This procedure is called *tamed frequency modulation* (TFM). Figure 5-4 shows the power spectra of MSK and TFM signals.[2]

The receiver structure for TFM signals is shown in Fig. 5-5. The system performance is about 1 dB poorer than for a quadriphase system (to be discussed later). Again phase coherency of the local reference oscillator is required.

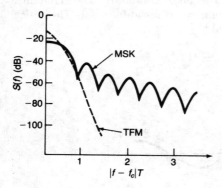

Figure 5-4 Comparison of power spectra of MSK and TFM signals.

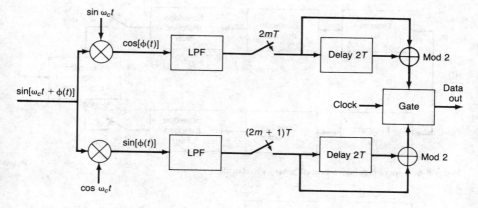

Figure 5-5 Receiver structure for TFM^2.

5-2 NONCOHERENT DETECTION

A number of schemes have been developed to eliminate the need for generating a phase synchronous reference at the receiver. Consider a binary system that uses signals of the following form.

$$s_1(t) = e_1(t)\cos(\omega_1 t + \theta) \qquad 0 \leq t \leq T \tag{5-35}$$

$$s_0(t) = e_0(t)\cos(\omega_0 t + \theta) \qquad 0 \leq t \leq T \tag{5-36}$$

Assume that $e_i(t) \geq 0$ for all t in the interval $(0, T)$. For $\omega_1, \omega_0 \gg 2\pi/T$, the signal energy is

$$E = \frac{1}{2} \int_0^T e_i^2(t) \, dt \tag{5-37}$$

These signals can be detected in a suboptimum manner using two bandpass filters followed by envelope detectors and matched filters as shown in Fig. 5-6. The decoding is done by determining which of the two receiver channels has the signal with the largest envelope. Phase is not used.

The decoder of Fig. 5-6 can be improved by combining the envelope matched filter with the bandpass filter as shown in Fig. 5-7.

Because of the nonlinear operation of the envelope detector, the statistics of samples that are compared in the decision element are no longer Gaussian. The error probability is given by

$$P_e = \Pr[\text{sample from correct channel} < \text{sample from incorrect channel}] \tag{5-38}$$

$$= \int_0^\infty \int_{r_1}^\infty p(r_1, r_2) \, dr_2 \, dr_1 \tag{5-39}$$

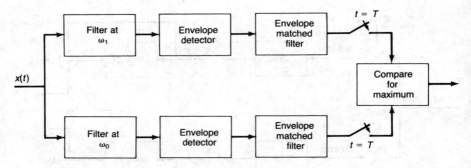

Figure 5-6 Noncoherent detection of binary waveforms.

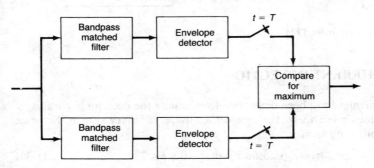

Figure 5-7 Simplified noncoherent detector.

where $p(r_1, r_2)$ is the joint PDF of the sampled envelope values of r_1 and r_2 in the correct and incorrect channels respectively. It has been shown that the error probability is minimized when the signals are orthogonal, and since in this case the sample values are independent, Eq. (5-39) becomes[3]

$$P_e = \int_0^\infty p(r_1) \int_{r_1}^\infty p(r_2)\, dr_2\, dr_1 \qquad (5\text{-}40)$$

In the channel without the signal, the envelope follows a Rayleigh PDF:

$$p(r_2) = \frac{r_2}{\sigma_n^2} \exp\left[\frac{-r_2^2}{2\sigma_n^2}\right] \qquad r_2 \geq 0 \qquad (5\text{-}41)$$

where

$$\sigma_n^2 = \frac{N_0}{2} \int_{-\infty}^\infty h^2(t)\, dt \qquad (5\text{-}42)$$

and for

$$e_i(t) = \text{rect}\left(\frac{t}{T} - \frac{1}{2}\right)$$

this becomes

$$\sigma_n^2 = \frac{N_0}{2} \int_0^T \cos^2[\omega(T-t)+\theta]\, dt = N_0 T/4 \tag{5-43}$$

In the channel with the signal, the envelope follows a Rician PDF:

$$p(r_1) = \frac{r_1}{\sigma_n^2} \exp\left[\frac{-(r_1^2 + A^2)}{2\sigma_n^2}\right] I_0\left(\frac{r_1 A}{\sigma_n^2}\right) \qquad r_1 \geq 0 \tag{5-44}$$

where $I_0(\cdot)$ is the modified Bessel function of zero order and A is the signal amplitude at the envelope detector input. Substituting these PDFs into Eq. (5-40) gives

$$\begin{aligned}
P_e &= \int_0^\infty \frac{r_1}{\sigma_n^2} \exp\left[\frac{-(r_1^2 + A^2)}{2\sigma_n^2}\right] I_0\left(\frac{r_1 A}{\sigma_n^2}\right) \int_{r_1}^\infty \frac{r_2}{\sigma_n^2} \exp\left[\frac{-r_2^2}{2\sigma_n^2}\right] dr_2\, dr_1 \\
&= \frac{1}{2} \exp\left[\frac{-A^2}{4\sigma_n^2}\right]
\end{aligned} \tag{5-45}$$

When a matched filter is used, the envelope detector input at $t = T$ is $A = \sqrt{ET/2}$ and the error probability is given by

$$P_e = \frac{1}{2} \exp\left[\frac{-E}{2N_0}\right] \tag{5-46}$$

This is the error probability for orthogonal binary signals with optimum noncoherent detection.

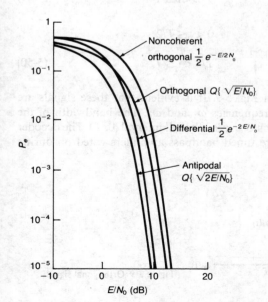

Figure 5-8 Comparison of several detectors.

A comparison of the error performance of several modulation schemes is shown in Fig. 5-8. It is seen that noncoherent processing of orthogonal signals is about 4 dB worse than antipodal signals at $P_e = 10^{-3}$ and 3 dB worse at $P_e = 10^{-5}$. For low P_e, coherent and noncoherent processing of orthogonal signals give equivalent performance.

The orthogonality requirement is

$$\int_0^T s_1(t)s_0(t)\, dt = \int_0^T e_1(t)\cos(\omega_1 t + \theta)e_0(t)\cos(\omega_0 t + \theta)\, dt = 0 \qquad (5\text{-}47)$$

Neglecting terms involving $\omega_1 + \omega_0$, this can be written as

$$\frac{1}{2}\int_0^T e_1(t)e_0(t)\cos(\omega_1 - \omega_0)t\, dt = 0 \qquad (5\text{-}48)$$

The signals can be orthogonal because of $e_1(t)$ and $e_0(t)$ or because of ω_1 and ω_0.

Pulse-Position Modulation

If the modulating waveforms $e_1(t)$ and $e_0(t)$ do not overlap in time, the waveforms $s_1(t)$ and $s_0(t)$ will be orthogonal. This is a type of pulse-position modulation. Consider modulating pulses of the form

$$e_1(t) = \left(\frac{4E}{T}\right)^{1/2} \qquad 0 < t \le T/2$$

$$= 0 \qquad T/2 < t \le T \qquad (5\text{-}49)$$

$$e_0(t) = 0 \qquad 0 < t \le T/2$$

$$= \left(\frac{4E}{T}\right)^{1/2} \qquad T/2 < t \le T \qquad (5\text{-}50)$$

The resulting signals are shown in Fig. 5-9. It is evident that these signals are orthogonal independently of the frequencies ω_1 and ω_0. The bandwidth of the signals is twice as great as if they each occupied the full interval $(0, T)$. The decoder matched filters for these signals are tuned bandpass amplifiers gated on during the time of signal occurrence.

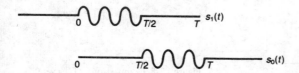

Figure 5-9 Orthogonal signals.

Frequency Shift Keying

For this type of modulation, the envelopes are constant and the frequencies are different.

$$e_1(t) = e_0(t) = \left(\frac{2E}{T}\right)^{1/2} \qquad 0 < t \le T \tag{5-51}$$

These signals are identical to coherent FSK signals except that no knowledge of phase is assumed. By separating the frequencies an exact multiple of the bit period, they will be orthogonal.

Matched filters for these signals consist of bandpass amplifiers tuned to the signal frequencies ω_1 and ω_0. These frequencies must be separated enough so that each is passed only by its own bandpass amplifier. In coherent FSK the signals were kept close together to take advantage of the slight negative correlation that could be obtained. This is not done in noncoherent FSK and the ease of encoding and detecting binary signals in a noncoherent manner has led to its wide usage. One problem that must be overcome is the need to know the frequencies very closely. Doppler shifts can cause difficulties and are usually handled by widening the bandwidth of the filters. With a wider bandwidth, the filters are no longer matched and the input noise to the decoder is increased. The PDF for the matched filter is no longer valid and each channel now measures the energy in its bandwidth and the decision is made as to which channel has the largest energy. The performance is slightly poorer ($1-2\,\text{dB}$) than would be predicted from the decrease in SNR resulting from the bandwidth increase alone.

An alternate detection procedure is shown in Fig. 5-10. In this detector the bandpass filter must pass the FSK signal with no band limiting. The bandwidth of the filter determines the system noise. An approximate analysis can be made assuming that the discriminator operates above threshold. Let the amplitude of the FSK signal be C at the system input. The discriminator input can then be represented as

$$\tilde{C} = C|\tilde{H}_{\text{BP}}(\Delta\omega)| \tag{5-52}$$

where $\tilde{H}_{\text{BP}}(\omega)$ is the low-frequency equivalent of the bandpass filter and $\Delta\omega$ is the frequency deviation.

$$\Delta\omega = \frac{\omega_1 - \omega_0}{2} \tag{5-53}$$

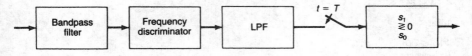

Figure 5-10 Alternate detector for noncoherent FSK.

The output signal is the sum of the PCM signal with amplitude corresponding to frequency shifts of $\pm \Delta\omega$ and the frequency discriminator noise. Assuming that the low-pass filter is ideal over a bandwidth of $1/T$ Hz, the noise at the sampler has a variance of

$$\sigma_n^2 = \frac{2N_{b0}(2\pi)^2}{3(C)^2 T^2} \tag{5-54}$$

Decision processing based on the sign of the sampler voltage will give an error probability of

$$P_e = Q\left\{ \left[3\sqrt{2}\beta^2 \frac{C^2 T^2}{2N_0} |\tilde{H}_{BP}(\Delta\omega)|^2 \right]^{1/2} \right\} \tag{5-55}$$

where

$$\beta = \frac{\Delta\omega T}{2\pi} = \Delta f T \tag{5-56}$$

is the FM modulation index with respect to the bit rate. As β increases the detection performance improves. However, as $\Delta\omega$ continues to increase, the detection band-pass filter reduces the carrier amplitude and some type of trade-off is required. For a given filter characteristic, there will be a $\Delta\omega$ that minimizes the error probability. This value for several filter functions is given in Table 5-1. Figure 5-11 shows a comparison of three methods of detecting FSK signals. It is seen that discriminator detection is slightly poorer than noncoherent processing.

Table 5-1 Optimum $\Delta\omega$ for noncoherent FSK detection

BPF function	Optimum β	Optimum $\Delta\omega$
Gaussian:		
$\exp\left[\dfrac{-\omega^2}{16\pi B_n^2}\right]$	$1.13B_n T$	$1.13B_n$
Butterworth:		
$\dfrac{1}{1+\left(\dfrac{\omega}{2\pi B_n}\right)^{2k}}$	$0.40B_n T$	$(k-1)^{1/2k}B_n$
Ideal:		
$1 \quad |\omega| \leq 2\pi B_n$ 0 elsewhere	$B_n T$	B_n

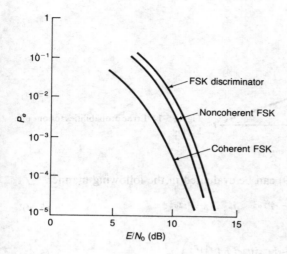

P_e axis labeled with 10^{-1}, 10^{-2}, 10^{-3}, 10^{-4}, 10^{-5}; horizontal axis E/N_0 (dB) from 0 to 15.

FSK discriminator

Noncoherent FSK

Coherent FSK

Figure 5-11 Comparison of FSK detectors.

On–Off Keying

The signals for on–off keying are of the form

$$e_1(t) = \left(\frac{2E}{T}\right)^{1/2} \qquad 0 \le t \le T$$

$$e_0(t) = 0 \qquad 0 \le t \le T \tag{5-57}$$

$s_1(t)$ and $s_0(t)$ are always orthogonal. The bandwidth is half that of PPM signaling since the pulse extends over the full time interval $(0, T)$. Since the signals corresponding to the 2 bits do not have equal energy, the error probability expression must be altered. The PDF is different when a one is sent from that when a zero is sent, and, therefore, the error probability calculations are different in the two cases.

The detector makes decisions by comparing the magnitude of the envelope of the received signal with a threshold η. For a zero bit (i.e., $e_0(t) = 0$), the error probability is obtained by integrating the Rayleigh PDF of the envelope of noise only.

$$P_0 = \text{PE}|0 \text{ transmitted} = \int_\eta^\infty p(r_2)\, dr_2 \tag{5-58}$$

where η is the threshold value. The error probability when a one bit is transmitted (i.e., $e_1(t) = (2E/T)^{1/2}$) is obtained by integrating the Rician PDF of the envelope of signal plus noise.

$$P_1 = \text{PE}|1 \text{ transmitted} = \int_0^\eta p(r)\, dr_1 \tag{5-59}$$

These error probabilities are indicated by the shaded areas in Fig. 5-12. The

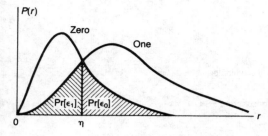

Figure 5-12 Error probabilities for on–off keying.

integrals of Eqs. (5-58) and (5-59) can be evaluated in the following manner.

$$P_0 = \int_\eta^\infty \frac{r_2^2}{\sigma_n^2} e^{-r_2^2/2\sigma_n^2} \, dr_2 = e^{-\eta^2/2\sigma_n^2} \tag{5-60}$$

$$P_1 = \int_0^\eta \frac{r_1}{\sigma_n^2} e^{-(r_1^2 + A_1^2)/2\sigma_n^2} I_0\left(\frac{A_1 r_1}{\sigma_n^2}\right) dr_1 \tag{5-61}$$

where A_1 is the amplitude of the signal out of the bandpass filter. The integral in Eq. (5-61) leads to a quantity called the Marcum Q function and is not expressible in closed form. However, for large SNR it can be approximated as follows:

$$I_0(v) \simeq \frac{e^v}{\sqrt{2\pi v}} \qquad v \gg 1 \tag{5-62}$$

$$P_1 \doteq 1 - \int_\eta^\infty \frac{r_1}{\sigma_n^2} e^{-(r_1^2 + A_1^2)/2\sigma_n^2} \frac{e^{A_1 r_1/\sigma_n^2}}{\sqrt{2\pi A_1 r_1/\sigma_n^2}} \, dr_1 \tag{5-63}$$

$$= 1 - \int_\eta^\infty \frac{\sqrt{r_1/A_1}}{\sqrt{2\pi}\sigma_n} e^{-(r_1 - A_1)^2/2\sigma_n^2} \, dr_1 \tag{5-64}$$

The integrand in Eq. (5-64) is sharply peaked at $r_1 = A_1$ and the slowly varying factor $\sqrt{r_1/A_1}$ can be replaced by its value at the peak without introducing significant error. This gives

$$P_1 = 1 - \int_\eta^\infty \frac{1}{\sqrt{2\pi}\sigma_n} e^{-(r_1 - A_1)^2/2\sigma_n^2} \, dr_1 \tag{5-65}$$

$$= 1 - Q\left(\frac{\eta - A_1}{\sigma_n}\right) \tag{5-66}$$

The optimum threshold occurs at the intersection of the two PDFs. This is a function of the SNR and for large SNRs is given by

$$\eta \doteq \frac{A_1}{2} \tag{5-67}$$

The error probabilities are then

$$P_0 = e^{-A_1^2/8\sigma_n^2} \tag{5-68}$$

$$P_1 = 1 - Q\left[\frac{A_1/2 - A_1}{\sigma_n}\right] = Q\left(\frac{A_1}{2\sigma_n}\right) \tag{5-69}$$

$$\simeq \sqrt{\frac{\sigma_n}{\pi A_1}}\, e^{-A_1^2/8\sigma_n^2} \tag{5-70}$$

Using this last approximation for the Q function, the total error probability can be written as

$$P_e = \frac{1}{2} P_0 + \frac{1}{2} P_1 \tag{5-71}$$

$$= \frac{1}{2} e^{-A_1^2/8\sigma_n^2}\left[1 + \sqrt{\frac{\sigma_n}{\pi A_1}}\right] \tag{5-72}$$

For large SNR, $A_1 \gg \sigma_n$ and the error probability is given closely by

$$P_e \simeq \frac{1}{2} e^{-A_1^2/8\sigma_n^2} \tag{5-73}$$

The term in Eq. (5-72) that was dropped for the large SNR case corresponded to the errors occurring when detecting ones. Clearly virtually all of the errors occur when detecting zeros.

The amplitude of the signal out of the bandpass filter is approximately A_1 at the sampling instant and the variance of the noise is $\sigma_n^2 = N_0/T$. Substituting these

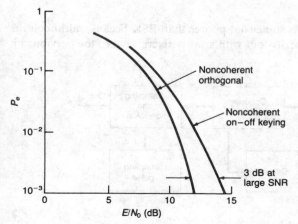

Figure 5-13 Error probabilities for noncoherent on–off keying and noncoherent orthogonal signaling.

values into Eq. (5-73) gives

$$P_e = \frac{1}{2} \exp\left[\frac{-A_1^2 T}{8N_0}\right] = \frac{1}{2} \exp\left[\frac{-E}{4N_0}\right] \tag{5-74}$$

where E is the signal energy of $s_1(t)$. In terms of the average signal energy E_a, the error probability is

$$P_e = \frac{1}{2} \exp\left[\frac{-E_a}{2N_0}\right] \tag{5-75}$$

Error curves for noncoherent on–off keying and noncoherent orthogonal signaling are given in Fig. 5-13.

5-3 DIFFERENTIAL PHASE SHIFT KEYING (DPSK)

It is possible to generate a type of coherent detection processor without using a coherent local oscillator. This is done by the use of differential phase shift keying in which consecutive bits are given the same phase for a one and opposite phase for a zero. The local oscillator is replaced by the signal delayed an amount equal to the bit spacing. This is illustrated in Fig. 5-14.

Differential encoding starts arbitrarily with the first bit and then indicates message bits by phase transition or no phase transition. As an example, consider the following bit sequence and encoded message.

Message	1 1 0 1 0 1 1 0 0 1
Encoded message	1 1 1 0 0 1 1 1 0 1 1
Phase	0 0 0 π π 0 0 0 π 0 0

The performance of DPSK is somewhat poorer than PSK because, although the detection is coherent, it is carried out with a noisy reference (i.e., the previous bit

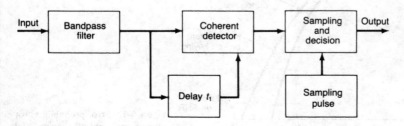

Figure 5-14 DPSK detector.

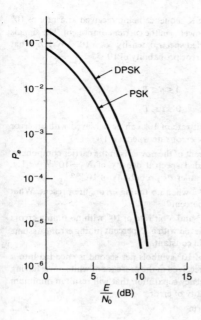

Figure 5-15 Error probabilities for PSK and DPSK.

consisting of signal plus noise). The error probability is[4]

$$P_e = \frac{1}{2} e^{-A^2/2\sigma_n^2} = \frac{1}{2} e^{-E/N_0}$$

The error performances of PSK and DPSK are shown in Fig. 5-15.

5-4 REFERENCES

1. Bennett, W. R., and J. R. Davey: *Data Transmission*, McGraw-Hill Book Company, New York, 1965.
2. De Jaeger, F., and C. B. Dekker: Tamed Frequency Modulation, A Novel Method to Achieve Spectral Economy in Digital Transmission, *IEEE Trans. Comm.*, vol. COM-26, no. 6, pp. 534–42, May 1978.
3. Weber, C.: *Elements of Detection and Signal Theory*, McGraw-Hill Book Company, New York, 1968.
4. Downing, J. J.: *Modulation Systems and Noise*, Prentice-Hall, Inc., Englewood Cliffs, N.J., 1964.

5-5 PROBLEMS

5.1-1 A binary PSK signal is decoded coherently in the presence of white noise having a one-sided spectral density of 10^{-6}. If the average signal power is 1 watt, what is the highest bit rate at which a probability of error of 10^{-5} can be achieved?

5.1-2 A matched filter is employed to decode a binary PSK sequence being received at a rate of 10^6 symbols per second. In the PSK signals, 10 percent of the power is in the carrier component. The signals are received in the presence of white noise having a one-sided spectral density $N_0 = 10^{-6}$ W/Hz. What average transmitted power level is required to achieve an error probability of 10^{-6}?

5.1-3 A binary PSK system employs signals of the form

$$s_1(t) = A \sin(\omega_c t + \cos^{-1} \alpha) \qquad 0 \leq t \leq T$$

$$s_0(t) = A \sin(\omega_c t - \cos^{-1} \alpha) \qquad 0 \leq t \leq T$$

Assume that one cycle occurs in the interval $(0, T)$. If a bit rate of 10^6 can be achieved with an error rate of 10^{-6} for $\alpha = 0$, what bit rate is possible for the same error rate when $\alpha = 1/\sqrt{2}$?

5.1-4 A PSK signal has bit durations of 100 ns and 20 percent of the power is in the carrier component. The signals are received in white noise having a one-sided spectral density of $N_0 = 10^{-10}$ V^2/Hz. What average received power level is required to achieve an error probability of 10^{-6}?

5.1-5 A PSK system operates with a bit error rate of 10^{-5} when no timing errors are present. What error rate would result for bit timing errors of 10 and 20 percent?

5.1-6 A PSK system operates with a bit error rate of 10^{-5} and a bit rate of 10^6 with no timing errors present. For the same error rate, what bit rate could be achieved with a 15 percent timing error? Assume the signal duration is $2\pi/\omega_c$ and the energy and noise remain constant.

5.1-7 A sequence of binary symbols produced at a rate of 10^6 symbols per second is encoded into a coherent FSK signal with an average power of 1 watt. It is received in the presence of white noise having a one-sided spectral density of 10^{-6} W/Hz. Find the frequency separation that will result in minimum probability of error and the value of the resulting probability of error.

5.1-8 A binary FSK system uses two waveforms of the form

$$s_1(t) = \sin(2\pi f_1 t + \theta_1) \qquad 0 \leq t \leq T$$

$$= 0 \qquad \text{elsewhere}$$

and

$$s_2(t) = \sin(2\pi f_2 t + \theta_2) \qquad 0 \leq t \leq T$$

$$= 0 \qquad \text{elsewhere}$$

Find the minimum frequency difference $(f_2 - f_1)$ such that these signals will be orthogonal for any relative phases.

5.1-9 A random binary sequence with a bit rate of 100 kb/s is encoded into a coherent FSK signal and is received with an average power level of 10^{-3} watts in the presence of white noise having a one-sided spectral density of $N_0 = 10^{-9}$ W/Hz.

(a) Find the frequency separation that will give the minimum probability of error and calculate the corresponding error probability.

(b) Find the minimum frequency separation that will make the signals orthogonal and calculate the corresponding error probability.

5.1-10 Two equally likely signals are transmitted as rectangular RF pulses with the same frequency but different durations. One signal has a duration of 24 μs and the other a duration of 48 μs. Assuming that the pulse amplitudes are 1 μV and the receiver noise is white with a one-sided spectral density of $N_0 = 10^{-18}$ V^2/Hz, find the minimum probability of error for coherent detection.

5.2-1 A noncoherent FSK system, with equally probable message states and orthogonal signals, operates with $E_b/N_0 = 20$ and an equivalent noise bandwidth that is twice the bit rate. Find the probability of error.

5.2-2 A binary communication system uses signals of the form

$$s_1(t) = \cos(\omega_1 t + \theta_1) \qquad 0 \leq t \leq \frac{3T}{4}$$

$$s_0(t) = \cos(\omega_0 t + \theta_0) \qquad \frac{T}{4} \le t \le T$$

and the two signals are equally probable. The bit rate is 30,000 b/s.

(a) Find the smallest frequency separation, in hertz, for which these signals will be orthogonal for any phase angles θ_0 and θ_1.

(b) The signals are to be detected noncoherently in the presence of white noise having a one-sided spectral density of 2×10^{-6} W/Hz. Find the resulting probability of error.

5.2-3 Find the optimum frequency separation of Δf for discriminator detection of FSK signals using a second-order Butterworth bandpass filter having a power transfer function of

$$|H_{BP}(\omega)|^2 = \frac{1}{1 + \left(\dfrac{\omega}{2\pi B_n}\right)^4}$$

5.2-4 An on–off keying system uses a signal of the form

$$s_1(t) = \left(1 - \cos\frac{2\pi t}{T}\right)\cos\omega_1 t \qquad 0 \le t \le T$$

The bit rate is 30,000 b/s. This signal is received in the presence of white noise having a one-sided spectral density of 4×10^{-6} V^2/Hz. This signal is detected noncoherently in a receiver whose threshold is set at one half of the maximum output signal value. Find the approximate probability of error.

5.2-5 Two equally likely signals are transmitted using a noncoherent FSK system with orthogonal signals. The equivalent noise bandwidth is twice the bit rate, the received signal amplitudes are 1 μV, the signal durations are 10 μs, and the one-sided noise spectral density is 2.5×10^{-19} V^2/Hz.

(a) Find the probability of error for noncoherent detection.

(b) What would be the probability of error for an optimum noncoherent FSK system?

5.2-6 A sequence of binary symbols produced at a rate of 10^6 symbols per second is encoded into a coherent FSK signal with an average power of 1 watt. The signal is received in the presence of white noise having a one-sided spectral density of 10^{-6} W/Hz. Find the frequency separation required for the signals to be orthogonal and the error probability for noncoherent detection.

5.2-7 Repeat Problem 5.2-6 assuming the frequency separation is that required for minimum-shift keying.

5.3-1 A binary sequence is encoded using differential PSK. This signal is decoded in the presence of white noise having a one-sided spectral density of 10^{-6} W/Hz. Assuming the average signal power is 1 watt, find the highest bit rate at which a probability of error of 10^{-5} can be achieved. Compare this result with the bit rate that could be achieved using coherent detection of the PSK signal.

SIX

M-ARY SIGNALS

6-1 ORTHOGONAL REPRESENTATIONS[1,2]

It is often convenient to represent a set of signals in terms of an arbitrary set of orthonormal (ON) basis functions. There are several reasons why this is so. In the first place, it is mathematically convenient to do so since many of the calculations involving signals can be carried out more conveniently in terms of orthonormal functions. In the second place, it makes it possible to visualize the signals by means of graphical representations that place in evidence the relationships among the various types of signals that will be considered.

An orthonormal set of basis functions must have the properties that every member of the set is orthogonal to every other member and every member must have unit energy. A mathematical definition of an orthonormal set $\phi_q(t)$, $q = 1, 2, \ldots, N$ over the interval $(0, T)$ is

$$\int_0^T \phi_i(t)\phi_k(t)\,dt = \delta_{ik} = 1 \qquad i = k$$

$$= 0 \qquad i \neq k \qquad (6\text{-}1)$$

These basis functions may be used to define an arbitrary set of M finite-energy waveforms by

$$s_i(t) = \sum_{j=1}^N s_{ij}\phi_j(t) \qquad 0 \leq t \leq T, i = 1, 2, \ldots, M \qquad (6\text{-}2)$$

where

$$s_{ij} = \int_0^T s_i(t)\phi_j(t)\,dt \qquad (6\text{-}3)$$

The coefficients in Eq. (6-2), s_{ij}, can be represented by a vector

$$\mathbf{s}_i = s_{i1}, s_{i2}, \ldots, s_{iN})$$

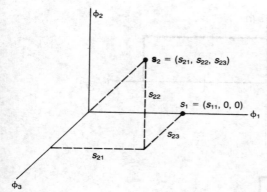

Figure 6-1 Three-dimensional signal space.

If the vector of coefficients is known and the basis set is known, the corresponding waveform is completely specified.

All possible linear combinations of the basis functions form a linear space known as a signal space. The coordinate axes in the signal space are the basis functions themselves, $\phi_1(t), \phi_2(t) \ldots \phi_n(t)$. These are usually abbreviated as $\phi_1, \phi_2 \ldots \phi_n$. Any signal formed from the basis functions can be represented as a point in the signal space, the point being the tip of the signal vector, $\mathbf{s}_i$. As an illustration consider a three-dimensional signal space—that is, one having three basis functions—and consider two different signal vectors in that space as shown in Fig. 6-1. The signal vector $\mathbf{s}_1$ is one whose ϕ_2 and ϕ_3 components are both zero (ϕ_1 being the only nonzero component). On the other hand, the signal vector $\mathbf{s}_2$ has all of its components nonzero. It is clear that by appropriately picking the components of the vector, any signal in the signal space can be represented in this geometrical manner.

Example of an Orthonormal Representation

To illustrate this method of signal representation explicitly, consider a set of three signals having the form shown in Fig. 6-2. It is clear that each of these signals has an energy of 2, but that they are not orthogonal and they occupy different time intervals. However, a possible set of orthogonal basis functions might be the two shown in Fig. 6-3, since combinations of these two basis functions can be used to represent any of the three signals of Fig. 6-2.

The coefficients that represent signal $s_1(t)$ can be obtained by using the defining relationship in Eq. (6-3). Thus in this case s_{11} and s_{12} become

$$s_{11} = \int_0^2 s_1(t)\phi_1(t)\,dt = \int_0^1 (1)(1)\,dt = 1$$

$$s_{12} = \int_0^2 s_1(t)\phi_2(t)\,dt = \int_1^2 (-1)(1)\,dt = -1$$

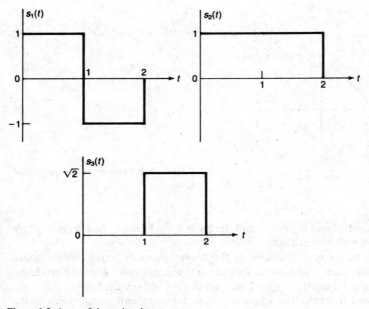

Figure 6-2 A set of three signals.

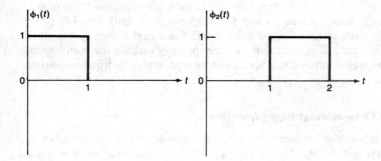

Figure 6-3 A possible set of basis functions for the signals of Fig. 6-2.

In a similar way, one can calculate the coefficients s_{21}, s_{22}, s_{31}, and s_{32}, and these become

$$s_{21} = 1 \qquad s_{31} = 0$$
$$s_{22} = 1 \qquad s_{32} = \sqrt{2}$$

Finally the three vectors that represent the three signals become

$$\mathbf{s}_1 = (1, -1)$$

$$\mathbf{s}_2 = (1, 1)$$

$$\mathbf{s}_3 = (0, \sqrt{2})$$

These signals can be represented graphically in a two-dimensional plane since there are only two basis functions required in this case. This graphical representation is as shown in Fig. 6-4.

The choice of basis functions shown in Fig. 6-3 is not unique and many other possibilities exist. To illustrate this consider a different set of basis functions as shown in Fig. 6-5. These functions are also orthogonal and they have unit energy, as is required. The coefficients can now be determined for signal $s_1(t)$ as

$$s_{11} = \int_0^2 s_1(t)\phi_1(t)\, dt = \int_0^1 \left(\frac{1}{\sqrt{2}}\right)(1)\, dt + \int_1^2 \left(\frac{-1}{\sqrt{2}}\right)(-1)\, dt = \sqrt{2}$$

$$s_{12} = \int_0^2 s_1(t)\phi_2(t)\, dt = \int_0^1 \left(\frac{1}{\sqrt{2}}\right)(1)\, dt + \int_1^2 \left(\frac{-1}{\sqrt{2}}\right)(1)\, dt = 0$$

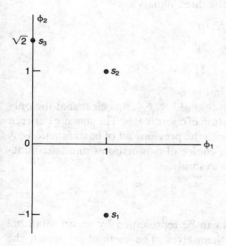

Figure 6-4 Graphical representation of the signals of Fig. 6-2.

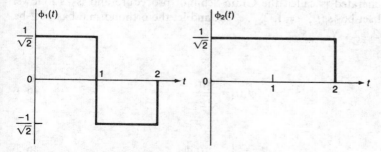

Figure 6-5 A different set of basis functions for the signals of Fig. 6-2.

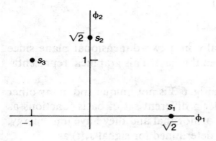

Figure 6-6 Graphical representation when the basis functions are those of Fig. 6-5.

In a similar way the coefficients for the other two signals become

$$s_{21} = 0 \qquad s_{31} = -1$$
$$s_{22} = \sqrt{2} \qquad s_{32} = 1$$

Therefore, the vector representations for the three signals are

$$\mathbf{s}_1 = (\sqrt{2}, 0)$$
$$\mathbf{s}_2 = (0, \sqrt{2})$$
$$\mathbf{s}_3 = (-1, 1)$$

These signals are also shown graphically in Fig. 6-6.

In comparing this representation with that of Fig. 6-4, it is clear that the only change that has taken place has been a rotation of coordinates. The spacing between all pairs is still exactly the same as it was with the previous set of basis functions. A similar result would occur with any other choice of two orthonormal basis functions that can represent the signals in this example.

Gram-Schmidt Procedure

Any finite set of finite energy waveforms can be represented by an orthonormal basis set derived from the waveforms themselves. The method by which the basis set is generated is called the Gram-Schmidt procedure and is as follows. Let the signal set be $\{s_i(t)\}$, $i = 1, 2, \ldots, M$, and let the orthonormal basis set be $\{\phi_j(t)\}$, $j = 1, 2, \ldots, N$.

1. Choose

$$\phi_1(t) = \frac{s_1(t)}{\sqrt{E_1}} \tag{6-4}$$

where

$$E_1 = \int_0^T s_1^2(t)\, dt \tag{6-5}$$

The signal $s_1(t)$ can now be represented as

$$s_1(t) = s_{11}\phi_1(t) \tag{6-6}$$

and from Eq. (6-3)

$$s_{11} = \int_0^T \frac{1}{\sqrt{E_1}} s_1^2(t)\, dt = \sqrt{E_1}$$

2. Define an auxiliary function $\Theta_2(t)$ as

$$\Theta_2(t) = s_2(t) - s_{21}\phi_1(t) \tag{6-7}$$

where

$$s_{21} = \int_0^T s_2(t)\phi_1(t)\, dt$$

The second basis function is then

$$\phi_2(t) = \frac{\Theta_2(t)}{\sqrt{E_{\Theta_2}}} \tag{6-8}$$

where

$$E_{\Theta_2} = \int_0^T \Theta_2^2(t)\, dt \tag{6-9}$$

The signal $s_2(t)$ can then be represented as

$$s_2(t) = s_{21}\phi_1(t) + s_{22}\phi_2(t) \tag{6-10}$$

The basis functions $\phi_1(t)$ and $\phi_2(t)$ each have unit energy by virtue of the way they were derived. They are also orthogonal to each other, as can be seen from the following.

$$\begin{aligned}
\int_0^T \phi_1(t)\phi_2(t)\, dt &= \frac{1}{\sqrt{E_{\Theta_2}}} \int_0^T \phi_1(t)\Theta_2(t)\, dt \\
&= \frac{1}{\sqrt{E_{\Theta_2}}} \int_0^T \phi_1(t)[s_2(t) - s_{21}\phi_1(t)]\, dt \\
&= \frac{1}{\sqrt{E_{\Theta_2}}} \left[\int_0^T s_2(t)\phi_1(t)\, dt - s_{21} \int_0^T \phi_1^2(t)\, dt \right] \\
&= \frac{1}{\sqrt{E_{\Theta_2}}} [s_{21} - s_{21}] = 0
\end{aligned}$$

3. The procedure is continued in the same manner by defining at each step an auxiliary function that is the signal minus the components of the signal that can be represented by the basis functions already derived. This auxiliary

function is then normalized by its energy. For example, at the jth step,

$$\Theta_j(t) = s_j(t) - s_{j1}\phi_1(t) - s_{j2}\phi_2(t) \cdots - s_{j,j-1}\phi_{j-1}(t)$$

$$\phi_j(t) = \frac{\Theta_j(t)}{\sqrt{E_{\Theta_j}}}$$

In using this procedure, the number of basis functions will always be less than or equal to the number of signals in the set.

As an example of this procedure consider the signals of Fig. 6-2. Starting with $s_1(t)$, it is found that

$$\phi_1(t) = \frac{s_1(t)}{\sqrt{E_1}} = \frac{1}{\sqrt{2}} s_1(t) \tag{6-11}$$

This waveform is the first waveform of Fig. 6-5. The auxiliary function is

$$\Theta_2(t) = s_2(t) - s_{21}\phi_1(t) \tag{6-12}$$

$$s_{21} = \int_0^T s_2(t)\phi_1(t)\, dt = 0 \tag{6-13}$$

$$\phi_2(t) = \frac{\Theta_2(t)}{\sqrt{E_{\Theta_2}}} = \frac{s_2(t)}{\sqrt{E_2}} = \frac{1}{\sqrt{2}} s_2(t) \tag{6-14}$$

This is the second waveform shown in Fig. 6-5. The final auxiliary function is

$$\Theta_3(t) = s_3(t) - s_{31}\phi_1(t) - s_{32}\phi_2(t) \tag{6-15}$$

$$s_{31} = \int_0^T s_3(t)\phi_1(t)\, dt = \frac{1}{\sqrt{2}} \int_1^2 -\sqrt{2}\, dt = -1 \tag{6-16}$$

$$s_{32} = \int_0^T s_3(t)\phi_2(t)\, dt = \frac{1}{\sqrt{2}} \int_1^2 \sqrt{2}\, dt = 1 \tag{6-17}$$

$$\Theta_3(t) = s_3(t) + \phi_1(t) - \phi_2(t) \tag{6-18}$$

Examination of the waveforms and basis functions shows that $\Theta_3(t) = 0$. Therefore, the entire signal set can be represented in terms of two basis functions. The vector representation of the three signals relative to the above basis set is

$$\mathbf{s}_1 = (\sqrt{2}, 0) \tag{6-19}$$

$$\mathbf{s}_2 = (0, \sqrt{2}) \tag{6-20}$$

$$\mathbf{s}_3 = (-1, 1) \tag{6-21}$$

If the procedure had been started with a different signal initially [e.g., $s_3(t)$], the basis set would have been different. However, the geometrical relationship between the signal vectors would be the same. Starting with a different signal is equivalent to a rotation of the coordinate system.

Signal Space Concepts

There are many characteristics of signals that can be obtained more conveniently from the vectorial representation than from analytic expressions. For example, the signal energy is readily determined as

$$E_i = \int_0^T s_i^2(t)\, dt = \int_0^T \sum_{j=1}^N \sum_{k=1}^N s_{ij} s_{ik} \phi_j(t) \phi_k(t)\, dt$$

$$= \sum_{j=1}^N \sum_{k=1}^N s_{ij} s_{ik} \delta_{jk} = \sum_{j=1}^N s_{ij}^2 \tag{6-22}$$

The signal energy is seen from Eq. (6-22) to be equal to the sum of the squares of the coefficients that are used to represent the signal. Geometrically this is the square of the distance of the signal from the origin. An alternative representation for this is the dot product or

$$E_i = \mathbf{s}_i \cdot \mathbf{s}_i = \|\mathbf{s}_i\|^2 \tag{6-23}$$

When binary signals were being considered, it was found that the correlation coefficient between the signals was an important quantity in determining how well one signal could be distinguished from the other. In the case of a larger number of signals, this is still true. Therefore, it is convenient to define the correlation coefficient between the ith signal and the jth signal and this is given by

$$\rho_{ij} = \frac{\displaystyle\int_0^T s_i(t) s_j(t)\, dt}{\sqrt{E_i E_j}} = \frac{\displaystyle\sum_{k=1}^N s_{ik} s_{jk}}{\sqrt{E_i E_j}} \tag{6-24}$$

$$= \frac{\mathbf{s}_i \cdot \mathbf{s}_j}{\sqrt{E_i E_j}} \tag{6-25}$$

An alternative representation is

$$\rho_{ij} = \cos \theta_{ij} \tag{6-26}$$

where θ_{ij} is the angle between the vector $\mathbf{s}_i$ and the vector $\mathbf{s}_j$. Even though θ_i is an angle measured in N-dimensional space, it has the same significance as it does in two or three dimensions.

Another quantity that is of considerable utility in determining how well one signal can be distinguished from another is the concept of distance. The distance between two signals in the signal space representation is the Euclidean distance between the tips of the vectors representing them. The corresponding (and exactly equal) quantity in the time domain is a measure of how different the signals are.

With this in mind, the square of the distance between two signals is defined as

$$d_{ij}^2 = \int_0^T [s_i(t) - s_j(t)]^2 \, dt$$

$$= \sum_{k=1}^N s_{ik}^2 + \sum_{k=1}^N s_{jk}^2 - 2 \sum_{k=1}^N s_{ik}s_{jk}$$

$$= E_i + E_j - 2\rho_{ij}\sqrt{E_i E_j} \tag{6-27}$$

Note that the distance between two signals depends only upon the energy of the signals and the correlation coefficient between them. Note also that the distance between two signals will always be greater if the correlation coefficient between them is negative. On the other hand, if the correlation coefficient between two signals is one, and the signals have equal energy, then the distance between them is zero.

It is useful to illustrate these concepts with the signals defined in Fig. 6-2. Using the values of the coefficients determined there, the three signal energies, the three correlation coefficients, and the three distances can be determined as

$$E_1 = 1^2 + (-1)^2 = 2$$

$$E_2 = 1^2 + 1^2 = 2$$

$$E_3 = 0^2 + (\sqrt{2})^2 = 2$$

$$\rho_{12} = \frac{1 \cdot 1 + (-1) \cdot 1}{\sqrt{2 \cdot 2}} = 0$$

$$\rho_{13} = \frac{1 \cdot 0 + (-1) \cdot \sqrt{2}}{\sqrt{2 \cdot 2}} = -0.707$$

$$\rho_{23} = \frac{1 \cdot 0 + 1 \cdot \sqrt{2}}{\sqrt{2 \cdot 2}} = \frac{1}{\sqrt{2}} = 0.707$$

$$d_{12} = [2 + 2 - 2 \cdot 0 \cdot \sqrt{2 \cdot 2}]^{1/2} = 2$$

$$d_{13} = \left[2 + 2 - 2 \cdot \frac{-1}{\sqrt{2}} \cdot \sqrt{2 \cdot 2}\right]^{1/2} = 2.613$$

$$d_{23} = \left[2 + 2 - 2 \cdot \left(\frac{1}{\sqrt{2}}\right) \cdot \sqrt{2 \cdot 2}\right]^{1/2} = 1.082$$

The distance between signals is an indication of how easily they can be distinguished from one another in the presence of noise. It is usually desirable to make this distance as large as possible. This point is considered in more detail when the detection of signal sets containing more than two signals is discussed.

6-2 *M*-ARY ORTHOGONAL SIGNALS

Signal sets containing M signals are said to be M-ary signal sets. There are many different classes of M-ary signal sets, but only a few of them are useful for communication purposes. These will be discussed in this and subsequent sections. One very important class of M-ary signals is the orthogonal signal set. The requirement for a set of M signals to be orthogonal is that the correlation coefficient between any pair of signals be zero. That is,

$$\rho_{ij} = 0 \qquad i \neq j, i, j = 1, 2, \ldots, M \tag{6-28}$$

It can be shown readily that the dimensionality of the signal space used to represent M orthogonal signals must be

$$N \geq M$$

It can also be shown (e.g., by the Gram-Schmidt procedure) that N need not be larger than M. Therefore, in most cases it is convenient to let

$$N = M$$

A possible form for M orthogonal signals is to let the signals be proportional to the basis functions since the basis functions are necessarily orthogonal. That is, let

$$s_i(t) = \sqrt{E}\,\phi_i(t) \qquad i = 1, 2, \ldots, M \tag{6-29}$$

where the coefficient $\sqrt{E}$ is used to ensure that the signal has an energy of E, since the basis functions themselves have unit energy. Note that in this method of defining M-ary orthogonal signals, all of the signals have equal energy. This assumption will be retained throughout the discussion, although it is not necessary to do so. Orthogonal signal sets may have different energies, and in some cases do have.

Multiple Frequency Shift Keying (MFSK)

A common form of orthogonal signal set is the frequency-shift-keyed set. This class of signals consists of sinusoids having a duration of T and in this interval having one frequency selected from a set of M possible frequencies. This can be represented mathematically as

$$s_i(t) = \sqrt{\frac{2E}{T}}\,\sin(\omega_i t + \theta_i) \qquad 0 \leq t \leq T \quad i = 1, 2, \ldots, M \tag{6-30}$$

where ω_i is equal to $\omega_0 + i\,\Delta\omega$, $\Delta\omega$ being the smallest frequency difference between any two signals in the set. For the signals to be orthogonal, it is necessary that $\Delta\omega$ be sufficiently large. This requirement is

$$\Delta\omega = 2\pi\,\Delta f; \quad \Delta f = \frac{1}{T}$$

Although some choices of $\Delta\omega$ smaller than this may produce orthogonality under special conditions, this is the minimum frequency spacing that is necessary for the signals to be orthogonal when there is an arbitrary set of phases θ_i. Since each signal in the signal set is a sinusoidal pulse, its energy spectrum is a sinc function in frequency. The energy spectrum of the entire set then is the super-position of these, displaced by frequencies of $1/T$. This is illustrated in Fig. 6-7. The bandwidth for the entire set is usually defined as the difference in frequency between the closest nulls of the lowest frequency and the highest frequency members of the set. From the figure it is readily seen that this bandwidth is approximately $(M + 1)/T$.

Quantized Pulse-Position Modulation (QPPM)

The MFSK signal set is orthogonal by virtue of the signals being displaced in frequency. An alternative procedure is to cause the signals to be displaced in time. This is illustrated in Fig. 6-8, in which there are pulses $g(t)$, identical in shape but appearing at different time intervals within the overall time interval of T.

To transmit a message, one would transmit one of these pulses within each interval of T duration. Hence the ith signal can be defined as

$$s_i(t) = \sqrt{E}g\left[t - (i-1)\frac{T}{M}\right] \qquad i = 1, 2, \ldots, M \qquad (6-31)$$

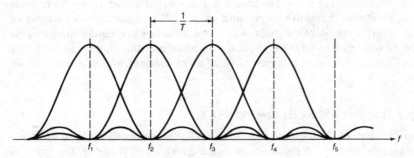

Figure 6-7 Energy spectrum for an MFSK signal.

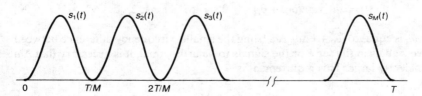

Figure 6-8 Quantized pulse-position modulation signals.

where

$$g(t) = 0 \quad \text{outside of } 0 \le t \le T/M$$

and

$$\int_0^{T/M} g^2(t)\, dt = 1$$

Although the quantized pulse-position modulation signals shown here are an illustration of orthogonal signals, in most applications of QPPM, they are not orthogonal. Instead the signals are closer together than T/M and a significant positive correlation exists between any pair of signals. This is done to conserve bandwidth.

Orthogonal Binary Vectors

A communication channel that is basically binary can also be used to transmit a set of M orthogonal signals by defining binary signals that are orthogonal. An illustration of one such signal is shown in Fig. 6-9. It consists of M bits having a particular sequence and existing over the time interval from 0 to T.

One can then define other sequences of M bits in the same time interval, such that all pairs of signals are mutually orthogonal. One way of achieving this orthogonality is to define the signal vectors as coming from the rows of a Hadamard matrix. A Hadamard matrix is a square $(n \times n)$ matrix all of whose rows have elements that are either $+1$ or -1, and each row has an arrangement of $+1$'s and -1's such that it is orthogonal to every other row. Hadamard matrices of large dimensionality can be obtained iteratively from those of smaller dimensionality. The sequence starts with the matrix $\mathbf{H}_1$, which is

$$\mathbf{H}_1 = \begin{bmatrix} 1 & 1 \\ 1 & -1 \end{bmatrix} \tag{6-32}$$

The nth matrix in the sequence is related to the $(n-1)$th matrix by

$$\mathbf{H}_n = \begin{bmatrix} \mathbf{H}_{n-1} & \mathbf{H}_{n-1} \\ \mathbf{H}_{n-1} & -\mathbf{H}_{n-1} \end{bmatrix} \tag{6-33}$$

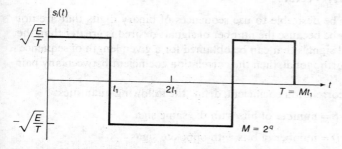

Figure 6-9 A binary signal from an orthogonal set.

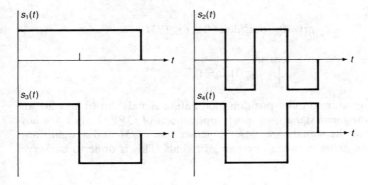

Figure 6-10 The orthogonal binary vectors for $M = 4$.

Note that the dimensionality of the matrix will always be a power of 2. In particular $N = 2^n$. As a simple example of orthogonal binary vectors, consider the second-order Hadamard matrix:

$$\mathbf{H}_2 = \begin{bmatrix} 1 & 1 & 1 & 1 \\ 1 & -1 & 1 & -1 \\ 1 & 1 & -1 & -1 \\ 1 & -1 & -1 & 1 \end{bmatrix}$$

The rows of this matrix define the state of the bits in each of four different signals, as illustrated in Fig. 6-10. These signals are all mutually orthogonal and they all have the same signal energy.

6-3 SIGNALS RELATED TO ORTHOGONAL SIGNALS

Nonorthogonal Binary Vectors

In some cases it may be desirable to use sequences of binary digits that are not orthogonal. This may be because the number of signals desired is greater than the number of orthogonal signals that can be obtained for a given length of sequence. If the signals are not orthogonal, then the correlation coefficient between any pair of signals is not zero.

To evaluate this correlation coefficient, define the following quantities:

$$S = \text{number of bits with the same sign}$$

$$D = \text{number of bits with opposite signs}$$

$$N = S + D$$

Then the correlation coefficient is given by

$$\rho_{ij} = \frac{1}{E} \sum_{k=1}^{N} s_{ik} s_{jk} = \frac{1}{E} \sum_{k=1}^{N} \pm \frac{E}{N}$$

$$= \frac{S - D}{N} = \frac{S - D}{S + D} \tag{6-34}$$

The distance between two binary vectors can be obtained from Eq. (6-27), modified by the condition that all signals have the same energy. Thus the distance squared is

$$d_{ij}^2 = 2E(1 - \rho_{ij}) = 2E\left(1 - \frac{S - D}{S + D}\right) = \frac{4ED}{N} \tag{6-35}$$

It may be noted from Eq. (6-34) that the required condition for orthogonal signals is that $S = D$. Since all signals in the set are to be mutually orthogonal, this must be true for every pair of signals in the set.

Biorthogonal Signals

A very important class of M-ary signals is one in which the signals are not all mutually orthogonal. In fact there are really two sets of signals. There is a set containing $M/2$ signals that are mutually orthogonal, and another set containing $M/2$ signals that are mutually orthogonal; but the two sets are antipodal to one another. To create a set of biorthogonal signals, one would start with a set of $M/2$ orthogonal signals and add to it the negative of each of these signals. This gives a signal set of size M, but the required dimensionality of the space is only $M/2$. This may be an advantage in some applications. One way of defining biorthogonal signals is to use the basis functions just as was done in the case of orthogonal signals. In this case, however, each basis function is used for two signals as shown here:

$$s_1(t) = \sqrt{E}\phi_1(t)$$

$$s_2(t) = -\sqrt{E}\phi_1(t)$$

$$\vdots$$

$$S_{M-1}(t) = \sqrt{E}\phi_{M/2}(t)$$

$$s_M(t) = -\sqrt{E}\phi_{M/2}(t)$$

The dimensionality of the signal space must be greater than or equal to $M/2$ and in general N would be equal to $M/2$. A typical example of biorthogonal signals for $M = 4$ is illustrated in Fig. 6-11. Several particular types of biorthogonal signals are discussed in greater detail subsequently.

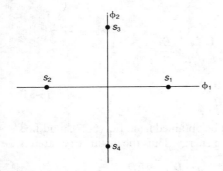

Figure 6-11 A biorthogonal signal set for $M = 4$.

Transorthogonal or Simplex Signals

A simplex signal set is one in which all signal pairs have the same correlation coefficient. Note that the orthogonal signal set is a special case of the transorthogonal set in which all correlation coefficients are zero. For equal-energy signals and with equal correlation coefficients, the distance between any two pairs of signals is

$$d^2 = 2E(1 - \rho) \tag{6-36}$$

To make this distance as large as possible, we would like to make ρ as negative as possible. The question, then, is to find out how negative ρ can be and still be the same value for all signal pairs. This question can be answered by considering the integral of the square of the sums of the signals. That is,

$$\int_0^T \left[\sum_{i=1}^M s_i(t) \right]^2 dt \geq 0 \tag{6-37}$$

since this integral must always be greater than zero. The square of the sum can be written as a double sum

$$\int_0^T \sum_{i=1}^M \sum_{j=1}^M s_i(t)s_j(t) \, dt \geq 0 \tag{6-38}$$

and this can be separated into two parts as

$$\int_0^T \left[\sum_{i=1}^M s_i^2(t) + \sum_{i=1}^M \sum_{\substack{j=1 \\ j \neq i}}^M s_i(t)s_j(t) \right] dt \geq 0 \tag{6-39}$$

The integral of s_i^2 becomes

$$\int_0^T s_i^2(t) \, dt = E \tag{6-40}$$

since all signals have the same energy. Likewise the integral of the product of the two signals becomes

$$\int_0^T s_i(t)s_j(t) \, dt = \rho E \tag{6-41}$$

because all signals also have the same correlation coefficient. Therefore, this result can be written as

$$ME + M(M-1)\rho E \geq 0 \tag{6-42}$$

It is straightforward to solve this for ρ and obtain the inequality

$$\rho \geq \frac{-1}{M-1} \tag{6-43}$$

This result says that the correlation coefficient can never be more negative than the quantity given above. Note that for $M = 2$, ρ is equal to -1, and hence we have an antipodal signal set. However, for larger values of M, ρ can never be as negative as -1. It must have some value closer to zero, although it can still be negative.

The correlation coefficient given by Eq. (6-43) can be used in the expression for the distance squared to obtain

$$d^2 = 2E(1-\rho) = 2E\left(1 - \frac{-1}{M-1}\right) = \frac{2EM}{M-1} \tag{6-44}$$

Since ρ has its most negative value, this represents the greatest distance that can be obtained between all pairs of signals. Thus simplex signals provide the minimum probability of error for a given signal energy.

The term simplex is derived from the geometrical concept that the distance d is the distance between the vertices of a regular simplex.

Construction of Simplex Signals

There is a straightforward procedure whereby one may construct a set of M equal-energy simplex signals. To do this start with a set of M equal-energy orthogonal signals, and define the average signal vector as

$$\mathbf{a} = \frac{1}{M} \sum_{q=1}^{M} \mathbf{s}_q \tag{6-45}$$

The simplex signal set is obtained by subtracting this average signal vector from each signal vector of the original orthogonal set. That is,

$$\mathbf{s}'_i = \mathbf{s}_i - \mathbf{a} \qquad i = 1, 2, \ldots, M \tag{6-46}$$

The energy of each signal in the simplex set can be obtained readily from this relationship. Specifically,

$$E'_i = (\mathbf{s}_i - \mathbf{a}) \cdot (\mathbf{s}_i - \mathbf{a}) = \mathbf{s}_i \cdot \mathbf{s}_i - 2\mathbf{s}_i \cdot \mathbf{a} + \mathbf{a} \cdot \mathbf{a}$$

$$= E - 2\frac{E}{M} + \frac{1}{M^2} \cdot ME$$

$$= E\left(1 - \frac{1}{M}\right) \tag{6-47}$$

It is also of interest to consider the sum of all of the vectors in the simplex signal set. That is, let

$$\sum_{q=1}^{M} \mathbf{s}_q' = \sum_{q=1}^{M} (\mathbf{s}_q - \mathbf{a}) = M\mathbf{a} - M\mathbf{a} = \mathbf{0} \qquad (6\text{-}48)$$

Since this sum is zero, this implies that any signal in the simplex signal set is a linear combination of the other signals in the set. Therefore, the dimensionality of the signal set is $M - 1$, instead of M.

As an illustration of the application of this technique to the generation of simplex signals, consider a very simple case for which $M = 3$. Start the design with the three orthogonal signals shown in Fig. 6-12. The vector representation of these three signals and the corresponding average signal vector is shown in Fig. 6-13.

Subtracting the average signal vector from each of the original orthogonal vectors leads to the simplex signal vectors:

$$\mathbf{s}_1' = (1, 0, 0) - (1/3, 1/3, 1/3) = (2/3, -1/3, -1/3)$$

$$\mathbf{s}_2' = (0, 1, 0) - (1/3, 1/3, 1/3) = (-1/3, 2/3, -1/3)$$

$$\mathbf{s}_3' = (0, 0, 1) - (1/3, 1/3, 1/3) = (-1/3, -1/3, 2/3)$$

The time function representation of these signals is shown in Fig. 6-14 and the vector representation in Fig. 6-15.

The energy of the simplex signal set is only 2/3, compared with the energy of unity for the original orthogonal set. However, the distance between any pair of

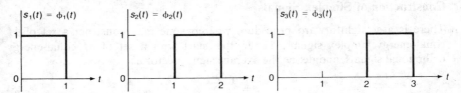

Figure 6-12 An orthogonal signal set for $M = 3$.

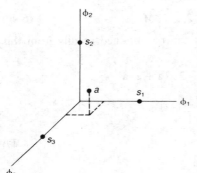

Figure 6-13 Vector representation of the signal set of Fig. 6-12.

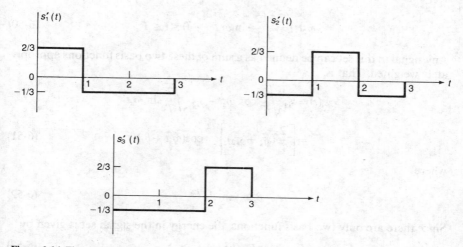

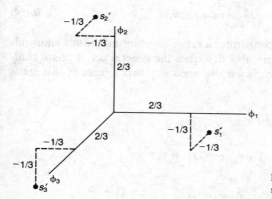

Figure 6-14 The simplex signal set derived from the signals of Fig. 6-12.

Figure 6-15 Vector representation of the simplex signal set of Fig. 6-14.

vectors in the simplex set is the same as it was in the original orthogonal set, and, therefore, the probability of error in detecting this signal in the presence of noise will be the same as it was for the orthogonal set. This point is discussed in more detail subsequently.

Signals from Quadrature Functions

A very important class of signals is one in which there are only two basis functions defined as

$$\phi_1(t) = \sqrt{\frac{2}{T}} \cos \omega_s t \qquad 0 \le t \le T \tag{6-49}$$

$$\phi_2(t) = \sqrt{\frac{2}{T}} \sin \omega_s t \qquad 0 \le t \le T \tag{6-50}$$

Any signal in this set can be defined as a sum of these two basis functions appropriately weighted. That is,

$$s_i(t) = s_{i1} \sqrt{\frac{2}{T}} \cos \omega_s t - s_{i2} \sqrt{\frac{2}{T}} \sin \omega_s t$$

$$= \left[\frac{2}{T} (s_{i1}^2 + s_{i2}^2) \right]^{1/2} \cos(\omega_s t + \theta_i) \tag{6-51}$$

where

$$\theta_i = \tan^{-1} \frac{s_{i2}}{s_{i1}} \tag{6-52}$$

Since there are only two basis functions, the energy in the signal set is given by

$$E = s_{i1}^2 + s_{i2}^2 \tag{6-53}$$

Therefore, Eq. (6-51) can be written as

$$s_i(t) = \sqrt{\frac{2E}{T}} \cos(\omega_s t + \theta_i) \tag{6-54}$$

It is clear that this set of signals constitutes a set of constant-amplitude sinusoids that differ only in phase. Therefore, this describes the general set of phase-shift-keyed signals. The correlation coefficient between any two signals in this set is obtained readily from

$$\rho_{ij} = \frac{1}{E} \int_0^T s_i(t) s_j(t) \, dt$$

$$= \frac{2}{T} \int_0^T \cos(\omega_s t + \theta_i) \cos(\omega_s t + \theta_j) \, dt$$

$$= \frac{1}{T} \int_0^T \left[\cos(\theta_i - \theta_j) + \cos(2\omega_s t + \theta_i + \theta_j) \right] dt$$

$$\simeq \cos(\theta_i - \theta_j) \tag{6-55}$$

When the phase angles are equally spaced, that is,

$$\theta_i = \frac{2\pi i}{M}$$

the signal set is known as a multiple-phase-shift-keyed (MPSK) or polyphase signal set. The correlation coefficient between any pair of polyphase signals is given by

$$\rho_{ij} = \cos \frac{2\pi(i-j)}{M} \tag{6-56}$$

Some of the signal classes that have already been discussed also are polyphase signals. This is observed by looking at values of $M = 2$, 3, and 4 as shown in the following.

$$M = 2 \quad \rho_{12} = \cos \pi = -1 \qquad \text{(antipodal)}$$

$$M = 3 \quad \rho_{12} = \cos 2\frac{\pi}{3} = -0.5$$

$$\rho_{13} = \cos 4\frac{\pi}{3} = -0.5 \qquad \text{(simplex)}$$

$$\rho_{23} = \cos 2\frac{\pi}{3} = -0.5$$

$$M = 4 \quad \rho_{12} = \rho_{23} = \rho_{34} = \rho_{41} = 0$$

$$\rho_{13} = \rho_{24} = -1 \qquad \text{(biorthogonal) (QPSK)}$$

6-4 COHERENT DETECTION OF *M*-ARY SIGNALS

The problem of detecting *M*-ary signal sets will now be considered. There are many ways in which this might be done, but it is basically a problem of deciding which of the *M* possible signals was transmitted given the signal actually received during the time interval of $(0, T)$. Furthermore, it is desired to make this decision in such a way as to minimize the probability of obtaining an incorrect decision. One of the possible techniques is illustrated by the matched-filter receiver shown in Fig. 6-16.

In this receiver there are *M* filters, each matched to one of the possible *M* signals. The outputs of the *M* filters are all sampled at the same time, these sample values are compared, and the largest one is selected. This leads to the so-called

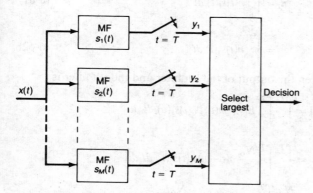

Figure 6-16 Matched-filter receiver for *M*-ary signals.

maximum-likelihood (ML) receiver for M-ary signal sets. To analyze the performance of such a receiver, let the input signal be

$$x(t) = s_i(t) + n(t) \qquad i = 1, 2, \ldots, M \tag{6-57}$$

where $n(t)$ is white noise that has a one-sided spectral density of N_0. The response of the ith matched filter to the signal s_i is

$$y_i = \int_0^T x(t)s_i(t) \, dt \qquad i = 1, 2, \ldots, M \tag{6-58}$$

On the other hand, the response of the ith filter, given that signal $s_j(t)$ was transmitted, plus noise, is given by

$$
\begin{aligned}
y_i|s_j &= \int_0^T [s_j(t) + n(t)]s_i(t) \, dt \\
&= \int_0^T s_j(t)s_i(t) \, dt + \int_0^T n(t)s_i(t) \, dt
\end{aligned}
\tag{6-59}
$$

This response can be separated into two parts—one that results from the signal component of the input and one that results from the noise component of the input. The signal component of the input is just the correlation between signals s_i and s_j and hence is

$$\int_0^T s_j(t)s_i(t) \, dt = E\rho_{ij} \tag{6-60}$$

This is seen to be proportional to the signal energy and to the correlation coefficient between the ith signal and the jth signal.

Next we need to examine the response for the matched filters to the input white noise. Since the same noise is going into all filters, the outputs of the filters will in general be correlated. To determine the correlation between the noise at the outputs of the various filters, define

$$n_i = \int_0^T n(t)s_i(t) \, dt \tag{6-61}$$

$$n_j = \int_0^T n(t)s_j(t) \, dt \tag{6-62}$$

Then the correlation between the output of the ith filter and the jth filter is

$$
\begin{aligned}
E[n_i n_j] &= \int_0^T \int_0^T E[n(t)n(u)]s_i(t)s_j(u) \, dt \, du \\
&= \int_0^T \int_0^T \frac{N_0}{2} \delta(t - u)s_i(t)s_j(u) \, dt \, du \\
&= \frac{N_0}{2} \int_0^T s_i(t)s_j(t) \, dt = \frac{N_0 E}{2} \rho_{ij}
\end{aligned}
\tag{6-63}
$$

Note that this correlation is also proportional to the energy of the signal and to the correlation coefficient between the ith signal and the jth signal, as well as being proportional to the noise spectral density.

It is now clear that the output of the ith filter, given that signal $s_j(t)$ is received, is a Gaussian random variable having a mean value of $E\rho_{ij}$, a variance of $EN_0/2$, and a covariance with all other outputs of $E\rho_{ij}N_0/2$. Since there are M different filter outputs, one can define a vector $\mathbf{y}$ whose elements are the various y_i. The probability density function of this vector can now be written as

$$p(\mathbf{y}|s_j) = \frac{1}{(2\pi)^{M/2}|\mathbf{Q}|^{1/2}} \exp\left[\frac{-1}{2}\tilde{\mathbf{y}}^T\mathbf{Q}^{-1}\tilde{\mathbf{y}}\right] \tag{6-64}$$

where

$$\mathbf{y} = \begin{bmatrix} y_1 \\ y_2 \\ \vdots \\ y_M \end{bmatrix} \quad \tilde{\mathbf{y}} = \begin{bmatrix} y_1 - E\rho_{1j} \\ y_2 - E\rho_{2j} \\ \vdots \\ y_M - E\rho_{Mj} \end{bmatrix} \quad \mathbf{Q} = \frac{N_0 E}{2}\Gamma = \frac{N_0 E}{2}\begin{bmatrix} 1 & \rho_{12} & & \rho_{1M} \\ \rho_{21} & 1 & & \vdots \\ \vdots & & - & \\ \rho_{M1} & & & 1 \end{bmatrix}$$

Note that the vector $\tilde{\mathbf{y}}$ is simply the vector $\mathbf{y}$ with the mean vector subtracted out. Hence it is a Gaussian random vector with zero mean. The matrix Γ is the correlation coefficient matrix of this zero mean Gaussian random vector.

The problem of interest is to find the probability of making an incorrect decision. This can be done by first finding the conditional probability of making an error given that the jth signal was transmitted, then multiplying this by the probability of the jth signal being transmitted, and finally summing this joint probability over all possible values of j. This leads to

$$P(\varepsilon) = \sum_{j=1}^{M} P(\varepsilon|j)P(j) \tag{6-65}$$

$$= \frac{1}{M}\sum_{j=1}^{M} P(\varepsilon|j) \tag{6-66}$$

in which the second form of this equation results from assuming that all signals are equally probable.

The conditional probability of making an error, given that signal j was transmitted, can be obtained by subtracting the probability that y_j is greater than y_k from one, for all values of k not equal to j. That is,

$$P(\varepsilon|j) = 1 - Pr[y_j \geq y_k|s_j] \quad \text{all } k \neq j$$

$$= 1 - \int_{-\infty}^{\infty} dy_j \int_{-\infty}^{y_j} \cdots \int_{-\infty}^{y_j} p(\mathbf{y}|s_j)\, d\mathbf{y}' \tag{6-67}$$

in which

$$d\mathbf{y}' = dy_1\, dy_2 \ldots dy_{j-1}\, dy_{j+1} \ldots dy_M$$

There are only a few specific cases in which Eq. (6-67) can be evaluated analytically. However, it is often possible to find an upper bound on this probability of error by using a result known as the union bound. This is a bound based on the property that a union of sets cannot be larger than the sum of the elements making up the sets and may be stated mathematically in a probability concept as

$$P(\varepsilon|j) \le \sum_{\substack{k=1 \\ k \ne j}}^{M} \Pr[y_j \le y_k|s_j] \tag{6-68}$$

Using Eq. (6-68) in Eq. (6-66) leads to an upper bound on the total probability of error. This is

$$P(\varepsilon) \le \frac{1}{M} \sum_{j=1}^{M} \sum_{\substack{k=1 \\ k \ne j}}^{M} \Pr[y_j \le y_k|s_j] \tag{6-69}$$

Some additional insight into the nature of this result can be obtained by pairing the terms in Eq. (6-69) to write it in the form

$$P(\varepsilon) \le \frac{1}{M} \sum_{j=1}^{M} \sum_{\substack{k=1 \\ k \ne j}}^{M} \left\{ \frac{1}{2} \Pr[y_j \le y_k|s_j] + \frac{1}{2} \Pr[y_k \le y_j|s_k] \right\} \tag{6-70}$$

It may now be recognized that each of the terms in the summation is simply a binary error probability. Thus the entire result can be written as

$$P(\varepsilon) \le \frac{1}{M} \sum_{j=1}^{M} \sum_{\substack{k=1 \\ k \ne j}}^{M} P_e(j, k) \tag{6-71}$$

in which $P_e(j, k)$ is simply the binary error probability for signals j and k. This union bound on the probability of error will be used in subsequent analyses of M-ary signal sets.

Correlation Receiver

An alternative approach to detecting M-ary signals is to use a set of M correlators as shown in Fig. 6-17. This receiver consists of M separate correlators, each correlator consisting of a multiplier that multiplies one of the M possible signals by the input signal and integrates the product over the signal duration T. This produces a set of outputs y_1 through y_M, which are then examined and the largest selected. The output of the ith correlator is

$$y_i = \int_0^T x(t)s_i(t) \, dt \tag{6-72}$$

It may be noted that this is identical to the result obtained for the matched-filter receiver as shown in Fig. 6-16. Therefore, the matched-filter receiver and the correlation receiver will produce identical outputs and the analysis of the probability of error is exactly the same in both cases.

Both the matched-filter receiver and the correlation receiver have M branches.

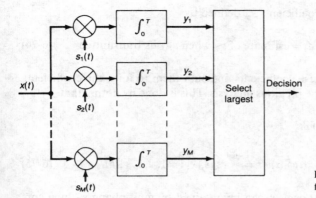

Figure 6-17 Correlation receiver for *M*-ary signals.

If M, the number of signals in the signal set, is less than N, the number of basis functions used to represent the signal set, then this implementation of the *M*-ary detector is probably the simplest form to use. On the other hand, if the number of basis functions, N, is less than the number of signals, M, there is an alternative implementation that may be preferable. This is based on the representation of each signal as

$$s_i(t) = \sum_{j=1}^{N} s_{ij} \phi_j(t) \tag{6-73}$$

This suggests that the ith signal can be detected by using filters that are matched to the basis functions and weighting the outputs of these filters by the corresponding coefficients s_{ij}. Such a receiver is shown in Fig. 6-18.

In this case there are N matched filters, and the output of each is sampled at

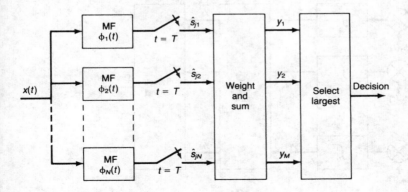

Figure 6-18 *M*-ary receiver with filters matched to the basis functions.

time T to produce a set of coefficients $\hat{s}_{jk}$ defined by

$$\hat{s}_{jk} = \int_0^T x(t)\phi_k(t)\,dt = \text{estimate of } s_{jk} \text{ when } s_j(t) \text{ is transmitted} \qquad (6\text{-}74)$$

It is necessary to weight these coefficients and sum them to form the coefficients y_i that were considered in the previous analysis. This is done by noting that

$$y_i = \int_0^T x(t)s_i(t)\,dt$$

$$= \sum_{k=1}^N s_{ik} \int_0^T x(t)\phi_k(t)\,dt = \sum_{k=1}^N s_{ik}\hat{s}_{jk} \quad i = 1, 2, \ldots, M \qquad (6\text{-}75)$$

It is clear from this that each y_i can be obtained by multiplying $\hat{s}_{jk}$ by a corresponding s_{ik} and summing over all values of k. Since the outputs from the summing circuit are exactly the same values of y_i that were obtained in the previous receiver structures, the analysis of the probability of the error will again be the same as in the matched-filter receiver.

An alternative version of this receiver, shown in Fig. 6-19, is to use N correlators, each correlator consisting of multiplying one of the N basis functions by the incoming signal and integrating the product over a signal interval T. This produces estimates of $\hat{s}_{jk}$ that can be weighted and summed to form the values of y_i. So again the analysis would be the same as for the original filter structure. When the basis functions are used, in the form of either matched filters or correlators, the number of branches in the receiver is always N, and if this is less than M, the receiver will be more economical to build. The correlator version of this receiver is most useful for MPSK signals, since in this case $N = 2$ for any value of M and the basis functions are simply sinusoids. In fact they can be steady-state sinusoids rather than pulsed sinusoids.

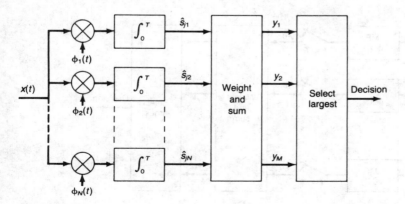

Figure 6-19 M-ary receiver correlating with the basis functions.

6-5 PERFORMANCE OF ORTHOGONAL SIGNALS

It was noted in the previous section that exact analytic evaluation of the probability of error is not possible in most circumstances. There are, however, certain cases in which this can be carried out—and the case of orthogonal signals is one such example. To review the general result, recall that the total probability of error was given by

$$P(\varepsilon) = \frac{1}{M} \sum_{j=1}^{M} P(\varepsilon|j) \tag{6-76}$$

where the conditional probability is

$$P(\varepsilon|j) = 1 - \int_{-\infty}^{\infty} dy_j \int_{-\infty}^{y_j} \cdots \int_{-\infty}^{y_j} p|\mathbf{y}|s_j)\, d\mathbf{y}' \tag{6-77}$$

Because of the Gaussian nature of the vector $\mathbf{y}$, the joint density function of $\mathbf{y}$ given s_j is

$$\rho(\mathbf{y}|s_j) = \frac{1}{(2\pi)^{M/2}|\mathbf{Q}|^{1/2}} \exp\left[-\frac{1}{2} \tilde{\mathbf{y}}^T \mathbf{Q}^{-1} \tilde{\mathbf{y}} \right] \tag{6-78}$$

where

$$\mathbf{Q} = \frac{N_0 E}{2} \mathbf{\Gamma} = \frac{N_0 E}{2} \begin{bmatrix} 1 & \rho_{12} & \cdots & \rho_{1M} \\ \rho_{21} & 1 & & \\ & & \vdots & \\ \rho_{M1} & & & 1 \end{bmatrix}$$

When the signals are mutually orthogonal, the correlation coefficient is $\rho_{ij} = 0$, $i \neq j$. Therefore, the matrix $\mathbf{\Gamma} = \mathbf{I}$. In this case the inverse of $\mathbf{Q}$ is readily obtained as

$$\mathbf{Q}^{-1} = \frac{2}{N_0 E} \mathbf{I} \tag{6-79}$$

The variance of each element of $\mathbf{y}$ is

$$\sigma^2 = \frac{N_0 E}{2} \tag{6-80}$$

and the mean value is given by

$$\begin{aligned} m_{ij} = E\rho_{ij} = 0 \qquad & i \neq j \\ = E \qquad & i = j \end{aligned} \tag{6-81}$$

Since the y_i are now independent, the joint probability density function can be written as the product of the marginal density functions of each element. Therefore,

$$p(\mathbf{y}|s_j) = \prod_{i=1}^{M} \frac{1}{\sqrt{2\pi}\sigma} e^{-(y_i - m_{ij})^2/2\sigma^2} \tag{6-82}$$

When $i \neq j$, the integrals from $-\infty$ to y_j can be expressed in terms of the Q function as

$$\int_{-\infty}^{y_j} \frac{1}{\sqrt{2\pi}\sigma} e^{-y_i^2/2\sigma^2} \, dy_i = 1 - \int_{y_j}^{\infty} \frac{1}{\sqrt{2\pi}\sigma} e^{-y_i^2/2\sigma^2} \, dy_i$$

$$= 1 - Q\left[\frac{y_i}{\sigma}\right] \tag{6-83}$$

Thus the conditional probability of error given j becomes

$$P(\varepsilon|j) = 1 - \int_{-\infty}^{\infty} \left[1 - Q\left(\frac{y_j}{\sigma}\right)\right]^{M-1} \cdot \frac{1}{\sqrt{2\pi}\sigma} e^{-(y_j-E)^2/2\sigma^2} \, dy_j \tag{6-84}$$

It is now convenient to make a change in variables by letting

$$y = \frac{y_j - E}{\sigma}$$

$$\frac{y_j}{\sigma} = y + \frac{E}{\sigma} = y + \sqrt{\frac{2E}{N_0}}$$

Upon doing this the conditional probability of error becomes

$$P(\varepsilon|j) = 1 - \int_{-\infty}^{\infty} \left\{1 - Q\left[y + \sqrt{\frac{2E}{N_0}}\right]\right\}^{M-1} \cdot \frac{1}{\sqrt{2\pi}} e^{-y^2/2} \, dy \tag{6-85}$$

Note that this conditional probability of error does not depend upon j. Therefore, every term in the summation is the same and the total probability of error of now becomes

$$P(\varepsilon) = \frac{1}{M} \sum_{j=1}^{M} P(\varepsilon|j) = P(\varepsilon|j) \tag{6-86}$$

This reduces the probability of error to a single integral that can be evaluated numerically. Tables and curves of the probability of error for orthogonal signals are available in the literature.[3]

It is desirable to modify the above result in two different ways. First, it is usually more convenient to express the energy in terms of the energy per bit, rather than the total signal energy. Second, it is usually desirable to determine the probability of bit error rather than the total probability of signal error. These two modifications are now made.

To relate the bit error probability to the probability of a signal error, note that given a particular transmitted signal, there are $M - 1$ possible erroneous signals. In M signals a given bit will be a 0 or 1 $M/2$ times. Therefore,

$$\frac{M}{2(M-1)} = \text{conditional probability of bit error given a signal error}$$

This leads immediately to

$$P_e = \frac{M}{2(M-1)} P(\varepsilon) \tag{6-87}$$

$$\simeq \frac{1}{2} P(\varepsilon) \quad \text{for large } M \tag{6-88}$$

Note that the result in Eq. (6-88) is intuitively obvious. When each signal represents a large number of binary digits, selecting the wrong signal will result in approximately half of the digits being incorrect.

The number of binary digits that can be represented by a set of M signals is $\log_2 M$. Therefore, the relationship between the energy per bit and the total signal energy is

$$E_b = \frac{E}{\log_2 M} = \frac{E}{k} \quad \text{if } M = 2^k \tag{6-89}$$

Hence

$$\frac{2E}{N_0} = \frac{2kE_b}{N_0} \tag{6-90}$$

A sketch of the probability of bit error for orthogonal signals, expressed in terms of the ratio of the energy to the noise spectral density, is shown in Fig. 6-20.

This is a conventional way of expressing probability of error curves. It is also possible to express the signal energy in terms of the duration of the signal and the bit rate. This relationship is

$$R = \frac{k}{T} = \frac{\log_2 M}{T} \tag{6-91}$$

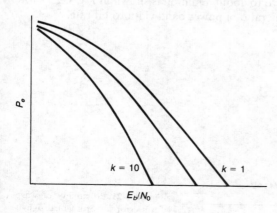

Figure 6-20 Probability of bit error for orthogonal signals.

For a given $R = R_0$, the duration of the signal is

$$T = \frac{\log_2 M}{R_0} \tag{6-92}$$

Therefore, the ratio of signal energy to noise spectral density becomes

$$\frac{E}{N_0} = \frac{P_s T}{N_0} \tag{6-93}$$

$$= \frac{P_s}{N_0 R_0} \log_2 M \tag{6-94}$$

where P_s is the average signal power. The probability of error as a function of the ratio of signal power to noise spectral density times bit rate is shown in Fig. 6-21.

A very interesting and unusual result takes place when the number of signals in the signal set is allowed to approach infinity. Specifically,

$$\lim_{M \to \infty} P(\varepsilon) = 1 \qquad \frac{P_s}{N_0 R_0} < \ln 2$$

$$= 0 \qquad \frac{P_s}{N_0 R_0} > \ln 2 \tag{6-95}$$

This result implies that if the ratio $P_s/(N_0 R_0)$ is less than the natural logarithm of 2, the probability of error is unity, while if this ratio is greater than the logarithm of 2, the probability of error is zero. The derivation of this unique result is beyond the scope of the present discussion, but it is basic to many communication system problems. In particular it may be noted that the ratio of P_s to N_0 has the dimensions of bandwidth, and is frequently referred to as the power bandwidth. This suggests that the ratio of power bandwidth to bit rate must be greater than some threshold value in order for communication to take place. Another viewpoint on this same result is discussed in a subsequent section. The probability of error for orthogonal signal sets ranging in size from two to about a million is shown in Fig. 6-22. Note that the abscissa in this case is the ratio of power bandwidth to bit rate.

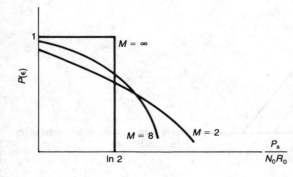

Figure 6-21 Probability of waveform error for orthogonal signals.

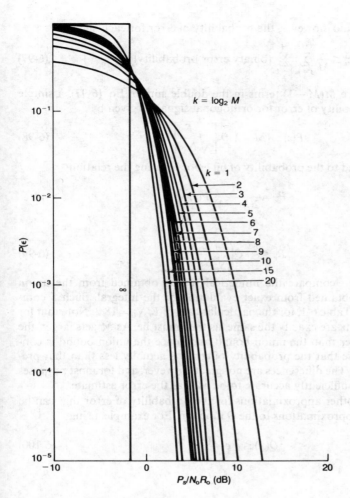

Figure 6-22 Probability of error for orthogonal signals.

Union Bound for Orthogonal Signals

Although the results given in the preceding section for orthogonal signals are exact, they do require numerical evaluation of an integral in order to obtain numerical values. For many analytical purposes, it is more convenient to have an explicit analytical result. This analytical result can be obtained from the union bound. It was noted that the probability of error for binary orthogonal signals is simply

$$P_e = P(\varepsilon) = Q\left[\sqrt{\frac{E}{N_0}}\right] \tag{6-96}$$

From the union bound, however, the probability of error for each signal is

$$P(\varepsilon) \le \frac{1}{M} \sum_{j=1}^{M} \sum_{\substack{k=1 \\ k \ne j}}^{M} \text{(binary error probability)} \qquad (6\text{-}97)$$

Noting that there are $M(M-1)$ terms in the double sum of Eq. (6-97), a simple bound on the probability of error for orthogonal signals is given by

$$P(\varepsilon) \le (M-1)Q\left[\sqrt{\frac{E}{N_0}}\right] \qquad (6\text{-}98)$$

This can be converted to the probability of bit error by using the relation

$$P_e = \frac{M}{2(M-1)} P(\varepsilon)$$

which yields

$$P_e \le \frac{M}{2} Q\left(\sqrt{\frac{E}{N_0}}\right) \qquad (6\text{-}99)$$

It is desirable to compare the numerical results obtained from the union bound with those obtained from exact evaluation of the integral. Such a comparison is shown in Table 6-1, for the particular case of $E/N_0 = 18.2$. Note that for $M = 2$, the two results are exactly the same, as they must be. As M gets larger, the exact result is smaller than the union bound and hence the union bound is conservative in the sense that the probability of error is actually less than that predicted by the bound. The differences are not great, however, and for most purposes the union bound is sufficiently accurate to be used as the error estimate.

There are also other approximations for the probability of error that can be obtained from the approximations to the Q function. For example, using

$$Q(x) \le \frac{1}{2} e^{-x^2/2} \qquad (6\text{-}100)$$

Table 6-1 Comparison of union bound with exact result for orthogonal signals

| M | $E/N_0 = 18.2$ | |
	Exact $P(\varepsilon)$	Union bound
2	10^{-5}	10^{-5}
4	2.9×10^{-5}	3×10^{-5}
8	6.9×10^{-5}	7×10^{-5}
16	1.45×10^{-4}	1.5×10^{-4}
32	2.70×10^{-4}	3.1×10^{-4}
64	5.1×10^{-4}	6.3×10^{-4}
128	1.1×10^{-3}	1.27×10^{-3}

the probability of bit error for orthogonal signals can be expressed as

$$P_e \leq \frac{M}{4} e^{-E/2N_0} \tag{6-101}$$

This is a particularly simple result that can be used in many analytical investigations. It is one that appears widely throughout the literature.

6-6 DETECTION OF NONORTHOGONAL *M*-ARY SIGNALS

The probability of error performance of orthogonal *M*-ary signal sets is a particularly easy case to consider because the correlation matrix Γ of the signal set is diagonal, and hence the inverse can be obtained readily. For nonorthogonal signals, however, the correlation matrix is not diagonal and the evaluation of the probability of error becomes more difficult. In general it is possible to obtain some bounds, although it may not be possible to obtain exact results. In this section some of the bounds that can be obtained for several of the classes of signals that have been mentioned previously are discussed.

Coherent Detection of Biorthogonal Signals

The biorthogonal signal set consists of $M/2$ pairs of antipodal signals, each pair of which is orthogonal to all other pairs. Because of the special form of this signal set, the detector must accomplish the detection in two steps. First it must look for the largest magnitude of the $M/2$ orthogonal signals, and then it must determine the polarity of that signal with the largest magnitude. The receiver structure for biorthogonal signals can take the same form as that for orthogonal signals. In particular the receiver may contain filters matched to the signals or it may consist of correlators that correlate the incoming signal with $M/2$ orthogonal reference signals. Alternatively it may contain filters that are matched to the basis functions, or it may contain correlators that correlate with the basis functions. Since all of these receiver structures are mathematically equivalent so far as the probability of error is concerned, only one will be considered here.

In Fig. 6-23 the matched-filter receiver for biorthogonal signals is illustrated. Note that it contains $M/2$ matched filters that are sampled at their outputs at time *T*. These outputs then pass through circuits that determine the magnitude of the signal without regard to the sign. These magnitudes are examined and the largest one is selected. Having determined which of the signals is the largest, that particular matched-filter output is examined again and the polarity of the signal determined. This is then used to produce the decision as to which of the signals is present.

The union bound can be used to obtain a bound on the probability of error for biorthogonal signals. This is[1]

$$P(\varepsilon) \leq (M - 2)Q\left[\sqrt{\frac{E}{N_0}}\right] + Q\left[\sqrt{\frac{2E}{N_0}}\right] \tag{6-102}$$

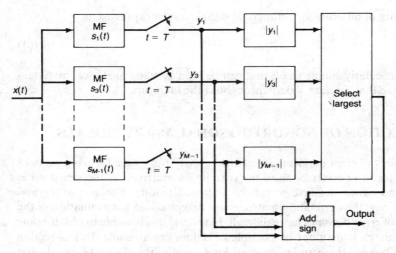

Figure 6-23 Matched-filter receiver for biorthogonal signals.

This bound is not good for all probabilities of error, but is reasonably close for $P(\varepsilon) < 10^{-2}$. It may also be noted that, under most circumstances, the first term of Eq. (6-102) is the one that contributes the most to the probability of error, the second term generally being quite small as compared with the first one.

The bit error probability of the biorthogonal signal set is not related to the signal error probability in the same way as it is for orthogonal signals. This is so because of the double-decision process that is necessary to establish which signal was received. However, a bound on the bit error probability can be obtained; this can be found in Lindsey and Simon.[3]

$$P_e \le \frac{M-2}{2} Q\left[\sqrt{\frac{E}{N_0}}\right] + Q\left[\sqrt{\frac{2E}{N_0}}\right] \qquad (6\text{-}103)$$

A numerical comparison of the probabilities of error for orthogonal and biorthogonal signals is useful. Consider the case in which $M = 8$, and $E/N_0 = 25$. In the case of orthogonal signals, the probability of error is given by

$$P_e \le \frac{M}{2} Q\left(\sqrt{\frac{E}{N_0}}\right) = \frac{8}{2} Q(5)$$

$$P_e \le 1.17 \times 10^{-6}$$

In the case of biorthogonal signals, the probability of error becomes

$$P_e \le \frac{8-2}{2} Q(5) + Q(5\sqrt{2})$$

$$P_e \le 8.74 \times 10^{-7}$$

It is seen that the probability of error is somewhat smaller for the biorthogonal case. For large values of M, the performances of orthogonal and biorthogonal signals approach one another. Otherwise the biorthogonal signal set is always somewhat better than the orthogonal signal set.

Coherent Detection of Simplex Signals

It may be recalled that simplex signals are obtained from an orthogonal signal set by translating the orthogonal set in signal space by an amount that minimizes the total energy. Because of this relationship to orthogonal signals, the receiver structure for simplex signals can be the same as that of orthogonal signals. Furthermore the probability of error is the same as orthogonal signals if an appropriate change in the energy is made. This is so because the distance between the signals has not been changed by virtue of translating the signal set in the space and it is the distance between signals that determines the probability of error. To make the appropriate correction for simplex signals, the energy E of the orthogonal set is replaced by $E[M/(M-1)]$. This leads to a result that would be exact if the integral expression for the probability of error were used. However, the result can also be used to obtain a bound on the probability of error by using the union bound result. Thus the probability of a signal error is given by

$$P(\varepsilon) \leq (M-1)Q\left[\sqrt{\frac{ME}{(M-1)N_0}}\right] \tag{6-104}$$

or the probabiliy of a bit error is

$$P_e \leq \frac{M}{2}Q\left[\sqrt{\frac{ME}{(M-1)N_0}}\right] \tag{6-105}$$

For purposes of comparison, consider the case in which there are eight signals and $E/N_0 = 25$. In this case the probability of error is

$$P_e \leq \frac{8}{2}Q\left[\sqrt{\frac{8 \times 25}{7}}\right] = 1.81 \times 10^{-7}$$

Note that this is smaller than the probability of error for the corresponding orthogonal signal set by almost an order of magnitude.

Coherent Detection of Polyphase Signals

Polyphase signals constitute another important class in which the signals are described in terms of sinusoids having a constant amplitude and different phase angles for each signal in the set. Thus the ith signal can be represented as

$$s_i(t) = \sqrt{\frac{2E}{T}}\cos(\omega_s t + \theta_i) \tag{6-106}$$

$$= \sqrt{\frac{2E}{T}}[\cos\theta_i\cos\omega_s t - \sin\theta_i\sin\omega_s t] \qquad 0 \leq t \leq T \tag{6-107}$$

Since every signal in the set can be represented by only two basis functions, regardless of the size of M, a receiver structure that correlates the received signal with the basis functions is the most convenient form. This is illustrated in Fig. 6-24.

Note that the correlators, consisting of the multipliers and the integrators, produce outputs that are indicated as X and Y and these outputs then go into a circuit that produces the inverse tangent of Y/X. This inverse tangent, designated as ψ, is an estimate of the phase of the incoming signal. To decide which of the possible phases is being received, this estimate of the phase is compared with all possible values of θ_i and the *smallest* difference between θ_i and ψ is selected as being the signal actually present. It is possible to obtain exact results for the probability of error of polyphase signals using the Γ matrix, but the results are very complicated. A simpler approach can be obtained from an approximate result that is valid for large values of M.[4] This is

$$P(\varepsilon) \simeq 2Q\left[\sqrt{\frac{2E}{N_0}}\sin\frac{\pi}{M}\right] \qquad M \geq 4 \qquad (6\text{-}108)$$

This result is valid for values of M greater than or equal to 4. It is not valid for $M = 2$, but for this particular case, the signals are antipodal and the probability of error is known to be

$$P(\varepsilon) = P_e = Q\left[\sqrt{\frac{2E}{N_0}}\right] \qquad (6\text{-}109)$$

For M greater than 16, it is possible to replace the sine by its angle. Thus

$$\sin\frac{\pi}{M} \simeq \frac{\pi}{M}$$

and this leads to a probability of error of the form

$$P(\varepsilon) \simeq 2Q\left[\sqrt{\frac{2\pi^2 E}{N_0 M^2}}\right] \qquad (6\text{-}110)$$

It is of interest to note from this expression that to maintain a given probability

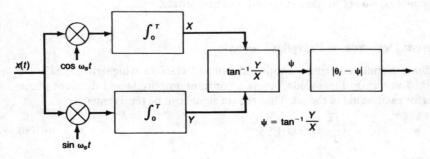

Figure 6-24 Receiver for polyphase signals.

of error as the number of signals in the set increases, the signal energy must increase as M^2. This points up one of the disadvantageous features of polyphase signals.

The relation between the bit error probability and the signal error probability is not the same for polyphase signals as it was for orthogonal signals. Nevertheless, for large values of M, an approximate result is

$$P_e \simeq \frac{1}{2} P(\varepsilon) \simeq Q\left[\sqrt{\frac{2E}{N_0}} \sin \frac{\pi}{M}\right] \tag{6-111}$$

To make a numerical comparison of the performance of polyphase signals and orthogonal signals, consider again the case in which $M = 8$ and $E/N_0 = 25$. The probability of error now becomes

$$P_e \simeq Q\left[\sqrt{2 \times 25} \sin \frac{\pi}{8}\right]$$

$$P_e \simeq 3.53 \times 10^{-3}$$

For an orthogonal signal set of this size and with the same signal energy, the probability of error was previously found to be bounded by 1.17×10^{-6}, which is more than three orders of magnitude smaller than it is for the polyphase signals. To obtain a probability of error equivalent to that for the orthogonal signals, the value of E/N_0 would have to be increased to 76.2, which represents an increase in signal energy of 4.8 dB.

Coherent Detection of Quadriphase Signals (QPSK)

The quadriphase signal is a special case of the polyphase signal in which $M = 4$. In this case the signals are explicitly defined by

$$s_i(t) = \sqrt{\frac{2E}{T}} \cos\left(\omega_s t + \frac{i\pi}{2} - \frac{\pi}{4}\right) \qquad i = 1, 2, 3, 4 \tag{6-112}$$

The receiver model is somewhat simpler in this case because it is not necessary to employ the inverse tangent operation. A possible receiver model is shown in Fig. 6-25. The outputs of the integrators are compared with zero and the decision

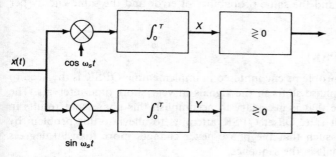

Figure 6-25 Receiver for QPSK signal.

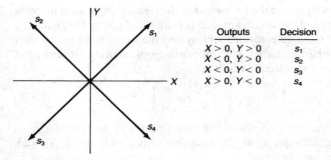

Outputs	Decision
$X > 0, Y > 0$	s_1
$X < 0, Y > 0$	s_2
$X < 0, Y < 0$	s_3
$X > 0, Y < 0$	s_4

Figure 6-26 QPSK signal vectors.

is made based on whether these outputs are greater or smaller than the threshold of zero.

The operation can be understood more clearly by referring to the vector diagram of Fig. 6-26. This diagram shows the four signals separated by 90 degrees. Since each signal occupies a different quadrant of the signal space, decisions can be made by determining which quadrant the observed signal is in. A table that illustrates this is shown with Fig. 6-26. It may also be noted that QPSK is a biorthogonal signal set. If we use the results obtained for biorthogonal signals for the probability of error for the case of $M = 4$, the union bound result leads to

$$P(\varepsilon) \leq 2Q\left[\sqrt{\frac{E}{N_0}}\right] + Q\left[\sqrt{\frac{2E}{N_0}}\right] \qquad (6\text{-}113)$$

The bit error probability is taken to be

$$P_e = \frac{1}{2} P(\varepsilon) \qquad (6\text{-}114)$$

From these results it is clear that QPSK achieves the same probability of error as a binary system with the same bit rate and the same energy per bit, but uses only one half the bandwidth; or, conversely, it can transmit at twice the bit rate with the same bandwidth and the same probability of error and the same energy per bit.

Offset QPSK (OQPSK)

One of the practical problems encountered in implementing QPSK is the need to produce 180-degree phase shifts in the signals in very short time intervals. The finite amount of time that is necessary to accomplish this phase shift results in slowing down the bit rate. Offset QPSK attempts to alleviate this problem by defining a signal set such that the phase never changes more than 90 degrees between any two signals in the sequence.

To see how this is done, consider the message to consist of two binary sequences

$$(a_1, a_2, a_3, \ldots.)$$

$$(b_1, b_2, b_3, \ldots.)$$

where $a_i = \pm 1$, $b_i = \pm 1$.

In normal QPSK the phase is determined by

$$\theta_i = \tan^{-1} \frac{b_i}{a_i} \tag{6-115}$$

Since both a_i and b_i may change sign, this can result in a total change in phase of 180 degrees at intervals of T seconds. This is illustrated in the top part of Fig. 6-27. In the case of offset QPSK, however, the sequence represented by the b symbols is displaced from the a sequence by one half of a period, as shown in the bottom of Fig. 6-27. This means that the phase θ_i can change at most ± 90 degrees every $T/2$ seconds since only one of the sequences can change sign at each time instant. However, since phase changes can occur every $T/2$ seconds, the bandwidth is increased unless the pulses are shaped. A typical shape is the half sinusoid.

A conceptual approach to implementing the transmitter for offset QPSK is shown in Fig. 6-28. Note that shaping filters are employed and the outputs of the shaping filters are then multiplied by $\sin \omega_s t$ and $\cos \omega_s t$ and added to produce the eventual output signal. The bit stream in one branch is delayed by $T/2$ seconds with respect to the other branch.

A receiver for offset QPSK is shown in Fig. 6-29. In this case the incoming signal is multiplied by $\sin \omega_s t$ and $\cos \omega_s t$, the products are filtered in low-pass filters, and the outputs of the low-pass filters are compared with a threshold of zero. A decision is then made as to whether these outputs are positive or negative.

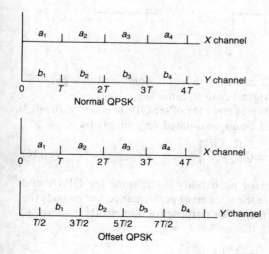

Figure 6-27 Comparison of normal and offset QPSK.

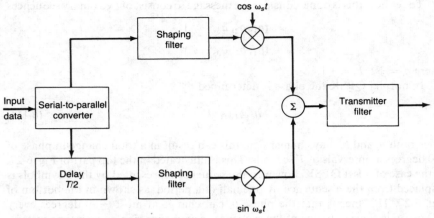

Figure 6-28 Transmitter for offset QPSK.

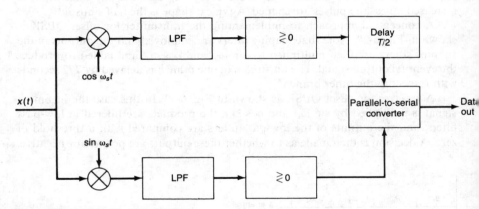

Figure 6-29 Receiver for offset QPSK.

One set of outputs is delayed by $T/2$ and then these are combined in a parallel-to-serial converter to produce the original data stream.

Exact evaluation of the probability of error for offset QPSK is very difficult.[3] However, the probability of error can be approximated very simply by

$$P_e \simeq Q\left[\sqrt{\frac{E}{N_0}}\right] \qquad (6\text{-}116)$$

In a white noise channel, the error probability is the same for QPSK and OQPSK. Thus it is possible to achieve the same error performance, but provide the capability of operating at higher bit rates. For purposes of numerical comparison, consider the case in which $E/N_0 = 25$. Then

$$P_e \simeq Q(5) = 2.91 \times 10^{-7}$$

Phase Reference Errors

The previous discussion of QPSK and offset QPSK has assumed that there was an accurate phase reference available at the receiver. In many cases this is not true. It is necessary, therefore, to investigate what happens to the probability of error when the phase reference is not exactly correct. To do this suppose that the phase reference has an error of θ_e. The bit error probability for QPSK can be approximated by

$$P_e \simeq \frac{1}{2} Q\left[\sqrt{\frac{E}{N_0}}(\cos \theta_e + \sin \theta_e)\right] + \frac{1}{2} Q\left[\sqrt{\frac{E}{N_0}}(\cos \theta_e - \sin \theta_e)\right] \quad (6\text{-}117)$$

To evaluate the significance of this result, consider the case in which $E/N_0 = 25$ and assume that the phase error is 15 degrees. From Eq. (6-117), then, the probability of error becomes

$$P_e \simeq \frac{1}{2} Q[5(\cos 15° + \sin 15°)] + \frac{1}{2} Q[5(\cos 15° - \sin 15°)] \simeq 1.05 \times 10^{-4}$$

Had the phase reference been exactly correct, the probability of the bit error would have been

$$P_e \simeq Q\left[\sqrt{\frac{E}{N_0}}\right] = Q(5) = 2.91 \times 10^{-7}$$

Note that the error in the phase reference has increased the probability of error by almost three orders of magnitude.

The situation is a little more involved for offset QPSK. In this case the probability of bit error can be approximated by[3]

$$P_e = \frac{1}{2} P_e(\text{QPSK}) + \frac{1}{2} P_e(\text{PSK}) \quad (6\text{-}118)$$

where

$$P_e(\text{PSK}) = Q\left[\sqrt{\frac{E}{N_0}} \cos \theta_e\right] \quad (6\text{-}119)$$

and in which $P_e(\text{QPSK})$ is given by Eq. (6-117).

Looking again at our example in which E/N_0 is equal to 25 and the phase reference error is 15 degrees, the probability of error for offset QPSK becomes

$$P_e(\text{QPSK}) = 1.05 \times 10^{-4}$$

$$P_e(\text{PSK}) = Q[5 \cos 15°] = 6.96 \times 10^{-7}$$

$$P_e = \frac{1}{2} \times 1.05 \times 10^{-4} + \frac{1}{2} \times 6.96 \times 10^{-7}$$

$$= 5.28 \times 10^{-5}$$

Note that the increase in probability of error is not as great for offset QPSK as it was for normal QPSK. This is another advantage of using the offset form of QPSK.

Differential Quadriphase (DQPSK)

In a previous discussion of binary PSK, the concept of differential encoding was presented. In this case the phase reference for decoding any particular bit in the sequence is the state of the preceding bit. A similar concept can be employed in quadriphase. Each signaling waveform becomes the phase reference for the next waveform. In the case of QPSK, it has been shown that to achieve the same probability of error as coherent QPSK, the signal energy must be increased by about 2.3 dB. In the binary case, it was noted that DPSK required only about 1 dB more energy for the same probability of error as binary PSK.

6-7 NONCOHERENT DETECTION OF M-ARY SIGNALS

In all of the examples considered so far, the detection process was considered to be either coherent or differentially coherent. There are some classes of M-ary signals, however, that can be detected noncoherently. These include multiple-frequency shift keying (MFSK), which was discussed previously, and multiple-amplitude shift keying (MASK). Both of these situations are considered here.

When orthogonal MFSK signals are employed, the ith signal can be described as

$$s_i(t) = \sqrt{\frac{2E}{T}} \sin\left[\omega_s t + \frac{2\pi i t}{T} + \theta\right] \qquad 0 \le t \le T, \qquad i = 1, 2, \ldots, M \qquad (6\text{-}120)$$

where θ is an arbitrary phase angle. Note that the M different frequencies in this signal set are all separated by $1/T$ Hz.

Since each signal in the set has a different center frequency, it is possible to obtain an indication of which signal is being transmitted by looking at the frequency of the signal on an energy basis. This can be accomplished in a noncoherent manner by the receiver shown in Fig. 6-30. Note that the receiver consists of M

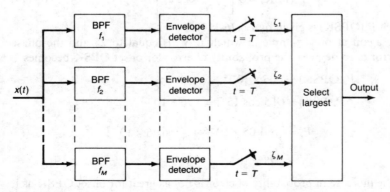

Figure 6-30 Noncoherent receiver for MFSK signals.

bandpass filters followed by envelope detectors. The outputs of the envelope detectors are sampled at $t = T$ and a decision is made based on which of these outputs is the largest. The bandpass filters may be matched filters for each particular signal, because this form would maximize the signal-to-noise ratio into the envelope detector. However, this would complicate the filter structure. Simple bandpass filters may well be adequate.

An exact analysis of the probability of error for noncoherent detection of MFSK signals is possible, but the analysis is very involved and the expression for the probability of error is very complicated.[4] The results can be portrayed graphically, however, and a curve of the probability of error as a function of the energy per bit is shown in Fig. 6-31. It is of interest to note in this figure that as the number of signals approaches infinity, the same phenomenon occurs that was observed in the case of orthogonal signals—namely, below a certain threshold the probability of error is unity, whereas above that threshold the probability of error is zero.

For the binary case, that is, $M = 2$, noncoherent detection of MFSK signals requires more energy than coherent detection. However, as M increases, the amount of energy required for noncoherent detection approaches coherent detection, and for very large values of M, the two modes of detection are essentially indistinguishable.

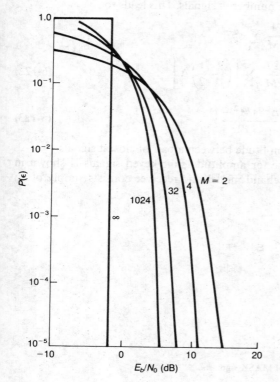

Figure 6-31 Probability of error for MFSK signals.

MFSK signaling is a common type for radio channels because of the ease with which transmitters and receivers can be implemented and because of its resistance to fading.

Amplitude Shift Keying (Baseband Model)

Amplitude shift keying describes a signal set in which the different signals have different amplitudes. This form of signaling is very commonly employed in wire transmission circuits where accurate control over signal amplitudes can be maintained. It is not as frequently employed in radio channels because of the difficulty in maintaining appropriate amplitude levels. The ith signal of the set can be described by

$$s_i(t) = s_i \qquad 0 \le t \le T \qquad i = 1, 2, \ldots, M \qquad (6\text{-}121)$$

The entire signal set is a one-dimensional set and can be represented graphically as shown in Fig. 6-32. It may be noted that this signal set differs from the previous ones that have been considered in that each signal has a different energy because of its difference in amplitude. In fact the energy for the ith signal is just $E_i = s_i^2 T$.

It is possible, however, to define an average signal energy by adding the energy of each signal and dividing by the number of signals. This leads to

$$E = \frac{1}{M} \sum_{i=1}^{M} s_i^2 T$$

$$= \frac{2T}{M} \sum_{i=1}^{M/2} \left[\left(i - \frac{1}{2} \right) \Delta \right]^2 \qquad (6\text{-}122)$$

$$= \frac{\Delta^2 T (M^2 - 1)}{12} \qquad (6\text{-}123)$$

in which Δ is the difference in amplitude between a pair of closest signals.

A possible receiver structure for amplitude shift keyed signals is shown in Fig. 6-33. Note that in this baseband model, the receiver consists simply of an

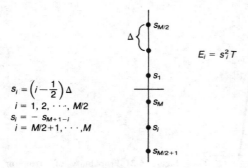

Figure 6-32 Graphical representation of MASK signals.

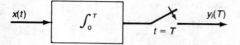

Figure 6-33 Receiver for MASK signals.

integrator followed by a sampler sampling the output at time T. The input to this receiver is represented as

$$x(t) = s_i(t) + n(t)$$

and the output becomes

$$y_i(T) = \int_0^T s_i(t)\, dt + \int_0^T n(t)\, dt$$

$$= s_i T + n_y \tag{6-124}$$

The mean square value of the noise out of the integrator is given by

$$E[n_y^2] = E\left[\int_0^T \int_0^T n(t)n(u)\, dt\, du\right] \tag{6-125}$$

If the noise at the input to the receiver is assumed to be white, with a one-sided spectral density N_0, then this mean square noise becomes

$$E[n_y^2] = \int_0^T \int_0^T \frac{N_0}{2}\, \delta(t - u)\, dt\, du$$

$$= \frac{N_0 T}{2} \tag{6-126}$$

Since the noise is Gaussian, the receiver output $y_i(T)$ is a Gaussian random variable with a mean value of $s_i T$ and a variance of $N_0 T/2$. Hence the probability density function of the ith signal is as shown in Fig. 6-34.

The amplitude levels at the output of the integrator are separated by an amount ΔT and, therefore, it is clear that errors occur if the noise exceeds $\Delta T/2$. However, conditions are slightly different at the end points. Therefore, it is convenient to

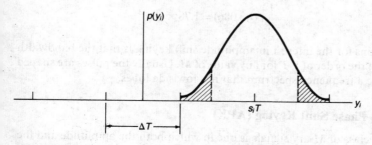

Figure 6-34 Probability density function for the ith received signal.

write the conditions under which errors occur as follows:

$$\text{(a)} \qquad |n_y| > \frac{\Delta T}{2} \qquad i \neq \frac{M}{2} \qquad i \neq \frac{M}{2} + 1 \tag{6-127}$$

$$\text{(b)} \qquad n_y > \frac{\Delta T}{2} \qquad i = \frac{M}{2} + 1 \tag{6-128}$$

$$\text{(c)} \qquad n_y < \frac{\Delta T}{2} \qquad i = \frac{M}{2} \tag{6-129}$$

If the M signals are equally probable, the probability of error now becomes

$$P(\varepsilon) = \frac{1}{M} \left\{ 2(M-2)Q\left[\frac{\Delta T/2}{\sqrt{N_0 T/2}} \right] + 2Q\left[\frac{\Delta T/2}{\sqrt{N_0 T/2}} \right] \right\} \tag{6-130}$$

$$= \frac{2(M-1)}{M} Q\left[\sqrt{\frac{\Delta^2 T}{2N_0}} \right] \tag{6-131}$$

This probability of error can also be expressed in terms of the average signal energy, which is

$$E = \frac{\Delta^2 T(M^2 - 1)}{12} \tag{6-132}$$

This leads to

$$P(\varepsilon) = \frac{2(M-1)}{M} Q\left[\sqrt{\frac{6E}{N_0(M^2-1)}} \right] \tag{6-133}$$

It may be noted that for a given probability of error, the energy required increases as M^2.

To make numerical comparisons, consider again the case in which $M = 8$ and $E/N_0 = 100$. The probability of error for this case is

$$P(\varepsilon) = \frac{2(8-1)}{8} Q\left[\sqrt{\frac{6 \times 100}{64 - 1}} \right]$$

$$= \frac{14}{8} Q(3.086) = 1.78 \times 10^{-3}$$

One of the reasons for the interest in amplitude shift keying is that the bandwidth of the signal is of the order of $1/T$ for any value of M. Usually the pulses are shaped so as to produce a frequency spectrum that has low side lobes.

Amplitude and Phase Shift Keying (APK)

A more general class of M-ary signals is one in which both the amplitude and the phase of the signal can be changed by discrete amounts. This leads to a much

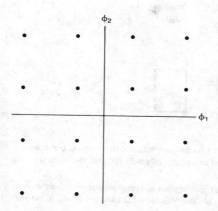

Figure 6-35 Signal space representation of a 16-ary APK signal set.

larger number of signals, since the number of signals is equal to the product of the number of amplitude levels and the number of phase shifts.

For example, if there are four amplitude levels and four phase shifts, the result will be a 16-ary signal set as illustrated in Fig. 6-35. For such a signal set, the ith signal can be described by

$$s_i(t) = s_i \cos(\omega_s t + \theta_i) \qquad 0 \le t \le T, \qquad i = 1, 2, \ldots, M \qquad (6\text{-}134)$$

The evaluation of the probability of error for APK signal sets is quite involved. Nevertheless results do appear in the literature. The main reason for the interest in amplitude and phase shift keying is that this class of signal conserves bandwidth. Again, the bandwidth will be of the order of $1/T$ for any value of M and side lobes can be controlled by shaping the pulses.

6-8 REFERENCES

1. Wozencraft, J. M., and I. M. Jacobs: *Principles of Communication Engineering*, John Wiley & Sons, Inc., New York, 1965.
2. Franks, L. E., *Signal Theory*, Prentice Hall, Inc., Englewood Cliffs, N.J., 1969.
3. Lindsey, W. C., and M. K. Simon: *Telecommunication Systems Engineering*, Prentice-Hall, Inc., Englewood Cliffs, N.J., 1973.
4. Gilbert, E. N.: A Comparison of Signalling Alphabets, *Bell Sys. Tech. J.*, vol. 31, pp. 504–522, May 1952.
5. Rhodes, S. A.: Effects of Noisy Phase Reference on Coherent Detection of Offset-QPSK Signals, *IEEE Trans. Com.*, vol. Com-22, no. 8, August 1974.
6. Stein, S., and J. J. Jones: *Modern Communication Principles*, McGraw-Hill Book Company, New York, 1967.

6-9 PROBLEMS

6.1-1 Starting with $s_1(t)$ and using the Gram-Schmidt procedure, find an orthonormal basis set for the signal set shown and express the signals as vectors of coefficients of the orthonormal representation.

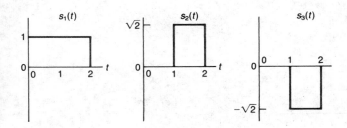

6.1-2 Starting with $s_2(t)$ and using the Gram-Schmidt procedure, find an orthonormal basis set for the signal set in Problem 6.1-1 and express the signals as vectors of coefficients of the orthonormal representation.

6.1-3 (a) Find a set of orthonormal basis functions that can be used to represent the signal set shown.

(b) Find the vector corresponding to each signal for the orthonormal basis set found in (a) and sketch the location of each signal in signal space.

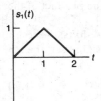 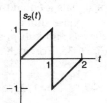 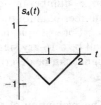

6.1-4 An equal-energy signal set has a correlation matrix of the form

$$\Gamma = \begin{bmatrix} 1 & \rho & \rho^2 \\ \rho & 1 & \rho \\ \rho^2 & \rho & 1 \end{bmatrix}$$

Find the smallest value of correlation coefficient ρ that can exist for this signal set.

6.1-5 A set of eight binary vectors has the form

$$s_i = (a_i, b_i, c_i) \qquad i = 1, 2, \ldots, 8$$

where a_i, b_i, and c_i are ± 1. Assuming the vectors are ordered in natural binary sequence (with -1 interpreted as 0), find the correlation matrix of the signal set. That is, s_i is the binary number for $(i-1)$ with 0 replaced by -1.

6.1-6 Write the normalized correlation matrix for the signal set shown.

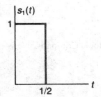 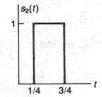 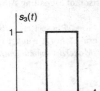 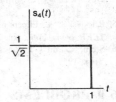

6.1-7 Express the signal set shown in terms of the orthonormal basis set shown. Compute the distances between each pair of signals using the time functions and using the vector representation.

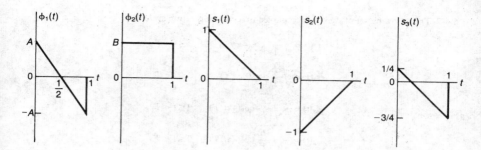

6.2-1 A 10-member MFSK signal set operating at a carrier of $f_0 = 10^7$ has signal durations of 10 μs. Find the bandwidth required to accommodate this signal set. How many QPPM signals can be generated in the same bandwidth?

6.2-2 Compute the third-order Hadamard matrix and sketch the corresponding orthogonal binary vectors.

6.3-1 A four-element signal set consists of the binary vectors shown. Find the correlation matrix of the signal set and determine the distances between all pairs of signals.

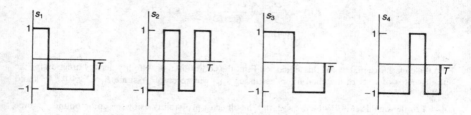

6.3-2 A set of orthogonal signals has the form

$$s_i(t) = x(t - 2i + 2) \qquad i = 1, 2, 3$$

where

$$x(t) = 1 - |t| \qquad |t| \le 1$$
$$= 0 \qquad \text{elsewhere}$$

(a) Sketch the waveforms of this signal set and calculate the signal energy.

(b) Sketch the waveforms of the corresponding simplex signal set (obtained by translation only) and calculate the signal energies of this set.

6.3-3 A signal set has a vector representation of

$$s_1 = (0, 5)$$
$$s_2 = (2, 2)$$
$$s_3 = (-2, 2)$$

(a) Assuming that the signals are equally likely, compute the average energy.

(b) Find the vector representation of the corresponding minimum energy signal set and compute its average energy.

6.4-1 A set of eight coherently detected orthogonal signals gives a bit error rate of 10^{-6}. If the transmitted power were reduced by 50 percent, what bit error rate would result?

6.4-2 The following set of signals is employed in a communication system.

$$s_1(t) = \sqrt{\frac{2E_s}{T}}\cos(4\pi t/T) \qquad 0 \leq t < T$$

$$s_2(t) = \sqrt{\frac{2E_s}{T}}\cos(8\pi t/T) \qquad 0 \leq t < T$$

$$s_3(t) = -\sqrt{\frac{2E_s}{T}}\cos(4\pi t/T) \qquad 0 \leq t < T$$

$$s_4(t) = -\sqrt{\frac{2E_s}{T}}\cos(8\pi t/T) \qquad 0 \leq t < T$$

Sketch the correlation receiver for this signal set utilizing two basis functions. Determine the weighting coefficients.

6.4-3 The following set of signals is employed in a communication system.

$$s_1(t) = \sqrt{\frac{2E_s}{T}}\sin\left(\frac{2\pi t}{T} + \frac{\pi}{3}\right) \qquad 0 \leq t < T$$

$$s_2(t) = \sqrt{\frac{2E_s}{T}}\sin\left(\frac{4\pi t}{T} + \frac{\pi}{3}\right) \qquad 0 \leq t < T$$

$$s_3(t) = \sqrt{\frac{2E_s}{T}}\sin\left(\frac{2\pi t}{T} - \frac{\pi}{3}\right) \qquad 0 \leq t < T$$

$$s_4(t) = \sqrt{\frac{2E_s}{T}}\sin\left(\frac{4\pi t}{T} - \frac{\pi}{3}\right) \qquad 0 \leq t < T$$

Find the covariance matrix for this signal set. Find the conditional probability density function for the output vector of a set of matched filters assuming $s_1(t)$ was received. Assume $E_s = 2 \times 10^{-6}$ V^2 s and $N_0 = 10^{-6}$ V^2/Hz.

6.4-4 For the signal set of Problem 6.4-3 find the bit error probability using the union bound. Assume $E_s = 2 \times 10^{-5}$ V^2 s and $N_0 = 10^{-6}$ V^2/Hz.

6.5-1 Four orthogonal signals are used in a communication system. For $E/N_0 = 25$:
   (a) Find the union bound on the bit error probability.
   (b) Find the bit error probability using the exponential approximation to the Q function.
   (c) Evaluate the integral in Eq. (6-85) using the exponential approximation to the Q function.

6.5-2 A binary message at a rate of 30,000 b/s is conveyed with an M-ary signal set having an average power of 0.25 watt and received in the presence of white noise having a one-sided spectral density of 10^{-6} W/Hz.
   (a) If eight orthogonal signals are used, what is the union bound on the bit error probability?
   (b) If M is made arbitrarily large, what is the smallest average signal power that could be used to achieve the same bit error probability?

6.5-3 Using the union bound, find the bit rate that can be achieved when 16 orthogonal signals are used with an average power of 2 V^2, a noise spectral density of 10^{-5} V^2/Hz, and a bit error probability of 10^{-7}.

6.6-1 A binary message at a rate of 30,000 b/s is conveyed with an M-ary signal having an average power of 0.25 watt and received in the presence of white noise having a one-sided spectral density of 10^{-6} W/Hz. If eight biorthogonal signals are used, what is the union bound on the bit error probability?

6.6-2 Repeat Problem 6.6-1 assuming eight simplex signals are used.

6.6-3 For the signal power and noise power of Problem 6.6-1 find the approximate word error probability when eight equally spaced polyphase signals are used.

6.6-4 Find the approximate bandwidths for the signals of Problems 6.5-2a and 6.6-3.

6.6-5 A QPSK signal set is used for signaling in the presence of white Gaussian noise. Using the union bound, find the value of E/N_0 required to obtain a bit error rate of 10^{-6}. If these signals were detected with a phase error of 20 degrees, what bit error rate would be achieved? What phase error could be tolerated if a bit error rate of 10^{-5} were required?

COMPARISON OF DIGITAL COMMUNICATION SYSTEMS

7-1 METHODS OF COMPARISON[1-3]

There are many ways in which digital communication systems might be compared. Only a few of these will be discussed here. One of the most important comparisons is based on how efficiently the system can utilize the available signal energy to transmit information. A useful measure of this efficiency is the energy per bit. Since all systems have noise in them, and since the probability of error in the system depends upon that noise, the energy utilization efficiency is defined as E_b/N_0, where E_b is the energy per bit and N_0 is the one-sided noise spectral density. It may be noted that the units of E_b/N_0 are

$$\frac{E_b(\text{watts} - \text{seconds/bit})}{N_0(\text{watts/cycle/second})} \rightarrow \frac{\text{cycle}}{\text{bit}}$$

Although this appears to be cycles per bit, it is really a dimensionless quantity.

To employ the energy utilization efficiency in comparing digital communication systems, it is necessary to relate this quantity to the signal power and bit rate in the system. The relation of energy to signal power is

$$E = E_b \log_2 M = P_s T \tag{7-1}$$

where M is the number signals, P_s is the signal power, and T is the signal duration. The relation between signal duration and bit rate R is

$$T = \frac{\log_2 M}{R} \tag{7-2}$$

Combining these results leads to

$$\frac{E_b}{N_0} = \frac{P_s}{R N_0} \tag{7-3}$$

Another quantity that is quite important in comparing digital communication systems is how efficiently the system utilizes the available bandwidth. This leads to the bandwidth utilization efficiency, which is defined as B/R, where B is the system bandwidth and R is the bit rate in the system. The ratio B/R has units of

$$\frac{B(\text{cycles/second})}{R(\text{bits/second})} \rightarrow \frac{\text{cycles}}{\text{bit}}$$

Note that this has the same units as the energy utilization efficiency.

Sometimes the quantity R/B is used for comparing communication systems. This quantity is usually referred to as the signaling speed.

It is also of interest to relate the energy utilization efficiency and the bandwidth utilization efficiency to the signal-to-noise ratio (SNR) in the communication channel. This relationship is

$$\text{SNR} = \frac{P_s}{N_0 B} = \frac{RE_b}{N_0 B} = \frac{E_b/N_0}{B/R} \qquad (7\text{-}4)$$

To make graphical comparisons among the various digital communication systems, one can set up a coordinate system as shown in Fig. 7-1, in which the abscissa is the bandwidth utilization efficiency and the ordinate is the energy utilization efficiency. Because the SNR is equal to the ratio of these two quantities, a line having a constant slope from the origin represents those points in the plane for which the SNR is constant. Such a line having a slope of unity indicates an SNR of one. Points below this line indicate SNRs of less than one, and points above the line represent SNRs greater than one.

A method of comparing communication systems is now evident. By determining the energy utilization efficiency and the bandwidth utilization efficiency of any particular communication system, that system can be represented as a point in the plane of Fig. 7-1. Because of the way these quantities are defined, it is desirable for this point to be as close to the origin as possible. By comparing points representing one system with those representing another system, it is possible to make visual comparisons of both of these quantities simultaneously.

However, there is still another parameter that has not yet been discussed. This is the output signal-to-noise ratio. Clearly, if we wish to compare communication systems, we should do so on the basis of comparable performance. By comparable performance is meant that two systems having the same input produce output

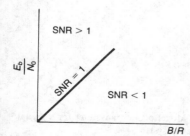

B/R   **Figure 7-1** Energy efficiency versus bandwidth efficiency.

SNRs that are the same. A basic definition of output SNR is

$$(SNR)_0 = \frac{\text{output message power}}{\text{output noise power}} \tag{7-5}$$

This measure is particularly useful when the message is an analog message. Such a message might be speech in which the system is expected to reproduce the analog waveform.

To transmit analog messages in a digital communication system, it is necessary that the message be digitized. This is normally done by sampling at an appropriate rate and quantizing the samples into a finite number of amplitude levels. As a consequence of doing this, the output noise power from the receiver consists of two separate and independent contributions. In the first place, there is the effect of quantizing the analog message. Even if the digital signal is received correctly, the resulting output message will not be exactly the same as the input because all that can be reproduced at the output are the quantized values of the input. Another source of noise in the analog output results from bit decision errors in the channel. Every time a bit is incorrectly decided, this leads to an incorrect amplitude level that is interpreted as noise in the resulting analog message. The operations that must be performed at the transmitter in order to transmit an analog signal over a digital communication system are illustrated in Fig. 7-2.

In constructing this model, certain assumptions have been made. First, it is assumed that the message is band-limited to W hertz. Second, it is assumed that the message is being sampled at the lowest possible rate, which is $2W$. Then the message samples are quantized into L levels, where L is a power of 2. In particular, if $L = 2^q$,

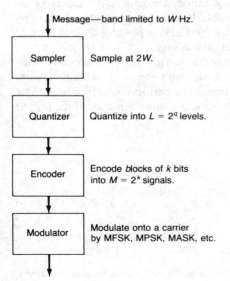

Message—band limited to W Hz.

Sampler — Sample at $2W$.

Quantizer — Quantize into $L = 2^q$ levels.

Encoder — Encode *blocks* of k bits into $M = 2^k$ signals.

Modulator — Modulate onto a carrier by MFSK, MPSK, MASK, etc.

Figure 7-2 Model of the transmitter for a digital communication system.

it requires q bits of information to define uniquely which level is being transmitted. Thus the bit rate is simply $2qW$ b/s. The resulting message bit stream is then broken up into blocks of length k. Since these blocks of length k can be arranged in 2^k possible combinations, it will require $M = 2^k$ signals to describe them. The resulting M binary sequences are then modulated onto a carrier with frequency shift keying, phase shift keying, or amplitude shift keying.

In considering the SNR of the output message, it is convenient to consider the two contributions separately. First, the SNR due to quantizing error only is

$$(\text{SNR})_q = L^2 = 4^q \tag{7-6}$$

This result assumes a uniform quantizer and a uniform error distribution between levels of the quantizer. Although it is not valid for all sources, it forms a reasonable basis for purposes of comparison.

Next it is possible to define an SNR due to bit decision errors. This is

$$(\text{SNR})_d = \frac{L^2}{4P_e(L^2 - 1)} \simeq \frac{1}{4P_e} \tag{7-7}$$

This result assumes that the ith level has been coded into the natural binary number for i, that the bit errors are independent, and that the signal level fully spans the range of the quantizer. Note that for large L, which is almost always the case, the SNR ratio due to bit decision errors is very nearly equal to $1/4P_e$. Since quantizing errors and bit decision errors are independent, the output noise powers resulting from them can be added directly. This leads to an equivalent output SNR of

$$
\begin{aligned}
(\text{SNR})_o &= \frac{1}{\dfrac{1}{(\text{SNR})_q} + \dfrac{1}{(\text{SNR})_d}} \\
&= \frac{L^2}{1 + 4P_e(L^2 - 1)}
\end{aligned} \tag{7-8}
$$

Since the bit error probability depends upon the SNR in the channel, one expects the output SNR to increase as the channel SNR increases. However, from Eq. (7-8) it is clear that the output SNR will never exceed L^2, even when the bit error probability goes to zero. Therefore, this represents an upper bound on the output signal-to-noise ratio. When the channel SNR becomes very small, the bit error probability approaches one half, and the output SNR then approaches a minimum value of $1/2$. The relationship between output SNR and channel SNR is illustrated in Fig. 7-3, for several different values of the parameter k. Recall that $M = 2^k$.

To make meaningful comparisons among digital communication systems, it is necessary to select a particular value of output SNR at which to make this comparison. A convenient value for this is what is called the PCM threshold, and is that value for which the SNR due to quantizing errors is equal to the SNR due

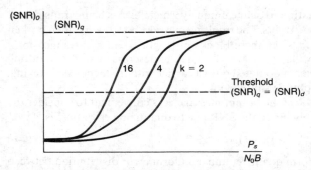

Figure 7-3 Output SNR for a digital communication system.

to decision errors. That is,

$$(\text{SNR})_q = (\text{SNR})_d \qquad (7\text{-}9)$$

For this particular case, the output SNR becomes

$$(\text{SNR})_o = \frac{1}{2}(\text{SNR})_q = \frac{1}{2} L^2 \qquad (7\text{-}10)$$

Note that this is an economical operating point in that if one increases the signal power, the SNR can never exceed that determined by the number of quantizing levels, while if one increases the number of quantizing levels, the SNR can never exceed that determined by the probability of error. Thus the situation in which these two contributions are approximately equal represents a reasonable compromise between power requirements and quantizing requirements.

It is now possible to see how various communication systems compare on the basis of their energy utilization efficiency and their bandwidth utilization efficiency. Such a comparison is shown in Fig. 7-4. All of the systems are compared on the basis of an output SNR of 40 dB, and 256 quantizing levels. This is not quite the PCM threshold, since the PCM threshold for this situation would be 45 dB. The types of systems compared are phase shift keying, amplitude shift keying, frequency shift keying, and differential phase shift keying. Also included are delta modulation, frequency modulation, and amplitude modulation, both double sideband and single sideband. Some general trends may be noted: In the case of frequency shift keying, increasing the values of M reduces the energy requirements but increases the bandwidth requirements, whereas in the case of phase shift keying, just the opposite is true. It may also be observed that if the point closest to the origin is the one that is most desirable, the point marked $P4$ occupies this position. This is quadriphase modulation (QPSK). Hence this is a very desirable form and one that is frequently used. Also shown in Fig. 7-4 is a curve marked "channel capacity." This is discussed subsequently.

There are other aspects of communication signals and channels that might be taken into account in comparing digital communication systems. One such aspect is the spectral characteristic of the transmitted signal. A method of measuring this

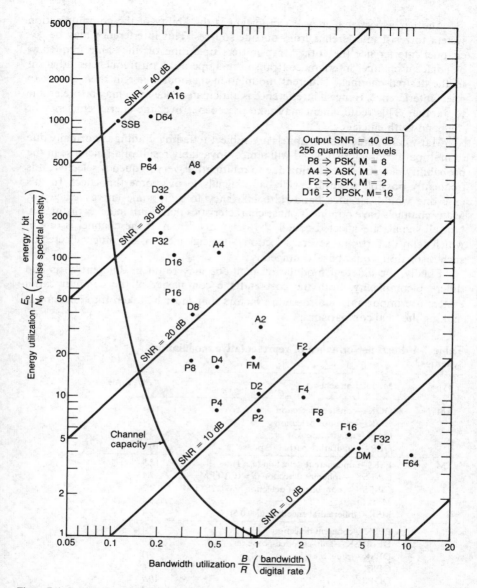

Figure 7-4 Comparison of digital communication systems.

spectral characteristic is to determine the attenuation of the power spectrum at a specified distance, say $8/T$, from the center frequency. To avoid interference with other transmitted signals, it is desirable for this portion of the spectrum to be as small as possible. Equally important may be the rate at which the spectrum falls off; that is, whether it falls off as f^{-2}, f^{-4}, or some higher power of f.

Another aspect that may be important is the ability of the communication system to resist interference from outside sources. This interference may be co-channel interference (i.e., other transmitters operating on the same frequency) or adjacent channel interference (transmitters operating on frequencies adjacent to the desired channel). Different modulation methods differ in their ability to reject interference. Hence if interference is anticipated, one form may be preferable to another. This requirement may take precedence over the energy efficiency or the bandwidth efficiency.

Many communication channels are subject to fading. Fading is normally due to multipath characteristics and will almost invariably result in an increase in the probability of error. Since fading due to multipath is very frequency selective, it is frequently possible to utilize wide-band signals to overcome this effect. In this case one would sacrifice bandwidth efficiency to gain some energy efficiency. Many channels have nonlinear phase characteristics that result in delay distortion. Not all signals are affected equally by delay distortion. If the channel is one in which delay distortion is severe, selection of the signal that is not much affected by such distortion would be advantageous.

Finally, the different modulation methods may result in different costs and different complexity. Both the cost and the complexity of the communication system are important, and these are factors that must be taken into account in making the final comparisons.

Table 7-1 Ideal performance of representative modulation schemes[3]

Type	Modulation scheme	E_B/N_0 (dB)*
AM	OOK—coherent detection	11.4
	OOK—envelope detection	11.9
	QAM—quadrature AM	8.4
	QPR—quadrature partial response	10.7
FM	FSK—noncoherent detection ($d = 1$)	12.5
	CP-FSK—coherent detection ($d = 0.7$) TFM	7.4†
	CP-FSK—noncoherent detection ($d = 0.7$) TFM	9.2†
	MSK ($d = 0.5$)	8.4
	MSK—differential encoding ($d = 0.5$)	9.4
PM	BPSK—coherent detection	8.4
	DE-BPSK—Diff.-encoded phase ref.	8.9
	DPSK—prev. bit ref.	9.3
	QPSK	8.4
	DQPSK	10.7
	OK-QPSK—offset	8.4
	8-ary PSK—coherent detection	11.8
	16-ary PSK—coherent detection	16.2
AM/PM	16-ary APK	12.4

d = FM modulation index.
*For bit error rate of 10^{-4}.
†Assumes 3-bit observation interval.

An extensive comparison based on many of the aspects listed is given in Tables 7-1 through 7-5 and in Fig. 7-5. Although these tables are almost self-explanatory, some of the terminology must be defined: OOK stands for on–off keying; QAM is quadrature AM, that is, two AM carriers 90 degrees out of phase; QPR is quadrature partial response, which is a result similar to that for partial

Table 7-2 Relative signaling speeds of representative modulation schemes[3]

Type	Modulation scheme	Speed (b/s per Hz)	E_B/N_0 (dB)*
AM	OOK—coherent detection	0.8	12.5
	OOK—envelope detection		
	QAM	1.7	9.5
	QPR	2.25	11.7
FM	FSK—noncoherent detection ($d=1$)	0.8	11.8‡
	CP-FSK—noncoherent detection ($d=0.7$)	1.0	10.7
	MSK ($d=0.5$)	1.9	9.4
	MSK—differential encoding ($d=0.5$)	1.9	10.4
PM	BPSK—coherent detection	0.8	9.4
	DE-BPSK	0.8	9.9†
	DPSK	0.8	10.6
	QPSK	1.9	9.9
	DQPSK	1.8	11.8
	8-ary PSK—coherent detection	2.6	12.8
	16-ary PSK—coherent detection	2.9	17.2
AM/PM	16-ary APK	3.1	13.4

d = FM modulation index
*For bit error rate of 10^{-4}.
†Calculated from results for BPSK.
‡Discriminator detection.

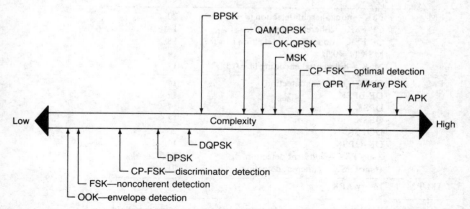

Figure 7-5 Relative complexity of representative modulation schemes.

Table 7-3 Performance of representative modulation schemes in the presence of CW interference[3]

Type	Modulation scheme	E_B/N_0 required*	
		$S/I = 10$ dB	$S/I = 15$ dB
AM	OOK—coherent detection		
	OOK—envelope detection	~20	14.5
	QAM		
	QPR		
FM	FSK—noncoherent detection ($d = 1$)	14.7	13.3
	CP-FSK—coherent detection ($d = 0.7$)		
	CP-FSK—noncoherent detection ($d = 0.7$)		
	MSK ($d = 0.5$)		
	MSK—differential encoding ($d = 0.5$)		
PM	BPSK—coherent detection	10.5	9.2
	DE-BPSK	11.0	9.7
	DPSK	12.0	10.3
	QPSK	12.2	9.8
	DQPSK	>20	14.0
	OK-QPSK		
	8-ary PSK—coherent detection	~20	15.8
	16-ary PSK—coherent detection		>24
AM/PM	16-ary APK		

d = FM modulation index.
*For bit error rate of 10^{-4}.

Table 7-4 Performance of representative modulation schemes on a Rayleigh fading channel[3]

Type	Modulation scheme	Average E_B/N_0 (dB)
AM	OOK—coherent detection	17†
	OOK—envelope detection	19†
	QAM	14
	QPR	
FM	FSK—noncoherent detection ($d = 1$)	20
	CP-FSK—coherent detection ($d = 0.7$)	13‡
	CP-FSK—noncoherent detection ($d = 0.7$)	18‡
	MSK ($d = 0.5$)	14
	MSK—differential encoding ($d = 0.5$)	17
PM	BPSK—coherent detection	14
	DE-BPSK	17
	DPSK	17
	QPSK	13.5
	DQPSK	20
	OK-QPSK	13.5
	8-ary PSK—coherent detection	16.5
	16-ary PSK—coherent detection	21
AM/PM	16-ary APK	18

d = FM modulation index   †Assumes optimum variable threshold.
*For bit error rate of 10^{-2}.   ‡Assumes 3-bit observation interval.

Table 7-5 Performance of representative modulation schemes in the presence of delay distortion $(d/T = 1)^3$

Type	Modulation scheme	E_B/N_0 (dB)*	
		Quadratic	Linear
AM	OOK—coherent detection	12.8	12.4
	OOK—envelope detection	13.3	16.9
	QAM	9.8	15.8
	QPR		>25
FM	FSK—noncoherent detection ($d = 1$)	13.5	16.0
	CP-FSK—coherent detection ($d = 0.7$)	8.8†	8.6†
	CP-FSK—noncoherent detection, ($d = 0.7$)	10.2†	12.7†
	MSK ($d = 0.5$)	9.8	15.8
	MSK—differential encoding ($d = 0.5$)	10.8	16.8
PM	BPSK—coherent detection	9.8	9.6
	DE-BPSK	10.3	10.1
	DPSK	11.8	
	QPSK	9.8	15.8
	DQPSK	16.3	
	OK-QPSK	9.8	15.8
	8-ary PSK—coherent detection	<25	~25
	16-ary PSK—coherent detection		
AM/PM	16-ary APK		

d = FM modulation index.
*For bit error rate of 10^{-4}
†Assumes 3-bit observation interval.

response signals. CPFSK is continuous-phase frequency shift keying. MSK is minimum shift keying, which is FSK with a frequency separation one half that of normal orthogonal FSK. BPSK is binary phase shift keying, DE-BPSK is differentially encoded binary phase shift keying, and OK-QPSK is offset QPSK.

7-2 THEORETICAL LIMITS ON PERFORMANCE

One of the most remarkable results of communication theory dates back to 1948 when Shannon published his paper on the mathematical theory of communication. In this paper he established a theoretical upper bound for the maximum rate at which information could be transmitted in a communication channel. This rate is called the channel capacity. Although general expressions for channel capacity that are applicable to any real communication channel are very difficult to come by, there are some results for ideal channels and systems that are quite simple, quite useful, and quite widely used. One of these is the result that applies to the band-limited channel with white noise and with a constraint on the average received power. This result says that the channel capacity is

$$C = B \log_2\left(1 + \frac{P_s}{N_0 B}\right) \quad \text{b/s} \tag{7-11}$$

In this result B is the bandwidth of the channel, P_s is the average received signal power, and N_0 is the one-sided spectral density of the white noise in the channel.

Although the channel capacity increases with channel bandwidth, it does not increase indefinitely. It may be shown that as the bandwidth approaches infinity, the channel capacity approaches a limiting value known as C_∞, which is related to the ratio of the average signal power and the noise spectral density. This result is illustrated in Fig. 7-6. Note also that when the bandwidth is numerically equal to C_∞, the channel capacity is $0.76\,C_\infty$, showing that it is getting relatively close to the ultimate limit.

When binary information is transmitted in the channel, the bit rate of this information must be less than or equal to the channel capacity. Furthermore, the average signal power can be related to the bit rate and energy per bit by

$$\frac{E_b}{N_0} = \frac{P_s}{N_0 R} \tag{7-12}$$

In the ideal case the bit rate in the channel would be equal to the channel capacity. For this situation

$$\frac{E_b}{N_0} = \frac{P_s}{N_0 C} = \frac{P_s}{N_0 B \log_2\left(1 + \dfrac{P_s}{N_0 B}\right)} \tag{7-13}$$

However, the SNR in the channel is

$$\frac{P_s}{N_0 B} = \text{SNR} \tag{7-14}$$

and from this the energy per bit becomes

$$\frac{E_b}{N_0} = \frac{\text{SNR}}{\log_2(1 + \text{SNR})} \tag{7-15}$$

This relationship is illustrated in Fig. 7-7. It may be noted that the greatest

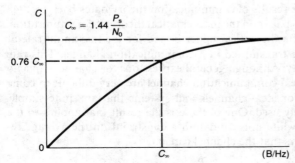

Figure 7-6 Channel capacity as a function of channel bandwith.

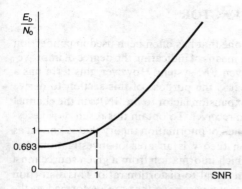

Figure 7-7 Energy efficiency for the ideal channel.

efficiency, that is, the smallest value of E_b/N_0, occurs when the SNR is the smallest, and in this limiting condition, the smallest value of E_b/N_0 is 0.693. This is an ultimate limit below which no physical channel can transmit without error. It is also possible to relate the bandwidth efficiency to channel capacity. For the condition in which the information rate is equal to the channel capacity, this relationship is

$$\frac{B}{R} = \frac{B}{C} = \frac{B}{B \log_2(1 + \text{SNR})} = \frac{1}{\log_2(1 + \text{SNR})} \tag{7-16}$$

Note that now both the energy efficiency and the bandwidth efficiency are expressed solely in terms of the signal-to-noise ratio. Thus it is possible to plot a curve of E_b/N_0 versus B/R and this curve is the channel capacity curve shown in Fig. 7-4. It is reproduced again in Fig. 7-8, with some particular values of SNR illustrated. No physically realizable communication system could occupy a point in this plane lying below this theoretical channel capacity curve.

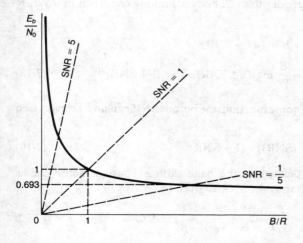

Figure 7-8 Energy efficiency versus the bandwidth efficiency for the ideal channel.

7-3 BANDWIDTH EXPANSION FACTOR

The term *bandwidth expansion factor* is one that has often been used in connection with frequency-modulation systems as a means of indicating the degree of improvement that can be accomplished by a given FM system. However, this term has a general application in any system and it is the purpose of this section to derive some results that relate the bandwidth expansion factor to the SNRs in the channel and at the output of the communication receiver. To obtain this result, it is necessary first to say a few words about a branch of information theory that is known as rate distortion theory. Rate distortion theory is an area of investigation that attempts to find the minimum rate at which information from a given source must be transmitted in order to obtain a given signal-to-distortion ratio. This distortion may result from quantizing, or from other operations that are performed in the modulation process. But in any event, when a continuous source is to be transmitted, there will always be some distortion at the receiver. The greater the information rate, the lower this distortion can be made. If one places a specification on an acceptable amount of distortion, then the rate distortion function indicates the rate at which information should be transmitted in order to achieve this level of distortion. The actual evaluation of rate distortion functions for practical information sources is extremely complicated.

If the source can be modeled as white, band-limited, and Gaussian, and if the distortion measure is the mean square error, then for this particular class of source the rate distortion function becomes

$$R(d) = W \log_2 \frac{P_m}{d} \tag{7-17}$$

where P_m is the average power of the message and d is the squared error, or the distortion power, or the mean square error. We can interpret P_m/d as the output SNR and $R(d)$ as the bit rate. The maximum output SNR occurs with the largest value of R, but R cannot be greater than C. For the limiting condition in which R is equal to C, it follows that

$$W \log_2(\text{SNR})_o = B \log_2(1 + \text{SNR})$$

$$\log_2(\text{SNR})_o = \frac{B}{W} \log_2(1 + \text{SNR}) = \log_2(1 + \text{SNR})^{B/W} \tag{7-18}$$

Since the logarithm is a monotonic function, the output SNR can now be expressed as

$$(\text{SNR})_o = (1 + \text{SNR})^{B/W} \tag{7-19}$$

In this expression B/W is defined as the bandwidth expansion factor and is represented by the symbol β; thus the SNR is given by

$$\text{SNR} = \frac{P_s}{N_0 B} = \frac{W}{B} \cdot \frac{P_s}{N_0 W} = \frac{1}{\beta} (\text{SNR})_W \tag{7-20}$$

where

$$(SNR)_W = \frac{P_s}{N_0 W}$$

Note that $P_s/N_0 W$ is the channel SNR in the message bandwidth. This is not an SNR that actually exists in the system, but simply a fictitious definition that turns out to be convenient not only in this problem, but in many other situations. It can be converted readily to the total SNR in the channel by dividing by β, the band-width expansion factor. Using the relation in Eq. (7-19), this leads to

$$(SNR)_0 = \left[1 + \frac{(SNR)_W}{\beta}\right]^{\beta} \tag{7-21}$$

The left side of Eq. (7-21) can now be interpreted as the maximum output SNR that can be obtained when the SNR in the message bandwidth is given and the bandwidth expansion factor is given. When the bandwidth expansion factor is very large, the output SNR approaches

$$\lim_{\beta \to \infty} (SNR)_0 = e^{(SNR)_W} \tag{7-22}$$

On the other hand, when the bandwidth expansion factor is 1,

$$(SNR)_0 = 1 + (SNR)_W \tag{7-23}$$

Figure 7-9 illustrates the dependence of the output SNR upon the SNR in the message bandwidth and the bandwidth expansion factor.

It is clear that for $\beta = 1$, these are linearly related. However, for $\beta = \infty$, the output SNR increases very rapidly as the SNR in the message bandwidth increases.

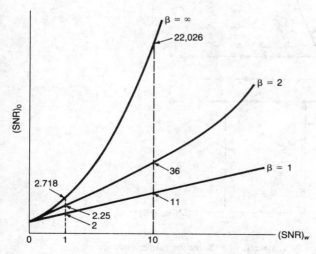

Figue 7-9 Dependence of output SNR upon bandwidth expansion factor.

A conclusion that can be drawn is that if the SNR in the message bandwidth is small, of the order of unity, even a very large value of β does very little to improve the output signal-to-noise ratio. However, when the SNR is large, say of the order of 10, then a large β can yield a very great improvement in signal-to-noise ratio. When the SNR in the message bandwidth is very large compared with β, it can be shown that each decibel increase in SNR in the message bandwidth yields a β-dB increase in output signal-to-noise ratio. The condition of SNR in the message bandwidth equal to β can be interpreted as the improvement threshold. It should be recalled that this result is for a band-limited message. When the messages are not band-limited, but have a spectral density that vanishes only at infinity, the improvement is not nearly as dramatic.

Table 7-6 indicates some typical values of β for practical communication systems. In this table NBFM is narrow-band frequency modulation and WBFM is wide-band frequency modulation. It may be noted that in most of these systems, the bandwidth expansion factor is relatively small.

Another comparison of practical with theoretical systems is shown in Fig. 7-10, in which the dots represent the performance of specific M-ary PCM systems having various numbers of bits per amplitude level and various numbers of messages. The curve is the theoretical result for band-limited messages with an infinite bandwidth expansion factor. It is clear from this graph that making the number of messages very large will produce results that approach the theoretical. However, even for as many as 256 messages, the practical system is 7 dB away from the theoretical result; in the binary case, the practical system is 14 dB away from the theoretical result.

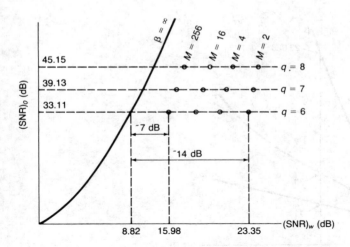

Figure 7-10 Comparison of M-ary PCM systems with the theoretical performance.

**Table 7-6 Bandwidth
expansion factors for
practical communication
systems**

System	β
SSB	1
DSB	2
NBFM	4–6
WBFM	10–20
7-bit PCM	$\simeq 28$

7-4 REFERENCES

1. Gagliardi, R. M.: *Introduction to Communications Engineering*, John Wiley & Sons, Inc., New York, 1978, pp. 381–388.
2. Viterbi, A.: *Principles of Coherent Communications*, McGraw-Hill Book Company, 1966, Chap. 9.
3. Oetting, J. D.: A Comparison of Modulation Techniques for Digital Radio, *IEEE Trans. on Comm.*, vol. COM-27, no. 12, pp. 1752–1762, December 1979.

7-5 PROBLEMS

7.1-1 A message band-limited to 4 kHz is sampled at the Nyquist rate and quantized to L levels where $L = 2^q$. The message is transmitted over a binary channel with $E_b/N_0 = 25$. What is the value of q that will provide approximately equal contributions to the SNR from quantizing and channel noise and what is the resulting SNR? What is the channel bit error probability? What is the bit rate for this channel?

7.1-2 A communication system employs eight orthogonal signals to transmit a message quantized to 8 bits. The signal error probability is 10^{-5}. Find the channel equivalent output signal-to-noise ratio.

7.2-1 A channel is band-limited to 25 kHz and has an average power limitation of 10 watts. The one-sided spectral density of the noise in the channel is 10^{-5} W/Hz.

(a) Find the maximum rate at which information can be sent over the channel without error.

(b) Find the limiting capacity of the channel if the bandwidth is not limited.

7.3-1 A message band-limited to 4 kHz is transmitted over a channel that is band-limited to 100 kHz. What is the smallest channel SNR in the message bandwidth that will yield an output SNR of 30 dB?

7.3-2 Approximately what channel SNR in the message bandwidth would be required to provide a 36-dB increase in SNR when the bandwidth expansion factor is 6?

7.3-3 Approximately what bandwidth expansion factor is required to obtain a 10-dB improvement in output SNR when the channel SNR in the message bandwidth is (a) 10 dB and (b) 20 dB?

EIGHT

FUNDAMENTALS OF SPREAD SPECTRUM

8-1 GENERAL CONCEPTS

The discussions of communication systems in previous chapters have been concerned with the efficiency with which these systems utilize signal energy and bandwidth. These are legitimate concerns, and in most communication systems are the concerns of paramount importance. There are situations, however, in which it is necessary for the system to resist external interference, to operate with a low-energy spectral density, to provide multiple-access capability without external control, or to make it difficult for unauthorized receivers to observe the message. In such a situation, it may be appropriate to sacrifice the efficiency aspects of the system in order to enhance these other features. Spread-spectrum techniques offer one way to accomplish this objective.

The use of spread-spectrum techniques originated in answer to the unique needs of military communications, and this application is still the predominant one. However, there are a number of nonmilitary applications that are under active consideration and it is reasonable to assume that these techniques will soon penetrate the civilian sector. Therefore, a discussion of modern communications would not be complete without a look at the fundamentals and the applications of spread spectrum. This discussion is brief, but the reader is referred to other sources for more complete details.[1-4]

For a communication system to be considered a spread-spectrum system, it is necessary that the transmitted signal satisfy two criteria. First, the bandwidth of the transmitted signal must be much greater than the message bandwidth. This by itself, however, is not sufficient because there are many modulation methods that achieve it. For example, frequency modulation, pulse code modulation, and delta modulation may have bandwidths that are much greater than the message bandwidth. Hence the second criterion is that the transmitted bandwidth must be determined by some function that is independent of the message and is known to the receiver. Methods of accomplishing this are discussed in a subsequent section.

Although it would appear that the bandwidth expansion factor of a spread-spectrum system is very large, the concept of such a factor as discussed in the preceding chapter does not really apply. In fact the bandwidth expansion does not combat white noise as it does in FM, PCM and other wide-band modulation methods. This is so because bandwidth expansion is achieved by something that is independent of the message, rather than being uniquely related to the message. Since the spread-spectrum system is not useful in combating white noise, it must have other applications that make it worth considering. These applications include:

1. Antijam capability—particularly for narrow-band jamming.
2. Interference rejection.
3. Multiple-access capability.
4. Multipath protection.
5. Covert operation or low probability of intercept (LPI).
6. Secure communications.
7. Improved spectral efficiency—in special circumstances.
8. Ranging.

These applications are discussed in more detail in subsequent chapters.

There are many different types of spread-spectrum systems and one way of classifying them is by concept. On this basis spread-spectrum systems may be considered to be either averaging systems or avoidance systems. An averaging system is one in which the reduction of interference takes place because the interference can be averaged over a large time interval. An avoidance system, on the other hand, is one in which the reduction of interference occurs because the signal is made to avoid the interference a large fraction of the time.

A second method of classifying spread-spectrum systems is by modulation. The most common modulation techniques employed are the following.

1. Direct sequence (pseudonoise).
2. Frequency hopping.
3. Time hopping.
4. Chirp.
5. Hybrid methods.

The relation between these two methods of classification may be made clearer by noting that a direct-sequence system is an averaging system, whereas frequency-hopping, time-hopping, and chirp systems are avoidance systems. On the other hand, a hybrid modulation method may be either averaging or avoidance, or both.

8-2 DIRECT SEQUENCE (DS) OR PSEUDONOISE (PN)

The terms *direct sequence* and *pseudonoise* are used interchangeably here and no distinction is made between them. A typical direct-sequence transmitter is illustrated in Fig. 8-1. Note that it contains a PN code generator that generates the pseudonoise sequence. The binary output of this code generator is added, modulo 2, to the binary message, and the sum is then used to modulate a carrier. The modulation in this case is biphase or phase reversal modulation so that the output is simply a phase shift keyed signal. The PN code is generated in a maximal-length shift register such as shown in Fig. 8-2. More details on these devices and the resulting sequences are considered in a subsequent chapter.

Pseudonoise code generators are periodic in that the sequence that is produced repeats itself after some period of time. Such a periodic sequence is portrayed in Fig. 8-3. The smallest time increment in the sequence is of duration t_1 and is known as a time chip. The total period consists of N time chips. When the code is generated by a maximal linear PN code generator, the value of N is $2^n - 1$, where n is the number of stages in the code generator. An important reason for using shift register codes is that they have very desirable autocorrelation properties. The auto-correlation function of a typical PN sequence is shown in Fig. 8-4. Note that on a normalized basis, it has a maximum value of one that repeats itself every period, but in between these peaks, the level is at a constant value of $-(1/N)$. If N is a very large number, the autocorrelation function will be very small in this region.

Another reason for using shift register codes is that the period of the PN sequence can easily be made very large. As a numerical example of this, suppose

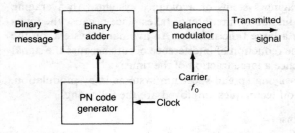

Figure 8-1 Direct-sequence transmitter.

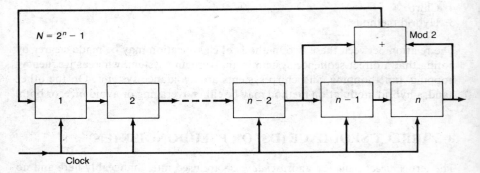

Figure 8-2 Maximal linear PN code generator.

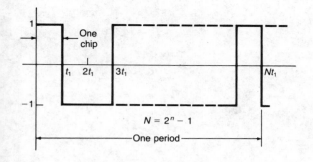

Figure 8-3 Periodic binary PN sequence.

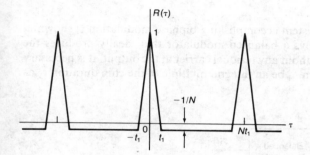

Figure 8-4 Autocorrelation function of PN sequence.

that the clock frequency is 10 MHz. This means that t_1 would be 10^{-7} seconds. If we use a shift register with 41 stages, the total length of the sequence is

$$N = 2^{41} - 1 = 2.199 \times 10^{12}$$

$$Nt_1 = 2.199 \times 10^{12} \times 10^{-7}$$

$$= 2.199 \times 10^5 \text{ seconds}$$

$$= 2.545 \text{ days}$$

Thus with a fairly modest amount of hardware, it becomes possible to make sequences that are almost arbitrarily long.

The modulation of the PN sequence on the spread-spectrum carrier can be either biphase or quadriphase. It is of interest to consider both of these methods.

Biphase Modulation

A phase-modulated carrier can be expressed in general as

$$s(t) = A \sin[\omega_0 t + \phi(t)] \tag{8-1}$$

where A is the constant carrier amplitude and $\phi(t)$ represents the phase modulation. In the case of biphase modulation, $\phi(t)$ will be either zero or π. The values of $\phi(t)$ for various combinations of the binary message, $m(t)$, and the PN sequence,

$b(t)$, are shown in Table 8-1.

Table 8-1 Truth table for $\phi(t)$

b(t)		m(t) 1	m(t) −1
	1	0	π
	−1	π	0

A block diagram of a system accomplishing biphase modulation is shown in Fig. 8-5. This system employs a balanced modulator that ideally produces the desired phase shift keying without any residual carrier at the output. It is necessary that the message bit duration t_m be an integral multiple of the chip duration t_1 as shown in Fig. 8-6.

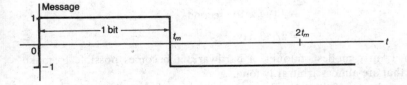

Figure 8-5 Block diagram for biphase modulation.

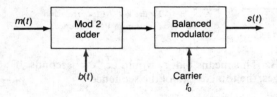

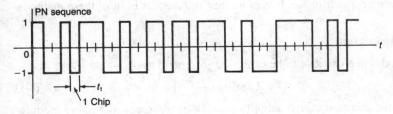

Figure 8-6 Relation between the code sequence and the binary message.

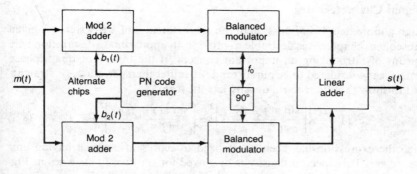

Figure 8-7 Block diagram for quadriphase modulation.

Quadriphase Modulation

A block diagram of a system producing quadriphase modulation is shown in Fig. 8-7. In this case two balanced modulators are used and the carriers to these two modulators are 90 degrees apart in phase. There are also two modulo-2 adders that add the message binary sequence to the PN code sequence, using alternate chips from the code sequence to do so. This means that each chip of the PN code is stretched to a duration of $2t_1$ before being added to the binary message. The quadriphase signal can again be represented as

$$s(t) = A \sin[\omega_0 t + \phi(t)] \tag{8-2}$$

in which A is the carrier amplitude and $\phi(t)$ is the phase modulation. The relation of $\phi(t)$ to the state of the message and the states of the PN code sequence is shown in Table 8-2.

Table 8-2 Truth table for
$\phi(t)$

		$m(t)$	
$b_1(t)$	$b_2(t)$	1	−1
1	1	$\pi/4$	$5\pi/4$
1	−1	$7\pi/4$	$3\pi/4$
−1	1	$3\pi/4$	$7\pi/4$
−1	−1	$5\pi/4$	$\pi/4$

It may be noted that the message modulation is still binary.

PN Signal Characteristics

Although a more detailed discussion of the characteristics of a PN signal is given in a subsequent chapter, it is desirable to provide an approximate description here as a means of introducing an important concept. If the binary or quadriphase PN sequence is considered to be purely random, rather than periodic, it is straightforward to show that its spectral density has the form[5]

$$S(f) = \frac{t_1}{2} \left\{ \left[\frac{\sin \pi(f - f_0)t_1}{\pi(f - f_0)t_1} \right]^2 + \left[\frac{\sin \pi(f + f_0)t_1}{\pi(f + f_0)t_1} \right]^2 \right\} \qquad (8\text{-}3)$$

in which the expression has been normalized to represent a signal having unit average power. This spectral density is displayed for positive frequencies in Fig. 8-8. It is customary to define the bandwidth of a PN signal as the frequency increment between the two zeros of the spectral density that are closest to the center frequency. It is clear from Fig. 8-8 that this signal bandwidth is $2/t_1$. For some analytical purposes, a more precise definition of bandwidth is required, but this is considered later.

Since the message is also binary, it will have a similar spectral density but centered around zero. Thus the message spectral density is

$$S_m(f) = t_m \left[\frac{\sin \pi f t_m}{\pi f t_m} \right]^2 \qquad (8\text{-}4)$$

The bandwidth of the message B_m is simply $1/t_m$ because it is customary to use only the positive frequency portion of the spectrum in defining bandwidth.

An important parameter that is sometimes useful in specifying the performance of a spread-spectrum signal in the presence of interference is known as the processing gain. This processing gain, PG, is frequently defined as the ratio of the signal bandwidth to the message bandwidth. Thus

$$PG = \frac{B_s}{B_m} = \frac{2t_m}{t_1} \qquad (8\text{-}5)$$

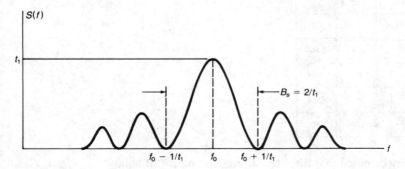

Figure 8-8 Spectral density of a random binary sequence.

Some authors prefer to define the processing gain of direct-sequence systems as the ratio of the chip rate to the message bit rate (see ref. 4, p. 348). This value is just one half of that given by Eq. (8-5).

As an illustration of the computation of processing gain, assume that we have the following conditions:

$$\text{PN chip rate} = 40 \times 10^6$$
$$B_s = 80 \times 10^6$$
$$\text{Message bit rate} = B_m = 10^4$$
$$\text{PG} = \frac{80 \times 10^6}{10^4} = 8000 = 39 \text{ dB}$$

Direct-Sequence Receiver

The receiver for a PN signal must perform three distinct functions: detect signal presence, despread the signal, and demodulate the message. The detection and despreading operations can be either active or passive, and both approaches are considered subsequently. Active methods involve searching for the signal presence in both time and frequency, tracking the sequence after it has been acquired, despreading the signal with a correlator, and then demodulating the message in the usual way.

Passive methods, on the other hand, only require that the signal be searched for in frequency, since the passive system will respond to the signal whenever it occurs. The despreading is accomplished in a matched filter rather than a correlator, and the demodulation is again performed in the usual manner.

The choice of methods, whether active or passive, depends upon the conditions that prevail. Active methods are preferred when the sequence is very long or when the processing gain is very large. On the other hand, passive methods may be preferred when the sequence is short or when used as an aid to acquisition. It is possible to combine these two approaches in a single receiver.

Figure 8-9 illustrates the essential parts of a direct-sequence receiver.[2] The

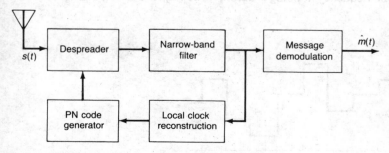

Figure 8-9 Direct-sequence receiver.

despreader, which is discussed in detail subsequently, accomplishes the task of multiplying the incoming signal with a locally generated PN sequence. When these are aligned in time, the output is simply the message, which is at a much lower frequency and thus can be filtered in a narrow-band filter. This is followed by the message demodulation. The output of the narrow-band filter can also be used as a means of controlling the clock that drives the PN generator to make sure that it stays in step with the incoming signal. The operation of this receiver is considered in more detail in a subsequent chapter.

8-3 FREQUENCY HOPPING

In a frequency-hopping signal, the frequency is constant in each time chip, but changes from chip to chip. This type of signal is illustrated in Fig. 8-10.

It is frequently convenient to categorize frequency-hopping systems as either "fast hop" or "slow hop," since there is a considerable difference in performance for these two types of systems. A fast-hop system is usually considered to be one in which the frequency hopping takes place at a rate that is greater than the message bit rate; in a slow-hop system, the hop rate is less than the message bit rate. There is, of course, an intermediate situation in which the hop rate and the message bit rate are of the same order of magnitude.

For purposes of illustration, a fast-hop system is considered here in which there are k frequency hops in every message bit. Thus the chip duration is

$$t_1 = t_m/k \qquad k = 1, 2, 3, \ldots \tag{8-6}$$

The number of frequencies over which the signal may hop is usually a power of 2, although not all of these frequencies are necessarily used in a given system. The reason for making the number of frequencies a power of 2 becomes apparent when methods of generating frequency-hopped signals are considered.

The Frequency-Hopping Transmitter

The block diagram of a frequency-hopping transmitter is shown in Fig. 8-11. Note that the frequency hopping is accomplished by means of a digital frequency

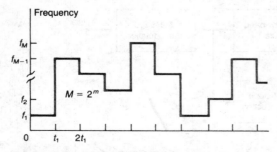

Figure 8-10 Frequency-hopping signal.

synthesizer, which in turn is driven by a PN code generator. The frequency synthesizer is controlled by m binary digits and produces one of $M = 2^m$ frequencies for each distinct combination of these digits. One of these m controlling digits comes from the message and the other $m - 1$ digits come from the PN code generator. If the digit from the message produced the smallest frequency change, then by itself it would produce a binary FSK signal. The $m - 1$ digits from the PN code generator then hop this FSK signal over the range of possible frequencies. More detail on the operation of the digital frequency synthesizer is provided in a subsequent section.

The message, prior to modulating the frequency synthesizer, normally will have error-correction coding applied to it. If any one hop is interfered with, all of the bits in that particular hop may be destroyed, and, therefore, it is necessary to be able to reconstruct the message by using error-correction techniques. It may also be noted that there is a frequency multiplier at the output of the system; its purpose is to increase the bandwidth and thereby increase the processing gain that is available. It also changes the shape of the spectrum, as discussed later.

If the M frequencies produced by the digital frequency synthesizer are separated by $f_1 = 1/t_1$, then the signal bandwidth is given by

$$B_s \simeq KMf_1 = \frac{KM}{t_1} \tag{8-7}$$

where K is the amount of frequency multiplication employed. When the chip duration is that shown in Eq. (8-6), the processing gain becomes

$$PG = \frac{B_s}{B_m} = \frac{KM/t_1}{1/t_m} = \frac{kKM/t_m}{1/t_m} = kKM \tag{8-8}$$

Note that the processing gain depends on the amount of frequency multiplication, the number of different frequencies employed, and the number of hops per message bit.

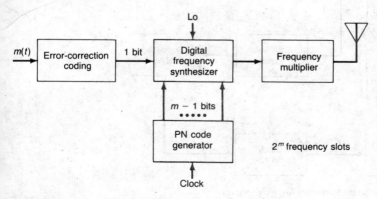

Figure 8-11 Frequency-hopping transmitter.

The Frequency-Hopping Receiver

Usually the reception of a frequency-hopping signal is done on a noncoherent basis. Coherent reception is possible, but it is more difficult to achieve and places some constraints on the nature of the transmitted signal and the transmission medium. A typical noncoherent, frequency-hopping receiver is shown in Fig. 8-12.[2] Note that this consists of a digital frequency synthesizer driven by a PN code generator and followed by a frequency multiplier. This locally generated frequency-hop signal is multiplied by the incoming signal in a mixer, and if the two are in step, the result will be a normal binary FSK signal, which is demodulated in the usual fashion. Error correction is then applied to produce the eventual message. The output of the mixer is also applied to early and late gates that produce an error signal to control the clock frequency. This keeps the locally generated frequency-hop signal in step with the incoming signal.

One of the advantages of frequency hopping is that the PN code generator can run at a considerably slower rate in this type of system than in a direct-sequence system. Specifically, when t_1 equals t_m/k, the PN code generator clock rate is simply $k(m-1)$ times the message bit rate, and is independent of the eventual signal bandwidth. Normally this clock rate will be considerably less than would be required if the clock had to run at half the signal bandwidth. The maximum rate at which the code generator can run in a frequency-hopping transmitter or receiver is generally determined by the switching speed in the frequency synthesizer, rather than by the code generator itself. The constraints on frequency synthesizer switching speed are considered in some detail in a subsequent chapter.

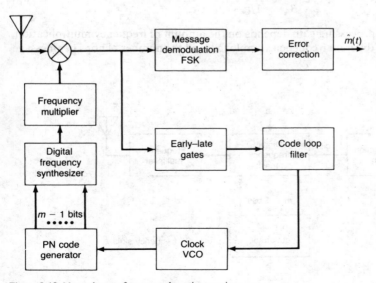

Figure 8-12 Noncoherent frequency-hopping receiver.

When several frequency-hopping signals occupy a common RF channel, simultaneous occupancy of a given frequency slot may result in message errors. This is the basic reason why error-correction coding is needed in a frequency-hop system since the error-correction capability of the code can restore the messages that are destroyed by mutual interference. This is particularly true in slow-hop systems in which there may be several message bits in each hop.

The "Near–Far" Problem

The "near–far" problem in spread-spectrum systems relates to the problem of very strong signals at a receiver swamping out the effects of weaker signals. This problem is particularly severe in the case of direct-sequence systems. A frequency-hop system is much less suceptible to the near–far problem because it is an avoidance system rather than an averaging system. Interference results whenever there is simultaneous occupancy of a given frequency slot, and in this case it does not matter much if the interfering signal is much stronger than the desired signal.

Another advantage of frequency-hopping signals is that acquisition is normally much faster than it is in the case of direct-sequence signals. This is so because the chip rate is considerably less in the frequency-hopping system. Acquisition is discussed in a later chapter.

A disadvantage of frequency-hopping systems is that one cannot easily use coherent message demodulation techniques. This results in poorer performance against thermal noise.

8-4 TIME HOPPING

A time-hopping waveform is illustrated in Fig. 8-13. The time axis is divided into intervals known as frames, and each frame is subdivided into M time slots. During each frame one and only one time slot will be modulated with a message by any

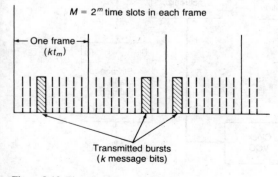

Figure 8-13 Time-hopping waveform.

reasonable modulation method. The particular time slot that is chosen for a given frame is selected by means of a PN code generator. All of the message bits accumulated in the previous frame are transmitted in a burst during the selected time slot. To quantify this concept, let

$$T_f = \text{frame duration}$$

$$k = \text{number of message bits in one frame}$$

$$T_f = kt_m$$

The width of each time slot in a frame is T_f/M and the width of each bit in the time slot is T_f/kM, which is simply t_m/M. This indicates that the transmitted signal bandwidth is $2M$ times the message bandwidth, and hence the processing gain of a time-hopping system is simply twice the number of time slots in each frame when biphase modulation is used, and half this when quadriphase modulation is used.

A typical time-hopping receiver is shown in Fig. 8-14.[2] It consists of an on–off switch that is driven by a PN code generator in order to do the switching at the proper time in each frame. The output of this switch is then demodulated by either a two-phase or four-phase demodulator, depending on the nature of the transmitted signal. Bit synchronization is also required and the output of the bit synchronizer is used to control the clock that drives the PN code generator to maintain synchronization. Since the message bits are coming out at a much greater rate than that at which they were originally produced, each burst must be stored and then retimed to the normal message rate. Time hopping is most often used in conjunction with other spread-spectrum techniques.

Interference among simultaneous users in a time-hopping system can be minimized by coordinating the times at which each user can transmit a signal. This also avoids the near–far problem. In a noncoordinated system, overlapping transmission bursts will result in message errors, and this will normally require the

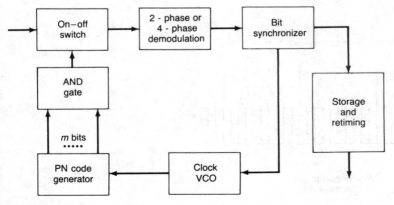

Figure 8-14 Time-hopping receiver.

use of error-correction coding to restore the proper message bits. The acquisition time is similar to that of direct-sequence systems for a given bandwidth. Implementation is simpler than that of a frequency-hop system.

8-5 COMPARISON OF MODULATION METHODS

It is desirable at this point to include a list of the advantages and disadvantages of the three types of spread-spectrum systems that have been discussed so far. The purpose of this comparison is to illustrate that each method has its advantages and disadvantages, and that it may be possible to overcome some of these disadvantages by the use of a combination of techniques. The advantages and disadvantages of the various systems are outlined below and more detailed discussion of some of these points is provided in following chapters.

Direct-Sequence (PN) Systems

Advantages
- Best noise and antijam performance.
- Most difficult to detect.
- Best discrimination against multipath.

Disadvantages
- Requires wide-band channel with little phase distortion.
- Long acquisition time.
- Fast code generator needed.
- Near–far problem.

Frequency-Hopping (FH) Systems

Advantages
- Greatest amount of spreading.
- Can be programmed to avoid portions of the spectrum.
- Relatively short acquisition time.
- Less affected by near–far problem.

Disadvantages
- Complex frequency synthesizer.
- Not useful for range and range-rate measurement.
- Error correction required.

Time-Hopping (TH) Systems

Advantages
- High bandwidth efficiency.
- Implementation simpler than FH.
- Useful when transmitter is average power limited but not peak power limited.

- Near–far problem avoided in a coordinated system.
Disadvantages
- Long acquisition time.
- Error correction needed.

8-6 HYBRID SPREAD-SPECTRUM SYSTEMS

The use of a hybrid system attempts to capitalize upon the advantage of a particular method while avoiding the disadvantages. Many different hybrid combinations are possible. Some of these are

PN/FH
PN/TH
FH/TH
PN/FH/TH

To illustrate how a hybrid system might operate, consider the case of a PN/FH hybrid system. This system might use a PN code to spread the signal to an extent limited by either code generator speed or acquisition time. Then frequency hopping would be used to increase the frequency spread. The difference between the frequencies in the frequency-hopping portion of the system would normally be equal to the bandwidth of the PN code modulation. Usually some form of non-coherent message modulation is used because of the frequency hopping, and differential phase shift keying is a typical example. Since there are fewer frequencies to be implemented, the frequency synthesizer is simpler for a given overall band-width. Thus this system gains some of the advantages of direct-sequence systems and of frequency-hop systems, and avoids some of the disadvantages of both. A typical example of a PN/FH system is shown in the following.

Example of PN/FH System

- PN rate of 250,000 chips per second.
- FH spreading to 2.048 GHz (8192 frequencies separated by 250 kHz).
- Error correction from one third rate convolutional coding.
- Message rate: 75 b/s or 2400 b/s.
- PG = 74.3 dB at 75 b/s = 59.2 dB at 2400 b/s.

8-7 CHIRP SPREAD SPECTRUM

A chirp spread-spectrum system utilizes linear frequency modulation of the carrier to spread the bandwidth. This is a technique that is very common in radar systems, but is also used in communication systems. The relationships between frequency and time are shown in Fig. 8-15, in which T is the duration of a given signal wave-

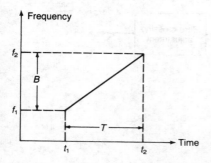

Figure 8-15 The linear chirp signal.

form and B is the bandwidth over which the frequency is varied. In this case the processing gain is simply BT. It is also possible to use nonlinear frequency modulation, and in some cases this may be desirable.

8-8 BASEBAND MODULATION TECHNIQUES

Digital Methods

An analog message is normally converted to binary form before being used to modulate a spread-spectrum signal. There are different ways to accomplish this. For purposes of illustration, assume that (PCM) is to be employed. Figure 8-16 illustrates the essential parts of a PCM system.

The input, which is an analog message, is sampled at a rate greater than twice the bandwidth of the message. These samples are then quantized into L different quantizing levels, where L is normally a power of 2. Each level is then encoded into q binary digits. This produces the output binary signal that is used to modulate the spread-spectrum signal. The duration of each message bit can be expressed in terms of the sampling rate and the number of bits per sample. Specifically $t_m = 1/qf_s$. When the chip duration is t_1, it is normally necessary to make the ratio t_m/t_1 an integer. This ratio is usually large in direct-sequence systems; typically it will be k in a frequency-hop system, and in a time-hopping system, it will be equal to the number of slots per frame.

There are several other techniques that might be used to convert an analog signal into a suitable digital form. These include differential PCM and delta modulation. All of these techniques are discussed in more detail in Sec. 1.6.

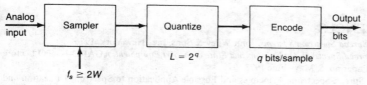

Figure 8-16 Pulse-code modulation.

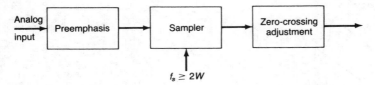

Figure 8-17 Pulse-duration modulation.

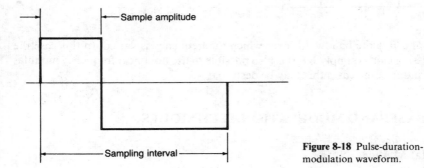

Figure 8-18 Pulse-duration-modulation waveform.

Sampled Analog Methods

Although true analog modulation of a spread-spectrum signal does not appear to be desirable, it is possible to use a sampled analog approach. A pulse-duration approach is illustrated in Fig. 8-17. The analog signal is normally passed through a preemphasis circuit in order to emphasize the high frequencies. It is then sampled at a rate that is twice the bandwidth of the message and passed through a device that adjusts the location of the zero crossing of the resulting sample. This results in a signal of the form shown in Fig. 8-18, in which the positive portion of the signal waveform has a duration that is proportional to the sample amplitude. The total duration of the waveform is simply the time interval between samples.

Although the pulse duration is a continuous quantity, it is subsequently quantized by the PN sequence. That is, it is terminated at the particular code zero crossing that is closest to the correct amplitude value. This is the reason for the original preemphasis, since this quantization introduces high-frequency noise that is reduced by the corresponding deemphasis circuit in the receiver.

8-9 REFERENCES

1. Dixon, R. C.: *Spread Spectrum Systems*, John Wiley & Sons, Inc., New York, 1976.
2. Cahn, C. R.: *Spread Spectrum Applications and State of the Art Equipment*, AGARD-NATO Lecture Series no. 58, Paper no. 5, May–June 1973.
3. Utlaut, W. F.: Spread-Spectrum: Principles and Possible Application to Spectrum Utilization and Allocation, *IEEE Com. Mag.*, pp. 21–30, September 1978.

4. Holmes, J. K.: *Coherent Spread Spectrum Systems*, John Wiley & Sons, Inc., New York, 1982.
5. Cooper, G. R., and C. D. McGillem: *Probabilistic Methods of Signal and System Analysis*, Holt, Rinehart and Winston, New York, 1971.

8-10 PROBLEMS

8.1-1 Given a message signal

$$f(t) = \sqrt{2} \cos \omega_m t$$

we form the FM signal

$$s(t) = A \cos\left[\omega_c t + k_f \int_0^t \sqrt{2} \cos \omega_m t\, dt \right]$$

$$= A \cos\left[\omega_c t + \frac{\sqrt{2}k_f}{2\pi W_m} \sin \omega_m t \right]$$

$$= A \cos[\omega_c t + \beta \sin \omega_m t]$$

where β was defined as in Sec. 1-5 as the modulation index and k_f as the frequency deviation constant in rad/V-s.

(a) With

$$W_m = \frac{2200}{\pi} \text{ Hz}$$

and

$$k_f = 44 \times 10^3$$

estimate the approximate signal bandwidth.

(b) The above signal bandwidth is then spread out to a wider bandwidth by passing it through a $\times 100$ multiplier and then translated back to ω_c as shown here:

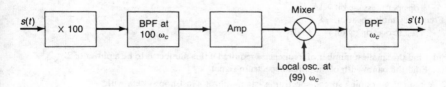

Now estimate the above approximate signal bandwidth. Can this method be a viable way to generate a spread-spectrum signal?

8.2-1 A recorded conversation is to be transmitted by a pseudonoise spread-spectrum system. Assuming the spectrum of the speech waveform is band-limited to 3 kHz, and using 128 quantization levels:

(a) Find the chip rate required to obtain a processing gain of 20 dB.

(b) Given that the sequence length is to be greater than 5 hours, find the number of shift register stages required.

8.2-2 A stationary random process is composed of an infinite sequence of independent binary digits. Each digit is a rectangular pulse of duration t_1 and has an amplitude of 0 or $+1$, with equal probability. Find the autocorrelation function of this random process.

8.2-3 The random process of the previous problem is to be added modulo-2 with the message bit sequence and the resulting sequence is used to phase-modulate a carrier (phase reversal keying), as shown.

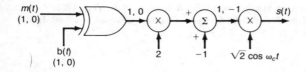

(a) Why is it necessary for the message bit duration t_m to be an integral multiple of the chip duration?

(b) If $m(t)$ has a bit rate of 1 kb/s and the logic gate (exclusive-or) above has a switching time of 30 ns, what is the maximum processing gain obtainable?

(c) Find the spectral density of the modulated signal, $s(t)$.

8.2-4 The binary process of Problem 8.2-2 is multiplied synchronously by another independent binary process in which the duration of the binary digits is $t_m = mt_1$, where m is an integer. Find the autocorrelation function of the product of the two binary sequences.

8.2-5 A direct-sequence system has a PN code rate of 192×10^6 chips per second and a binary message bit rate at 7500 b/s.

(a) If QPSK is used instead of biphase modulation, what is the processing gain?

(b) Assuming the received signal power is 4×10^{-14} watts and the one-sided noise spectral density level, N_0, is 1.6×10^{-20} W/Hz, find the signal-to-noise power ratio in the input bandwidth of the receiver.

8.2-6 Figure 8-8 shows the general shape of the spectral density of a random binary sequence centered at a carrier frequency, f_0. The usual convention for defining the bandwidth is the frequency increment between the two nulls or, rather, the width of the main lobe.

(a) If the chip duration is 0.1 μs and the signal used is normalized to unit average power, what percent of the total power is not contained in the bandwidth as defined above? Is the above bandwidth definition a realistic one in terms of containing a major portion of the power?

(b) What would be the 3-dB bandwidth of the above spectrum?

8.3-1 A frequency-hopping spread-spectrum system is to have the following parameters:

Message bit rate = 2400 bp (after error correction coding)

Hops per message bit = 16
Frequency multiplication = 8
Processing gain = ≥ 45 dB
$f_1 t_1 = 1$

(a) Find the smallest number of frequencies required if this number is to be a power of 2.

(b) Find the bandwidth of the spread-spectrum signal.

8.3-2 A frequency-hopping spread-spectrum system utilizes a fast-hop system with

$$k = 10 \text{ hops per message bit}$$

$$m = 1024 \text{ frequencies:}$$

$$\text{Message bit rate} = 2500 \text{ b/ps}$$

$$\text{Final RF multiplication of } K = 10$$

Find:

(a) The RF signal bandwidth.

(b) Processing gain in decibels.

(c) PN code generator clock rate.

(d) Frequency separation in kilohertz.

8.3-3 For the same processing gain as in the previous problem, what must an equivalent PN code rate be when $t_m = 1/42,000$?

8.3-4 With $k = 10$ hops per message bit and $K = 10$ as in Problem 8.3-2, how many hopping frequencies would be required to achieve the same processing gain as in the QPSK system of Problem 8.2-5?

8.3-5 In a frequency-hopping system, the switching speed of the synthesizer is 4 μs, the message bit rate is 5 kb/s after error-correction coding, there are five hops per message bit, and the final frequency multiplication is 8. What is the maximum processing gain obtainable?

8.4-1 A certain time-hopping spread-spectrum system is only allocated a maximum signal bandwidth of 8 Mhz. Assuming the message bit rate after error correction is 3200 b/s:

(a) Find the number of bits necessary from the code generator to control the on–off switching times.

(b) If biphase modulation is used, what is the processing gain?

8.4-2 The power amplifier in a time-hopping transmitter is rated at a maximum peak power of 50 kW, and is designed for a duty cycle of 0.1953 percent. Assuming the message bit rate after error-correction coding is 3200 b/s, and assuming biphase modulation:

(a) Find the average power that the power amplifier must handle.

(b) Find the minimum bandwidth for which the power amplifier must be designed.

(c) Find the processing gain achieved with the above system.

8.6-1 A hybrid FH/PN spread-spectrum system uses a PN code rate of 125 kb/s. The RF bandwidth is 286 MHz centered at 858 MHz. However, we are told that we are not allowed to transmit within the UHF television frequencies allocated to channels 55 through 59. How many frequencies must the synthesizer produce for it to cover the overall bandwidth of 286 MHz and still avoid these channels?

8.6-2 A PN/FH hybrid spread-spectrum system is to have the following parameters:

Message bit rate = 9600 b/s (after error-correction coding)

PN code rate = 15,3600 chips per second

Chips per hop = 16

Hops per message bit = 1

Number of frequencies = 4096

(a) Find the processing gain.

(b) Find the bandwidth of the spread-spectrum signal.

8.6-3 In a FH/PN hybrid spread-spectrum system, we are given the following design requirements:

Chips per message bit = 20

Message bit rate = 4800 b/s after coding

PG $\geq$ 41 dB

(a) Specify the requirements of the synthesizer in terms of the necessary switching speed and the number of frequencies it must produce.

(b) Specify the bandwidth requirements of the final RF amplifier if all the frequencies are used.

8.8-1 A FH/PN hybrid spread-spectrum system is specified by the following parameters:

B_s = 1.92 GHz

M = 512

PN code rate = 3.75 Mb/s

15 chips per hop

2 hops per message bit

32-level quantizer

If error-correction coding is not used, what is the maximum frequency the message could have?

8.8-2 It is known that if a signal has a bandpass spectrum of bandwidth B and has negligible energy above the frequency f_u, then it can be sampled at a rate $f_s = 2f_u/m$, where m is the largest integer not exceeding f_u/B.

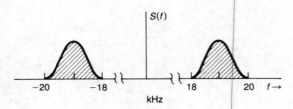

In our case we are interested in sampling an analog message whose spectral density is a bandpass spectrum, quantizing into L levels, and then spreading it out to a bandwidth of 500 MHz by an appropriate PN sequence. Given the analog message spectral density as shown, and if we desire a PG $\geq$ 43 dB, how many levels must we use to quantize?

ANALYSIS OF DIRECT-SEQUENCE SPREAD-SPECTRUM SYSTEMS

In this chapter we consider methods of analyzing the characteristics of spread-spectrum systems under a variety of conditions. This analysis is carried out explicitly for direct-sequence systems. However, many of the results are applicable to other systems also. Specific situations in which these results are not applicable are considered subsequently.

The mathematical properties of the pseudorandom sequences that are used in direct-sequence systems are of fundamental importance in determining the characteristics of the system. It is necessary, therefore, that we begin our study by looking at some of the properties of PN sequences.

9-1 PROPERTIES OF PN SEQUENCES

A pseudorandom (PN) sequence consists of a sequence of plus or minus ones that possess certain specified autocorrelation properties. As observed in the previous chapter, such sequences play an important role in almost all types of spread-spectrum systems. There are two general classes of PN sequences—aperiodic and periodic. An aperiodic sequence is one that does not repeat itself in a periodic fashion. It is usually assumed that the sequence has a value of zero outside its stated interval. The periodic sequence, however, is a sequence of plus or minus ones that repeats itself exactly with a specified period. Although periodic sequences are much more important in spread-spectrum systems, it is useful to review the situation briefly with regard to aperiodic sequences in order to identify their limitations.

Aperiodic Sequences

An aperiodic sequence is described analytically by a sequence of N plus or minus ones and is denoted by

$$(a_1, a_2, \ldots, a_N) \qquad a_i = \pm 1$$

To be a true pseudorandom sequence, this sequence must have certain properties.[4] One such property is autocorrelation, defined as

$$C(k) = \sum_{n=1}^{N-k} a_n a_{n+k} \qquad k = 0, 1, \ldots, N-1 \qquad (9\text{-}1)$$

As an example of this, consider a sequence consisting of four digits, a_1 through a_4, where each of the a's is either plus or minus one. The autocorrelation $C(1)$ is obtained by multiplying elements of this sequence by corresponding elements of the same sequence shifted one digit. This is illustrated as

$$
\begin{array}{cccc}
a_1 & a_2 & a_3 & a_4 \\
a_1 & a_2 & a_3 & a_4 \\
\hline
\end{array}
$$
$$C(1) = a_1 a_2 + a_2 a_3 + a_3 a_4$$

An ideal aperiodic sequence would have an autocorrelation given by

$$
\begin{aligned}
C(k) &= N & k = 0 \\
&= 0 \text{ or } \pm 1 & k \neq 0
\end{aligned}
\qquad (9\text{-}2)
$$

Such sequences are called Barker sequences, but they are known to exist for only a very few values of N. Specifically they have been found for $N = 1, 2, 3, 4, 5, 7, 11,$ and 13. No larger sequences have been found and it has been hypothesized that no larger Barker sequences exist. Thus such sequences are normally too short for use as a spreading function in a spread-spectrum system, although they may be used for synchronization purposes under some conditions.

Periodic Sequences

A periodic sequence consists of an infinite sequence of plus or minus ones divided up into blocks of length N, in which the particular sequence in each block is the same. This is illustrated by

$$\ldots a_{N-1}, a_N, a_1, a_2, \ldots, a_N, a_1, \ldots$$

Such a sequence is said to be pseudorandom if it satisfies the following conditions:[4]

1. In every period the number of plus ones differs from the number of minus ones by exactly one. Hence N is an odd number. Thus

$$N_+ + N_- = N$$
$$|N_+ - N_-| = 1$$

2. In every period half of the runs of the same sign have length 1, one fourth have length 2, one eighth have length 3, and so forth. Also the number of positive runs equals the number of negative runs.

3. The autocorrelation of a periodic sequence is two-valued. That is, it can be described by

$$C(k) = \sum_{n=1}^{N} a_n a_{n+k}$$

$$= N \qquad k = 0, N, 2N, \ldots$$

$$= -1 \qquad \text{otherwise} \tag{9-3}$$

where

$$a_{n+N} = a_n$$

As an example of this autocorrelation, consider a sequence consisting of five digits in each period and determine the autocorrelation for a shift of one digit. This is illustrated as

$$\ldots a_3\ a_4\ a_1\ a_2\ a_3\ a_4\ a_5\ a_1 \ldots$$

$$\ldots a_4\ a_1\ a_2\ a_3\ a_4\ a_5\ a_1 \ldots$$

$$C(1) = a_1 a_2 + a_2 a_3 + a_3 a_4 + a_4 a_5 + a_5 a_1$$

The two-valued nature of periodic correlations is extremely important in determining the characteristics of spread-spectrum systems employing such sequences. However, before investigating this in more detail, it is of interest to look at several classes of periodic sequences and to select one of these classes for further consideration.

9-2 CLASSES OF PERIODIC SEQUENCES

Many different classes of periodic PN sequences are known. Of these the ones that can be generated in shift registers are of the greatest practical importance.

Maximal-Length Linear Shift Register Sequences (m Sequences)[5]

The m sequences all have a length that is defined by

$$N = 2^m - 1 \qquad m = 1, 2, 3, \ldots$$

Thus some typical short-length sequences are

$$N = 7, 15, 31, 63, 127, 255, \ldots$$

The m sequences are most commonly used because they are easily generated in shift registers with a relatively small number of stages. As is demonstrated later, it requires only m stages of shift register to generate an m sequence. These sequences are considered in more detail later because they are by far the most important ones in direct-sequence spread-spectrum systems. They also have application in other types of spread-spectrum systems.

Another application of m sequences arises in connection with multiple access. As discussed in a later chapter, this application requires sequences that have good cross-correlation properties rather than the autocorrelation properties described above. An example of these are the Gold sequences that are obtained by combining two m sequences.

Quadratic Residue Sequences ($q-r$ Sequences)[4]

Quadratic residue sequences have lengths that are prime numbers of the form

$$N = 4k - 1 = \text{prime}$$

where k is an integer. Since prime numbers are fairly uniformly distributed, and since half of all prime numbers can form quadratic residue sequences, there are many more sequences of this type than there are m sequences. Hence if one needs a sequence of a length that is not readily obtainable from an m sequence, the quadratic residue sequence may fill the bill. However, they are not readily generated by shift registers and hence the implementation becomes more difficult. There are some values of m for which both quadratic residue and m sequences may exist. For example, for $N = 3, 7, 31, 127, 8191, 131071, \ldots$ except for $N = 3$ and 7, quadratic residue sequences and m sequences are distinct. That is, they are not the same sequence although they have the same correlation properties.

Hall Sequences[4]

Hall sequences also have a length that is a prime number described by

$$N = 4k - 1 = 4q^2 + 27 = \text{prime}$$

where both k and q are integers. Since both conditions must be satisfied, there are fewer Hall sequences than there are quadratic residue sequences. However, every Hall sequence has a quadratic residue sequence of the same length but distinct from it. Since Hall sequences have the same properties as quadratic residue sequences, there is no particular advantage to using this form.

Twin Primes[4]

The twin-prime sequences are ones in which the length of the sequence is defined by

$$N = p(p + 2)$$

where both p and $p + 2$ are prime numbers. Some examples of N for which this occurs are

$$N = 3 \times 5 = 15$$

$$N = 5 \times 7 = 35$$
$$N = 11 \times 13 = 143$$
$$N = 17 \times 19 = 323$$
$$N = 29 \times 31 = 899$$

The only advantage of twin-prime sequences is that it is possible to achieve some lengths that are not readily achieved by any of the other forms. However, the implementation is not simple.

9-3 PROPERTIES OF m SEQUENCES

In this and the following sections, we consider some of the properties of m sequences that are necessary in order to analyze the performance of a spread-spectrum system. Frequently it is necessary to use more than one m sequence in a complete system. Thus the number of m sequences of given length is an important property. This number can be shown to be[3]

$$\frac{1}{m} \phi(2^m - 1)$$

where $\phi(k)$ is the number of positive integers less than k and relatively prime to k. This is frequently referred to as the Euler totient function. Several examples of the number of sequences for sequences of different lengths are shown below.

m	N	Number of sequences
4	15	2
5	31	6
6	63	6
8	255	16
10	1,023	60
12	4,095	144
16	65,535	2,048

One point of interest may be noted by observing the results for $m = 5$ and $m = 6$. In both cases the number of sequences is six, even though sequences for $m = 6$ are twice as long as those for $m = 5$. This is so because $2^m - 1$ is a prime number in the case of $m = 5$, and is not a prime number in the case of $m = 6$. When the argument of the Euler totient function is prime, all of the integers less than the prime number are relatively prime to it, and, therefore, the number of sequences is much greater. Hence if one is looking for the largest number of sequences, it would be desirable to pick those values of m that lead to N values that are prime numbers.

Another property of m sequences that may be of interest in applications where the m sequence is used in a pseudorandom radar system has to do with the sum of two m sequences. Specifically, if an m sequence is added modulo 2 with a shifted replica of itself, the sum is another replica of the same sequence with a shift different from either one. As an example of this, consider the specific case for $m = 3$, $n = 2^3 - 1 = 7$ and consider the sequence shifted by one digit added to the original sequence. Thus

$$
\begin{array}{ccccccccccc}
+1 & -1 & +1 & -1 & -1 & +1 & +1 & +1 & -1 & +1 & -1 \\
+1 & +1 & -1 & +1 & -1 & -1 & +1 & +1 & +1 & -1 & +1 \\
-1 & +1 & +1 & +1 & -1 & +1 & -1 & -1 & +1 & +1 & +1
\end{array}
$$

In the case of modulo-2 addition, note that the sum of two digits with the same sign results in a minus one, while the sum of two digits of opposite sign results in a plus one. This would correspond in the more familiar case to associating minus one with a zero and plus one with a one in normal binary arithmetic. It may be noted from this example that adding a sequence shifted by one digit from its original sequence results in a third replica that is shifted by three digits from the original sequence. Thus if only the sum could be observed, it would be indistinguishable from a single sequence in this position.

9-4 PARTIAL CORRELATION

The subject of partial correlation is extremely important in spread-spectrum systems in which the duration of the sequence is much greater than the duration of a message bit. This is so because correlation in the receiver must take place over a time interval corresponding to the message bit duration, and as a consequence correlation over a full period of the sequence cannot occur. The particular two-value correlation function described in Eq. (9-3) does not result unless correlation is over a full period. Therefore, it is necessary to examine what happens when correlation is over only part of a period. The partial correlation may be defined as

$$
C_M(k) = \sum_{n=1}^{M} a_n a_{n+k} \qquad M < N, \, k = 0, 1, \ldots, N \qquad (9\text{-}4)
$$

Although this correlation is represented as being a function only of the delay k of one sequence with respect to the other, it also depends upon where in a sequence the summation actually starts. However, since we are going to be considering the a_n's as random variables from a stationary random process, this dependence upon the starting point will average out and so is not considered here.

When the sequences are short, it may be possible to evaluate the partial correlation explicitly for all combinations of k and relative starting points. But for very large sequences, this is clearly not possible and, therefore, modeling the sequence as a random process is really the only viable alternative. The randomness properties described earlier that are necessary to ensure that we have a pseudo-random sequence are also adequate to guarantee that modeling this sequence as a random process is reasonable. Since the a_n values in each sequence are assumed to be statistically independent and have values of only plus or minus 1, with equal probability, it follows that

$$E[a_n] = 0 \qquad E[a_n^2] = 1$$

Hence the mean value of the partial correlation is readily shown to be

$$E[C_M(k)] = \sum_{n=1}^{M} E[a_n a_{n+k}]$$

$$= \sum_{n=1}^{M} E[a_n^2] = M \qquad k = 0, N, 2N, \ldots$$

$$= \sum_{n=1}^{M} E[a_n]E[a_{n+k}] = 0 \qquad k \neq 0, N, \ldots \qquad (9\text{-}5)$$

It is clear from this result that the mean value of the partial correlation is zero when the sequences are not lined up in time and has a maximum value of M when they are. This, of course, is the desired result. The actual value of the partial correlation for some value shift other than zero may be quite different from zero, however. Therefore, it is necessary to evaluate the variance of the partial correlation. This is done by taking the difference between the expected value of the square of the partial correlation and the square of the expected value. Hence

$$\text{Var}[C_M(k)] = E[C_M^2(k)] - E^2[C_M(k)] = \sum_{n=1}^{M} \sum_{m=1}^{M} E[a_n a_{n+k} a_m a_{m+k}] - E^2[C_M(k)]$$

$$= M^2 - M^2 = 0 \qquad k = 0, N, 2N, \ldots$$

$$= M - 0 = M \qquad k \neq 0, N, 2N, \ldots \qquad (9\text{-}6)$$

From this it follows that the standard deviation of the partial correlation is simply the square root of M. In all future analysis it thus is assumed that the partial correlation is a random variable with a mean value of zero and a standard deviation of the square root of M.

It is also possible to obtain slightly more accurate results for the mean and variance of partial correlation by using the known properties of m sequences instead of assuming purely random sequences. To accomplish this the expectations in the above results are replaced by averages over all possible starting positions in one period. If this is done, Eqs. (9-5) and (9-6) become (see ref. 5, p. 546)

$$E[C_M(k)] = M \qquad k = 0, N, 2N, \ldots$$

$$= -\frac{M}{N} \qquad k \neq 0, N, 2N, \ldots \qquad (9\text{-}5a)$$

and

$$\text{Var}[C_M(k)] = 0 \qquad k = 0, N, 2N, \ldots$$

$$= M\left(1 + \frac{1}{N}\right)\left(1 - \frac{M}{N}\right) \qquad k \neq 0, N, 2N, \ldots \qquad (9\text{-}6a)$$

It is clear that when $N \gg M$, which is often the case, the mean and variance are not greatly different from the values obtained by assuming purely random sequences.

9-5 PN SIGNALS FROM PN SEQUENCES

So far we have only considered sequences of plus and minus ones. What actually is used in a spread-spectrum system, however, is a time function whose value may be plus or minus one. One period of such a time function is illustrated in Fig. 9-1 for the PN sequence corresponding to $N = 7$.

To analyze the correlation properties of such PN signals, it is convenient to define a basic chip function as

$$f(t) = 1 \qquad 0 \leq t \leq t_1$$

$$= 0 \qquad \text{elsewhere} \tag{9-7}$$

This is illustrated in Fig. 9-2.

The total PN signal $b(t)$ may be represented as an infinite sum of the basic chip functions. This results in

$$b(t) = \sum_{n=-\infty}^{\infty} a_n f(t - nt_1) \tag{9-8}$$

where $a_n = a_{n+N}$. First, we consider the correlation function for a PN signal when the averaging is done over a full period of Nt_1. This correlation function may be

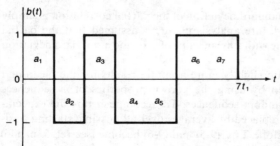

Figure 9-1 One period of a PN sequence for $N = 7$.

Figure 9-2 The basic chip function.

defined as

$$R_b(\tau) = \frac{1}{Nt_1} \int_0^{Nt_1} b(t)b(t+\tau)\, dt \qquad (9\text{-}9)$$

which can be written as

$$R_b(\tau) = \frac{1}{Nt_1} \sum_{n=0}^{N-1} \sum_m a_n a_m \int_0^{Nt_1} f(t - nt_1)f(t + \tau - mt_1)\, dt \qquad (9\text{-}10)$$

Note that the range of m values in the inner summation depends upon the value of τ, since the summation may start somewhere in the middle of a period and continue until the middle of the next period. The integral in Eq. (9-10) may be used to define the correlation function of the basic chip function. Thus let

$$R_f(\beta) = \frac{1}{t_1} \int_{-\infty}^{\infty} f(t)f(t + \beta)\, dt \qquad (9\text{-}11)$$

This correlation function for the rectangular chip function defined in Eq. (9-7) is simply a triangular function of the form shown in Fig. 9-3. Inserting Eq. (9-11) into Eq. (9-10) leads to

$$R_b(\tau) = \frac{1}{N} \sum_{n=0}^{N-1} \sum_m a_n a_m R_f[\tau + (n - m)t_1] \qquad (9\text{-}12)$$

From the two-valued autocorrelation property of the sum of the products of a_n and a_m, as defined in Eq. (9-3), it follows that the correlation function for $b(t)$ is

$$R_b(\tau) = R_f(\tau) - \frac{1}{N} R_f(\tau - t_1) \qquad 0 \le \tau \le t_1$$

$$= -\frac{1}{N} \qquad t_1 \le \tau \le (N-1)t_1 \qquad (9\text{-}13)$$

One period of this is illustrated in Fig. 9-4.

It may be noted that except for the triangular portion in the interval between $-t_1$ and $+t_1$, the autocorrelation function of the PN signal is at a constant value of $-(1/N)$. This is a desirable characteristic that we would like to achieve in most circumstances. However, as mentioned earlier, when the period of the sequence is longer than the integration time in the receiver, then the entire correlation function does not occur. It is necessary, therefore, to consider the partial correlation of PN signals.

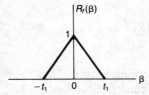

Figure 9-3 The basic chip correlation function.

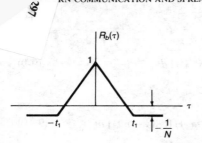

Figure 9-4 PN signal correlation function.

9-6 PARTIAL CORRELATION OF PN SIGNALS

If the time interval over which the correlation function can be averaged is less than Nt_1, then it becomes necessary to define the partial correlation as

$$R_M(\tau) = \frac{1}{Mt_1} \int_0^{Mt_1} b(t)b(t + \tau)\, dt \qquad M < N \qquad (9\text{-}14)$$

Using the same procedure as in the previous section, it may then be shown that the partial correlation can be represented as

$$R_M(\tau) = R_f(\tau) + \frac{C_M(1)}{M} R_f(\tau - t_1) \qquad 0 \le \tau \le t_1 \qquad (9\text{-}15)$$

in the interval for τ between 0 and t_1 and as

$$R_M(\tau) = \frac{1}{M}\left[C_M(1)R_f(\tau - t_1) + C_M(2)R_f(\tau - 2t_1) \right] \qquad t_1 \le \tau \le 2t_1 \qquad (9\text{-}16)$$

in the interval for τ between t_1 and $2t_1$. A similar result can be obtained for any larger value of τ by simply changing the argument of C_M in an obvious fashion. Since the correlation values C_M are treated as random variables, it follows that $R_M(\tau)$ is also a random variable. Thus we cannot specify its exact value for any value of M and τ, but we can specify the mean and variance. In particular, using the results that were developed in the earlier section, it follows that

$$E[R_M(\tau)] = R_f(\tau)$$

$$\qquad\qquad\qquad\qquad 0 \le \tau \le t_1 \qquad (9\text{-}17)$$

$$\text{Var}[R_M(\tau)] = \frac{1}{M} R_f^2(\tau - t_1)$$

and that

$$E[R_M(\tau)] = 0$$

$$\qquad\qquad\qquad\qquad t_1 \le \tau \le (N-1)t_1 \qquad (9\text{-}18)$$

$$\text{Var}[R_M(\tau)] = \frac{1}{M}$$

for values of τ between t_1 and $(N-1)t_1$. Because of the periodic nature of $b(t)$, the function then repeats itself with a period Nt_1, and is symmetrical around $\tau = 0$.

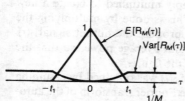

Figure 9-5 Mean and variance of partial correlation.

The mean and variance of the partial correlation are illustrated in Fig. 9-5.

It is of interest to note that the variance of the partial correlation is zero when the sequences are lined up exactly in time, but is at a constant value of $1/M$ when the sequences are shifted in time with respect to one another by any amount greater than t_1. This variance represents the self-noise of the signal or, as it is frequently called, the code noise. During the time interval when the system is attempting to synchronize itself, this code noise may be the predominant noise in the receiver and may be a limiting factor in system performance. Therefore, it is important to make the code noise as small as possible, and this, of course, is done by making M as large as possible. There are, however, limits as to how large M can become, as discussed subsequently.

9-7 THE PN SIGNAL

This section discusses in a little more detail the spectral characteristics of the PN signal as a prelude to discussing methods of receiving such signals. Figure 9-6 indicates the operations that must be performed in generating a PN signal. The quantities in this figure are defined as

$$m(t) = \pm 1, \text{ binary message}$$

$$b(t) = \pm 1, \text{ PN code}$$

$$P_t = \text{transmitted power}$$

$$s(t) = \text{normalized signal}$$

$$\lim_{T_s \to \infty} \frac{1}{2T_s} \int_{-T_s}^{T_s} s^2(t)\, dt = 1$$

$$s(t) = \sqrt{2} m(t) b(t) \cos \omega_0 t$$

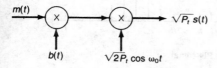

Figure 9-6 Generation of PN signals.

Note that the binary signals $m(t)$ and $b(t)$ are simply multiplied to create a new binary signal that phase-modulates the carrier. This is done by multiplying the new binary signal with a steady-state sinusoid to produce the output signal $s(t)$. Note also that $s(t)$ has been normalized to have unit average power so that the transmitted signal power is simply P_t as shown in the figure.

In the previous chapter, the spectral density of $s(t)$ was obtained by assuming $m(t)$ and $b(t)$ to be random quantities and taking the Fourier transform of the auto-correlation function. The result is

$$S_s(f) = \frac{t_1}{2} \left\{ \left[\frac{\sin \pi(f - f_0)t_1}{\pi(f - f_0)t_1} \right]^2 + \left[\frac{\sin \pi(f + f_0)t_1}{\pi(f + f_0)t_1} \right]^2 \right\} \tag{9-19}$$

This is not quite right, however, because $b(t)$ is really a periodic time function. To assess the effect of this periodicity, it is convenient to consider first the spreading sequence without any message modulation. Thus for a signal of the form

$$s'(t) = \sqrt{2}b(t)\cos \omega_0 t \qquad -\infty < t < \infty$$

with

$$b(t + nt_1) = b(t) = \pm 1$$

the continuous spectral density of Eq. (9-19) becomes a line spectrum of the form (see ref. 5, p. 379)

$$S_{s'}(f) = \frac{N+1}{2N^2} \left\{ \left[\frac{\sin \pi t_1(f - f_0)}{\pi t_1(f - f_0)} \right]^2 \sum_{k=-\infty}^{\infty} \delta\left(f - f_0 - \frac{k}{Nt_1} \right) \right.$$

$$\left. + \left[\frac{\sin \pi t_1(f + f_0)}{\pi t_1(f + f_0)} \right]^2 \sum_{k=-\infty}^{\infty} \delta\left(f + f_0 - \frac{k}{Nt_1} \right) \right\}$$

$$- \frac{1}{2N} [\delta(f - f_0) + \delta(f + f_0)] \tag{9-20}$$

This spectrum is illustrated in Fig. 9-7 for positive frequencies only. It may be noted that the relative magnitudes of the discrete frequency components follow the general shape described by Eq. (9-19) (except at $f = \pm f_0$).

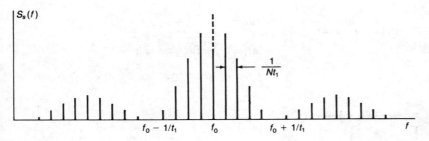

Figure 9-7 Spectral density of an unmodulated periodic PN signal.

When the spreading function is modulated by the message $m(t)$, each discrete frequency in Eq. (9-20) is spread out into a replica of the message spectrum. If $m(t)$ is a binary random process as shown in Fig. 9-8, its spectral density is

$$S_m(f) = t_m \left[\frac{\sin \pi f t_m}{\pi f t_m} \right]^2$$

and is displayed in Fig. 9-9. Hence each discrete line of Fig. 9-7 assumes the shape of Fig. 9-9. It is clear, therefore, that the resulting spectrum of the modulated signal will have a shape that depends upon the relative size of $1/t_m$ and $1/Nt_1$.

As a means of illustrating this conclusion, consider the two different cases shown in Fig. 9-10 for positive frequencies only. In Fig. 9-10a it is assumed that $1/t_m$ is less than $1/2Nt_1$. In this case the resulting spectrum has distinct peaks as shown. However, if $1/t_m$ is greater than or equal to $1/Nt_1$, the peaks will merge into a continuous smooth spectrum as shown in Fig. 9-10b. This spectrum is essentially identical to that of Eq. (9-19). Since this latter condition is the one that pertains most often, the continuous smooth spectrum of Eq. (9-19) is usually assumed to exist. It is important to be aware, however, that there are circumstances in which the spectral density of a modulated PN signal may be quite different.

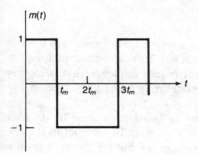

Figure 9-8 The binary message.

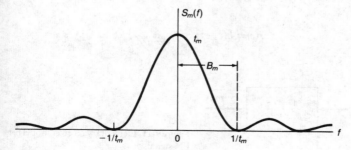

Figure 9-9 Spectral density of the binary message.

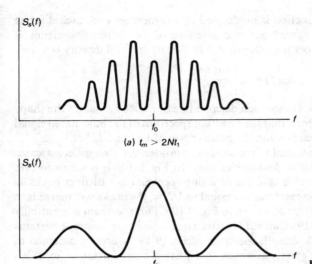

(a) $t_m > 2Nt_1$

(b) $t_m \leq Nt_1$

Figure 9-10 Spectral density of a modulated PN signal.

9-8 DESPREADING THE PN SIGNAL

One of the first functions that must be performed by the spread-spectrum receiver is to eliminate the effects of the spreading function $b(t)$. This can be accomplished by multiplying the received signal by an identical binary sequence that is in time synchronization with it. The basic receiver model may be as illustrated in Fig. 9-11, in which the received signal power is P_r.

The averaging operation is indicated here as being performed by an ideal finite time integrator with an integration time of T seconds. In general this integration time will be equal to the duration of a message bit, but for the moment it is left arbitrary. The reference signal is

$$v(t) = \sqrt{2}r(t)\cos(\omega_0 t + \theta)$$

where $r(t)$ is

$$r(t) = b(t - \tau)$$

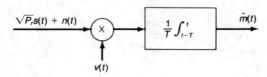

Figure 9-11 Elementary receiver model for PN signals.

Hence the output signal component is given by

$$\frac{1}{T} \int_{t-T}^{t} \sqrt{P_r} s(u) v(u) \, du = m_0(t) \tag{9-21}$$

The product of the received signal and the reference signal can be written in more complete form as

$$\sqrt{P_r} s(u) v(u) = \sqrt{2P_r} m(u) b(u) \cos \omega_0 u \sqrt{2} r(u) \cos(\omega_0 u + \theta)$$

$$= \sqrt{P_r} m(u) b(u) b(u - \tau)[\cos \theta + \cos(2\omega_0 u + \theta)] \tag{9-22}$$

Note that one of the terms in Eq. (9-22) involves the cosine of $2\omega_0 u$ and so is a frequency that is twice that of the carrier frequency. Since the integration time is very large compared with one period of this frequency, it is reasonable to assume that this term integrates essentially to zero. The output signal component thus becomes

$$m_0(t) = \sqrt{P_r} \int_{t-T}^{t} \frac{1}{T} m(u) b(u) b(u - \tau) \cos \theta \, du \tag{9-23}$$

It is apparent from Eq. (9-23) that the signal component in the integrator output is a function of time that depends upon the message state within the integration interval. To maximize this output for each message bit without interference from adjacent bits, it is clear that the integration time should be equal to the message bit duration (i.e., $T = t_m$) and that the output should be observed precisely at the end of each message bit (i.e., $t = kt_m$, $k = 0, \pm 1, \pm 2, \ldots$). It is appropriate, therefore, to choose the first bit as a typical one and let $t = t_m$. At this instant of time, the output signal is

$$m_0(t_m) = \pm \frac{\sqrt{P_r}}{t_m} \int_{0}^{t_m} b(u) b(u - \tau) \cos \theta \, du$$

$$= \pm \sqrt{P_r} R_b(\tau) \cos \theta \tag{9-24}$$

in which the $\pm$ sign comes from the sign of $m(u)$. Note that this result will be a maximum when τ is equal to zero and when θ is equal to zero. These are the conditions that would apply when the reference signal and the received signal are properly synchronized.

It is also necessary to evaluate the noise out of the receiver. The output noise component can be expressed as

$$\frac{1}{T} \int_{t-T}^{t} n(u) \cdot \sqrt{2} r(u) \cos(\omega_0 u + \theta) \, du = n_0(t) \tag{9-25}$$

This component has a mean value of zero since the expected value of $n(t)$ is zero. However, it does have a variance, and this variance can be obtained by looking at the noise component at $t = T = t_m$ and noting that the autocorrelation function of the noise is

$$R_n(\tau) = \frac{N_0}{2} \delta(\tau)$$

since white noise has been assumed. Thus the mean square value of the output noise becomes

$$\overline{n_0^2} = \frac{2}{t_m^2} \int_0^{t_m} \int_0^{t_m} \frac{N_0}{2} \delta(u-v)b(u-\tau)b(v-\tau)\cos(\omega_0 u + \theta)\cos(\omega_0 v + \theta) \, du \, dv$$

The product of cosines in this expression can be written as the sum of cosines as

$$\cos(\omega_0 u + \theta)\cos(\omega_0 v + \theta) = \frac{1}{2}\cos\omega_0(u-v) + \frac{1}{2}\cos[\omega_0(u+v) + 2\theta]$$

Because of the delta function in the double integral, it is possible to integrate over the variable v simply by replacing v by u. When this is done, one of the cosine terms becomes a constant and the other one becomes a double-frequency term whose integral can again be assumed to be zero. Thus the mean square value of the noise reduces to

$$\overline{n_0^2} = \frac{N_0}{2t_m^2} \int_0^{t_m} b^2(u-\tau) \, du = \frac{N_0}{2t_m} \tag{9-26}$$

Note that this mean square value depends only upon the noise spectral density and the duration of the message bit and is independent of the chip rate of the PN sequence. It is also independent of the relative time shift τ between the received signal and the reference signal. Thus it does not depend upon whether or not the signal is synchronized. Consider now the entire signal out of the correlator, $\hat{m}(t)$. At time $t = t_m$, this signal has a mean value given by Eq. (9-24), which may be written as

$$E[\hat{m}(t_m)] = \pm\sqrt{P_r}\left[1 - \frac{|\tau|}{t_1}\right]\cos\theta \qquad |\tau| \leq t_1$$

$$= 0 \qquad\qquad\qquad |\tau| > t_1 \tag{9-27}$$

since the correlation function of the message is a triangular shape as discussed previously. The noise component of the total output has a variance given by

$$\text{Var}[\hat{m}(t_m)] = \frac{N_0}{2t_m} + \frac{P_r\tau^2}{t_1 t_m} \qquad |\tau| \leq t_1$$

$$= \frac{N_0}{2t_m} + \frac{P_r t_1}{t_m} \qquad |\tau| > t_1 \tag{9-28}$$

in which the first term results from the receiver noise as given in Eq. (9-26) and the second term results from the code noise as described in Eqs. (9-17) and (9-18). Since the maximum component occurs at $\tau = 0$, and the code noise variance is zero at $\tau = 0$, it follows that the maximum SNR will occur at $\tau = 0$ and $\theta = 0$. This maximum SNR is simply the ratio of the square of the mean value of $\hat{m}(t)$ to the variance

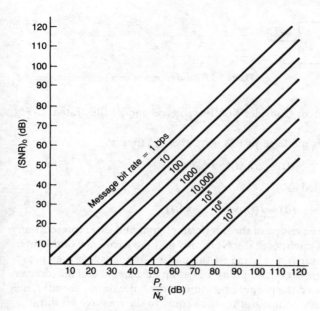

Figure 9-12 Correlator output signal-to-noise ratio.

of $\hat{m}(t)$. Thus it is

$$(\text{SNR})_0 = \frac{E^2[\hat{m}(t_m)]}{\text{Var}[\hat{m}(t_m)]}$$

$$= \frac{2P_r t_m}{N_0} \tag{9-29}$$

The SNR is an important parameter for a spread-spectrum system because it is a measure of signal detectability, as well as a measure of output signal quality. It may be noted that SNR depends only upon the ratio of received power to noise spectral density P_r/N_0 and t_m, which is the reciprocal of the message bit rate. This relationship is illustrated graphically in Fig. 9-12, which indicates how the output SNR can be increased by either increasing the received signal power or by reducing the message bit rate.

9-9 INTERFERENCE REJECTION

As noted previously one of the objectives in using spread spectrum is the ability of such a system to reject interference that otherwise might prohibit useful communication. The purpose of this section is to evaluate the degree to which this

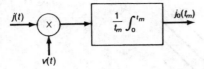

Figure 9-13 Receiver model with interference input.

interference can be rejected. Consider first the receiver model illustrated in Fig. 9-13.

The reference signal is modeled just as it was before; that is,

$$v(t) = \sqrt{2}r(t)\cos(\omega_0 t + \theta)$$

The interference is modeled as

$$j(t) = J(t)\cos[\omega_1 t + \psi(t)]$$

where $J(t)$ represents the envelope of the interfering signal and $\psi(t)$ represents any phase modulation that it might contain. Note also that the frequency of the interference, ω_1, is not necessarily the same as the carrier frequency of the spread-spectrum system, although, of course, it must be within the passband of the receiver. Having previously discussed the proper operation of the finite time integrator, it is shown here as being simply an integration time equal to the message bit duration t_m. No desired signal or noise components are shown in this model because these can be added in after the response of the receiver to interference has been obtained.

The output of the receiver in response to interference is simply

$$j_0(t_m) = \frac{1}{t_m} \int_0^{t_m} j(t)v(t)\, dt$$

If the interference is assumed to be random with a mean value of zero, then the mean value of $j_0(t_m)$ will also be zero. The mean square value is given by

$$\overline{j_0^2} = E[j_0^2(t_m)]$$

$$= \frac{1}{t_m^2} \int_0^{t_m} \int_0^{t_m} R_j(t-u)R_v(t-u)\, dt\, du$$

where $R_j(t-u)$ is the autocorrelation function of the interfering signal, and $R_v(t-u)$ is the autocorrelation function of the reference, which is, of course, the same as the autocorrelation function of the desired signal. Initially let us assume that the bandwidth of the interference is narrow compared with that of the spread-spectrum signal. In this case the autocorrelation function of the interference will be broad compared with that of the reference signal and the double integral can be reduced to a single integral as

$$\overline{j_0^2} = \frac{J}{t_m^2} \int_0^{t_m} (t_m - \tau)R_f(\tau)\cos(\omega_1 - \omega_0)\tau\, d\tau$$

where

$$J = R_j(0) = \text{interference power}$$

Note that the autocorrelation function of the reference has now been expressed in terms of the chip correlation function $R_f(\tau)$. Since $R_f(\tau)$ is zero, when the magnitude of τ is greater than t_1, and since t_m is very much greater than t_1, the integral reduces very nearly to

$$\overline{j_0^2} \simeq \frac{Jt_1}{2t_m} \tag{9-30}$$

This indicates that the interference at the output of the receiver depends upon the ratio of chip duration to message bit duration, which is, of course, a very small number if the spread-spectrum bandwidth is large. It may be recalled that the processing gain of a spread-spectrum system is defined as

$$\text{PG} = \frac{B_s}{B_m} = \frac{2/t_1}{1/t_m} = \frac{2t_m}{t_1}$$

Thus the mean square value of the output interference signal can be written as

$$\overline{j_0^2} = \frac{J}{\text{PG}} \tag{9-31}$$

Note that the interference power has been reduced by a factor equal to the processing gain. This is an average result, however, with the averaging taking place over all phases of the interference. For some fixed-phase relationships, the mean square value of the output interference may be twice the value given by Eq. (9-31).

It is also of interest to consider the situation in which the interference has a bandwidth that is greater than that of the spread-spectrum signal. That is, $B_j > B_s$. In this case the mean square value of the output of the correlator becomes

$$\overline{j_0^2} = \frac{J}{2B_j t_m} = \frac{B_s}{2B_j} \cdot \frac{J}{\text{PG}} \tag{9-32}$$

Since J is the interference power in its total bandwidth, the factor $(B_s/B_j)J$ in Eq. (9-32) represents simply that portion of the interference power that exists in the receiver bandwidth. This power is again reduced by a factor equal to the processing gain.

9-10 OUTPUT SIGNAL-TO-NOISE RATIO

To compute the output signal-to-noise ratio under general conditions, it is necessary to consider both the total output noise power and the output power due to interference. Since these are independent quantities, they may be added and the SNR expressed as

$$(\text{SNR})_0 = \frac{P_r}{n_0^2 + j_0^2} = \frac{P_r}{\dfrac{N_0}{2t_m} + \dfrac{J}{\text{PG}}} = \frac{(\text{PG})P_r}{\dfrac{N_0(\text{PG})}{2t_m} + J}$$

However, the denominator can be simplified because

$$\frac{PG}{t_m} = \frac{B_s}{B_m} \cdot B_m = B_s$$

Hence the output SNR becomes

$$(SNR)_0 = \frac{(PG)P_r}{\dfrac{N_0 B_s}{2} + J} \tag{9-33}$$

Note that the output SNR is proportional to the received signal power divided by the sum of the interference power and the total noise in half the spread-spectrum signal bandwidth. Since half the spread-spectrum signal bandwidth is really the equivalent noise bandwidth of the receiver, this term represents the receiver noise power. This ratio is then multiplied by the processing gain to obtain the output signal-to-noise ratio. This can be expressed more compactly by defining an input SNR as

$$(SNR)_i = \frac{P_r}{\dfrac{N_0 B_s}{2} + J} \tag{9-34}$$

the output SNR now becomes

$$(SNR)_0 = (PG)(SNR)_i \tag{9-35}$$

In the derivation so far, no mention has been made of losses that might occur in the system. These are discussed in a following section.

9-11 ANTIJAM CHARACTERISTICS

One of the important reasons for being interested in the interference rejection capabilities of a spread-spectrum system is to be able to evaluate the degree to which such a system can reduce the effects of intentional jamming. This ability is often expressed in terms of a quantity that is known as the jamming margin (or AJ margin), and it is usually expressed in decibels. From Eq. (9-35), it follows that the input SNR can be written in decibels as

$$(SNR)_i(dB) = (SNR)_0(dB) - (PG)(dB) \tag{9-36}$$

We now define the AJ margin as

$$M_j(dB) = -(SNR)_i(dB) - L(dB) \tag{9-37}$$

where the quantity L represents the losses in the system expressed in decibels. The AJ margin now becomes

$$M_j(dB) = (PG)(dB) - L(dB) - (SNR)_0(dB) \tag{9-38}$$

As an example of this result, consider the spread-spectrum system in which the chip rate is 10^7 chips per second and the message bit rate is 100 b/s. If it is desired to obtain an output SNR of 25, which is about 14 dB, and if the losses are determined to be 2 dB, then it follows that

$$\text{PG} = \frac{2 \times 10^7}{100} = 2 \times 10^5 \leftrightarrow 53 \text{ dB}$$

$$M_j = 53 - 2 - 14 = 37 \text{ dB}$$

This result indicates that the desired output SNR can be obtained if the jamming signal is less than 37 dB greater than the desired signal. It should be emphasized that this result has assumed some ideal conditions at the receiver, and under more practical circumstances, the actual jamming margin may be something less than this. This matter is discussed in more detail in a later section.

9-12 INTERCEPTION

As noted previously another objective in using spread-spectrum techniques is to make it more difficult for an unauthorized observer to detect the presence of the signal or to decode the signal even if it is known to be present. It is desirable, therefore, to evaluate the degree to which this objective can be accomplished. In order to do this, let us review the detection of known signals in white noise. As stated in an earlier chapter, this is normally accomplished by use of a matched filter such as shown in Fig. 9-14.

The output SNR of a matched filter for a known signal in the presence of white noise was shown to be

$$(\text{SNR})_0 = \frac{2E}{N_0} \tag{9-39}$$

where

$$E = \text{energy in } s(t)$$

$$N_0 = \text{one-sided spectral density of } n(t)$$

The significance of this result is twofold. First, the detectability of a *known* signal depends on the ratio of the signal energy to noise spectral density and is independent of the shape of the signal. Second, the output SNR is linearly related to the above ratio. As is seen in the following analysis, this is not true for an unknown signal.

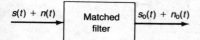

Figure 9-14 Matched filter for a known signal.

When an unknown spread-spectrum signal is to be detected, it is necessary to assume something about what is known. In particular it is assumed that the spread-spectrum code is not known, and that it is desired only to determine the existence of the spread-spectrum signal and not to decode it. This implies then that some form of noncoherent detection must be employed and that the existence of a spread-spectrum signal can be determined only by noting a change in energy between the signal present and the signal not present. Thus it is necessary to consider an energy detector. Such a detector is frequently referred to as a radiometer and might have a block diagram as shown in Fig. 9-15.

Note that the radiometer consists of a bandpass filter followed by a square-law detector and an averaging device or integrator. The output can be expressed as

$$y(t) = \int_{t-T}^{t} [s_1(u) + n_1(u)]^2 \, du \tag{9-40}$$

where $s_1(u)$ is the response of the bandpass filter to the spread-spectrum signal, $s(t)$. The quantity $n_1(u)$ is the noise from the bandpass filter and is assumed to have a variance of $\sigma_n^2 = N_0 B$, where B is the equivalent noise bandwidth of the bandpass filter. Assuming that the bandpass filter is tuned to a frequency that does, in fact, contain energy from the spread-spectrum signal, and that an observation is made at a time in which the spread-spectrum signal is actually present, then the output SNR can be shown to be[6,7]

$$(\text{SNR})_0 = \frac{(E/N_0)^2}{\xi^2 (TB)} \tag{9-41}$$

where ξ is a correction factor that depends upon the quantity TB. The nature of this correction factor is discussed in more detail subsequently. It may be noted, however, that the output SNR is now proportional to the square of the ratio E/N_0.

It is convenient to rewrite Eq. (9-41) as

$$\frac{E}{N_0} = \xi \sqrt{(\text{SNR})_0 (TB)} \tag{9-42}$$

in which E/N_0 can now be interpreted as a measure of the signal energy that is required to obtain a given signal-to-noise ratio. Since a given output SNR provides approximately the same detectability as the same output SNR with a known signal and a matched filter, it follows that this result indicates the degree to which a spread-spectrum signal can be detected.

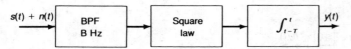

Figure 9-15 The radiometer or energy detector.

The correction factor ξ depends upon the value of the product TB. In particular it can be shown to be

$$\xi \simeq 1 \qquad TB \gg 1$$

$$\xi = \frac{(SNR)_0}{4}\left[1 + \sqrt{1 + \frac{18.4}{(SNR)_0^2}}\right] \qquad TB = 1$$

$$\xi > 1 \qquad TB < 1$$

As an example of the magnitude of this factor, for $(SNR)_o = 10$, $\xi = 5.22$ at $TB = 1$. From these results it is clear that the interceptor needs to make the detector TB small in order to be able to detect the spread-spectrum signal with the minimum E/N_0. On the other hand, the spread-spectrum communicator wishes to make the E/N_0 available to the interceptor as small as possible. The energy E in the spread-spectrum signal is that energy in time T and bandwidth B. Making B_s large reduces the fraction of the spread-spectrum signal energy in the detector bandwidth B. If the interceptor attempts to make the bandwidth of the detector larger, TB is made greater and the value of E/N_0 that is required is increased. The relation between these results and the results for matched filter detection of known signals is worth noting. Note from Eq. (9-41) that the output SNR again depends upon the ratio of signal energy to noise spectral density and not upon the form of the signal. However, in the case of noncoherent detection, the detectability depends upon the square of this ratio, rather than upon the first power; thus operating with small values of E/N_0 makes things more difficult for the interceptor than it does for the desired signal receiver. A more complete discussion of interception appears elsewhere.[8]

9-13 ENERGY AND BANDWIDTH EFFICIENCY

The output SNR from a spread-spectrum receiver in the presence of noise and interference is shown in the previous analysis to be

$$(SNR)_0 = \frac{P_r}{\dfrac{N_0}{2t_m} + \dfrac{J}{PG}} \tag{9-43}$$

where J is the interference power in the receiver bandwidth. However, the energy associated with a message bit is

$$E_b = P_r t_m = \text{energy per bit}$$

Therefore, the output SNR can be expressed as

$$(SNR)_0 = \frac{P_r t_m}{N_0/2 + J t_m/PG} = \frac{E_b}{N_0/2 + J/B_s} \tag{9-44}$$

since the processing gain is given by

$$PG = t_m B_s$$

This output SNR can be written as

$$(SNR)_0 = \frac{E_b/N_0}{1/2 + J/N_0 B_s} \tag{9-45}$$

Solving this for the ratio of energy per bit to noise spectral density yields

$$E_b/N_0 = \left[\frac{1}{2} + \frac{J}{N_0 B_s}\right] (SNR)_o \tag{9-46}$$

Two important conclusions can be reached from this result. First, when the interference is narrow band, increasing the spreading improves the energy efficiency. Second, when the interference is wide band, such as white noise, increasing the spreading does not improve the energy efficiency because the interference power J increases directly with the signal bandwidth B_s.

To obtain the bandwidth utilization efficiency, it may be noted that

$$R_m = \frac{1}{t_m} = \text{message bit rate}$$

The bandwidth utilization efficiency may now be written as

$$\frac{B_s}{2R_m} = \frac{B}{R}$$

Note that the quantity $B_s/2$ is used since this is the equivalent energy bandwidth of the signal. Note also that B/R increases linearly with spreading; thus the band-

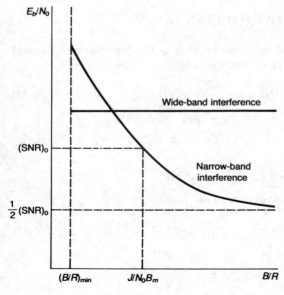

Figure 9-16 Energy and bandwidth utilization for a spread-spectrum system.

width utilization efficiency becomes poorer as the bandwidth of the spread-spectrum signal increases. Combining the results for energy and bandwidth utilization efficiencies leads to

$$\frac{E_b}{N_0} = (\text{SNR})_0 \left[\frac{1}{2} + \frac{J}{2N_0 R_m (B/R)} \right] = \frac{1}{2} (\text{SNR})_0 \left[1 + \frac{J/N_0 R_m}{B/R} \right] \qquad (9\text{-}47)$$

for the case in which the interference is narrow band compared with the bandwidth of the spread-spectrum signal and to

$$\frac{E_b}{N_0} = \frac{1}{2} (\text{SNR})_0 \left[1 + \frac{J_0}{N_0} \right] \qquad (9\text{-}48)$$

when the interference is wide band with respect to the spread-spectrum signal. The quantity J_0 is defined to be the spectral density of the interfering signal. These results can be displayed in terms of the chart of energy utilization efficiency versus bandwidth utilization efficiency, as has been done previously. This is shown in Fig. 9-16.

9-14 REFERENCES

1. Golomb, S. W.: *Shift Register Sequences*, Holden-Day, 1967. Republished by Aegean Park Press, Laguna Hills, CA, 1982.
2. Gold, R.: Optimal Binary Sequences for Spread Spectrum Multiplexing, *IEEE Trans. Infor. Theory*, pp. 619–621, October 1967.
3. Zierler, N., Linear Recurring Sequences, *J. Soc. Indust. Appl. Math.*, vol. 7, 1959.
4. Golomb, S. W., Ed.: *Digital Communications with Space Applications*, Prentice-Hall, Inc., Englewood Cliffs, N.J., 1964.
5. Holmes, J. K.: *Coherent Spread Spectrum Systems*, John Wiley & Sons, Inc., New York, 1982.
6. Urkowitz, H.: Energy Detection of Unknown Deterministic Signals, *IEEE Proc.*, vol. 55, pp. 523–531, April 1982.
7. Park, K. Y.: Performance Evaluation of Energy Detectors, *IEEE Trans. Aerospace Elec. Sys.*, vol. AES-14, no. 2, pp. 237–241, March 1978.
8. Torrieri, D. J.: *Principles of Military Communication Systems*, Artech House, Dedham, MA, 1981.

9-15 PROBLEMS

9.1-1 A seven-digit periodic m sequence has one period of the form

$$+1, +1, +1, -1, -1, +1, -1$$

Does a cyclic shift of this sequence have the same correlation properties?

9.1-2 A seven-digit aperiodic Barker sequence has the form $+1, +1, +1, -1, -1, +1, -1$.

(a) Determine the aperiodic correlation sequence for this code.

(b) Does a cyclic shift of this sequence (e.g., $+1, +1, -1, -1, +1, -1, +1$) have the same correlation properties?

9.1-3 A certain aperiodic PN waveform is defined for the interval as shown.

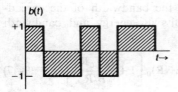

The sequence is $+1, -1, -1, +1, -1, +1, +1$.

    (a) Find the autocorrelation function of this process.

    (b) Is this an ideal aperiodic sequence? [See Eq. (9-2).]

9.1-4 Repeat the previous problem for the aperiodic sequence defined for the interval as shown.

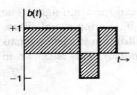

The sequence is $+1, +1, +1, -1, +1$.

9.1-5 One period of a certain sequence is as shown. Does this sequence satisfy the requirements for it to be a periodic pseudorandom sequence? Plot the autocorrelation.

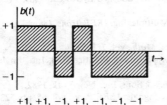

$+1, +1, -1, +1, -1, -1, -1$

9.2-1 A quadratic residue sequence of length N can be generated by squaring the integers from 1 to $(N-1)/2$, reduced modulo N, and inserting plus ones in the positions denoted by the resulting values. The remaining positions are filled with minus ones.

    (a) Find the q–r sequence obtained by this method for $N = 7$.

    (b) Determine the periodic correlation sequence for this code.

    (c) How is this q–r sequence related to the m sequence of Problem 9.1-1?

9.3-1 To aid in initial synchronization, a periodic PN sequence is sometimes used as a preamble, so that, upon reception, the receiver will be able to determine that transmission is ready to begin, and can then establish an accurate timing reference for synchronization.

    If it is desired to use a different preamble for each transmitter and receiver pair, how many such communication links can we have with length 127 ($m = 7$)? Why would it be desirable to use distinctly different m sequences?

9.3-2 Show that the modulo-2 addition of this sequence $(1, 1, 1, -1, 1, -1, -1)$ with itself shifted by two positions is still a shifted replica of itself. How many places is this replica shifted?

9.3-3 How many distinct m sequences can be generated by a linear shift register with 13 stages?

9.4-1 For an $m = 12$ sequence from a periodic pseudorandom signal where the ratio $t_m/t_1 = 20$, find the expected value and the variance of the partial correlation $C_m(K)$. Compare your results with what you would get if the m sequence were purely random. Comment.

9.4-2 Prove Eq. 9-6; that is,

$$\text{Var}[C_m(k)] = 0 \qquad k = 0, N, 2N, \ldots$$
$$= M \qquad k \neq 0, N, 2N, \ldots.$$

9.5-1 Given that the autocorrelation function of the basic chip function is as shown;

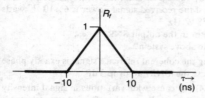

for a periodic m sequence, where $m = 4$, find and plot the autocorrelation function for the total PN signal $b(t)$ for the bipolar case ($a_n = 1, -1$ with equal probability). [See Eqs. (9-12) and (9-13).]

9.5-2 Show that the periodic PN signal normalized autocorrelation function corresponding to the on–off binary case ($a_n = 0, 1$ with equal probability) with $N = 7$ is of the form [see Eq. (9-12)]:

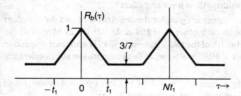

9.6-1 It is desired to send a high-performance reproduction of a singer's voice via a direct-sequence spread-spectrum system. Assuming the highest frequency to be recorded is 16,200 Hz, and it is to be quantized into 256 levels, specify the PN clock rate required if the processing gain is to be greater than 30 dB and the maximum rms code noise is ≤ -34 dB.

In light of your answer, is this a viable way to send this particular message?

9.6-2 A direct-sequence system has a PN code rate of 40×10^6 chips per second and a binary message at 75 b/s. The received signal power is 2×10^{-15} watts and the receiver noise spectral density at the same point is 4×10^{-20} W-s. For this system, in which the correlation time is exactly one message bit, find the ratio of code noise to receiver noise at the output of the correlator when the system is not correctly synchronized.

9.6-3 (a) Prove that

$$E[R_M(\tau)] = R_f(\tau)$$

and

$$\text{Var}[R_M(\tau)] = \frac{1}{M} R_f^2(\tau - t_1) \qquad 0 \leq \tau \leq t_1$$

(b) Plot the expected value and the variance for the random variable $R_M(\tau)$, where $R_f(\tau)$ is given as

$$R_f(\tau) = \left[1 - \frac{|\tau|}{10^{-2}} \right] \qquad |\tau| \leq 0.01 \text{ and } M = 5$$

(c) Find the mean square value of this random variable for $\tau = 7.5$ ms.

9.8-1 It was shown that the output of the correlator was $m_0(t_m)$ given by

$$m_0(t_m) = \frac{\pm \sqrt{P_r}}{t_m} \int_0^{t_m} b(u)b(u - \tau)\cos\theta \, du = \pm \sqrt{P_r} R_b(\tau)\cos\theta$$

Show that $E[m_0(t_m)]$ yields the result given in Eq. (9-27). What can then be said about the PN process $b(t)$?

9.8-2 A direct-sequence spread-spectrum system has the following parameters:

Message bit rate $= 4800$ b/s

PN clock rate $= 220$ Mb/s

(a) Given that the coherent reference signal is off by 10 degrees and the synchronization error is 30 percent, find the output signal-to-noise power ratio if the received signal power is 6×10^{-17} watts and the one-sided spectral density level, N_0, is 1×10^{-20} W/Hz.

(b) Is the code noise term above a determining factor in the output SNR?

(c) What is the maximum $(SNR)_o$ obtainable for the above system?

9.8-3 For the system of the previous problem, assuming the coherent reference signal is exactly phase coherent, find the synchronization error that would cause the $(SNR)_o$ to drop 3 dB.

9.9-1 The results of multipath propagation (see Sec. 1.4) can cause wide variations in signal intensity at the point of reception. Cancellation of one wave by another results in a loss of signal strength and is called "fading." Give a plausible reason why spread-spectrum techniques are able to combat this situation.

In addition, if the reflected signal is delayed (due to multipath) as compared with the direct signal path by more than one code bit, the reflected signal is treated exactly the same as any other uncorrelated input signal. If a direct-sequence system is used with a 5-Mb/s rate, how much longer than the direct path must the reflected path be in order for this multipath to have any effect?

9.9-2 A direct-sequence spread-spectrum system is operating under the conditions of two independent jammers whose center frequencies are very close to each other. The PN code rate is 125 Mb/s and the message bit rate is 2500 b/s. If the combined bandwidth of the two jammers is 50 kHz and their respective received signal powers are 5×10^{-7} watts and 1×10^{-6} watts, what is the approximate interference power at the output of the correlator.

9.9-3 A spread-spectrum system is subjected to a jamming signal of the form $j(t) = A \cos(\omega_1 t + \phi)$ whose phase, ϕ, is a random variable that is uniformly distributed between $(0, 2\pi)$:

$$P(\phi) = \frac{1}{2\pi} \qquad 0 \le \phi \le 2\pi$$

(a) Show that the mean value of the output $j_o(t_m)$ is zero.

(b) Show that the autocorrelation function of the jamming signal is

$$R_j(t - u) = \frac{A^2}{2} \cos \omega_1(t - u)$$

(c) Assuming that the phase, θ, in the reference signal $v(t)$ is zero

$$[v(t) = \sqrt{2} b(t - \tau) \cos \omega_0 t]$$

find the mean-square value of the interference $\overline{j_0^2}$ at the output of the correlator; that is,

$$\overline{j_0^2} = \frac{1}{t_m^2} \int_0^{t_m} \int_0^{t_m} R_j(t - u) R_v(t - u) \, dt \, du$$

9.9-4 If the interference bandwidth is much greater than the receiver input bandwidth of the spread-spectrum system, then it is possible to model the input interference as white noise. Specifically, if the interference power is J watts over a bandwidth B_j, then the two-sided spectral density of the interference can be written as $J/2B_j$ W/Hz. For this case show that the mean square value of the interference at the output of the correlator reduces to that of Eq. (9-32).

9.10-1 A direct-sequence system has a PN code rate of 40×10^6 chips per second and a binary message at 75 b/s. The received signal power is 2×10^{-15} watts and the receiver noise spectral density at the same point is 4×10^{-20} W-s. Assuming a sinusoidal jamming signal at the carrier frequency is introduced into the receiver input, find the ratio of jamming power to desired signal power (JSR) that will cause the correlator output SNR to be unity.

9.10-2 Given the following parameters of a spread spectrum communications system:

$N_0 = 2 \times 10^{-15}$ W/Hz
J = interference power = 5×10^{-7} watts
$P_r = 1.2 \times 10^{-8}$ watts
Equivalent receiver noise bandwidth = 50 MHz
(a) Find the input SNR, $(SNR)_i$.
(b) If the message bit rate is 10 kb/s, find the output SNR, $(SNR)_o$.

9.10-3 For a direct-sequence communication system, find the output SNR if the PN code generator is running at 48 Mb/s, the message bit rate is 2400 b/s, $P_r/N_o = 50$ dB, and $J/P_r = 30$ dB.

9.11-1 A simple direct-sequence receiver is shown in the following.

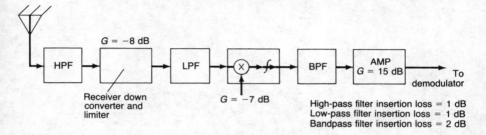

Given that the received signal power is 3×10^{-12} watts, the PN code rate is 100 Mb/s, and the one-sided spectral density, N_o, is 1×10^{-17} W/Hz:
(a) Find the jamming margin of the above system.
(b) What processing gain is necessary if we desire an output SNR of 125?
(c) What is the maximum jamming power the system can handle to achieve the desired output SNR?

9.11-2 It is desired to transmit high-priority information in a secure manner across part of an alien nation. However, it is known that this country has two jammers positioned approximately as shown. Intelligence reports indicate that the jamming powers of J_1 and J_2 are believed to be 25 and 40 watts respectively.

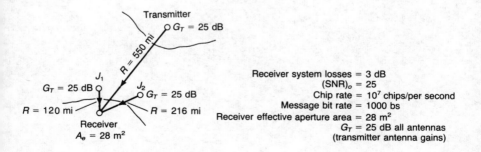

$$\text{Receiver system losses} = 3 \text{ dB}$$

$$(\text{SNR})_o = 25$$

$$\text{Chip rate} = 10^7 \text{ chips per second}$$

$$\text{Message bit rate} = 1000 \text{ b/s}$$

$$\text{Receiver effective aperture area} = 28 \text{ m}^2$$

$$G_T = 25 \text{ dB all antennas}$$

$$(\text{transmitter antenna gains})$$

(a) In this ideal condition, what is the jamming margin $M_j(\text{dB})$?

(b) Neglecting atmospheric losses, what is the absolute minimum transmitter power required?

ANALYSIS OF AVOIDANCE-TYPE SPREAD-SPECTRUM SYSTEMS

This chapter considers methods for analyzing avoidance-type spread-spectrum systems such as frequency-hopping and time-hopping systems. Because of the nature of the signals and the receivers, an exact analysis is considerably more complicated than the analysis of direct-sequence systems. Therefore, only a few fundamental points are discussed here and no attempt at a complete analysis is made.

10-1 THE FREQUENCY-HOPPED SIGNAL

A time-frequency representation of a typical frequency hopped-signal is shown in Fig. 10-1. If the frequency is constant during each hop, in the qth frequency slot the signal may be expressed as

$$s_i(t) = \sqrt{2} \cos 2\pi(f_0 + q f_1)t \qquad it_1 \le t \le (i+1)t_1 \qquad (10\text{-}1)$$

Note that this is the normalized version of the signal, which has an average power of one, and must be multiplied by the square root of the transmitted power in order to obtain the total amplitude during that particular frequency interval. Since the frequencies in different time slots are different and noncoherent, the total energy

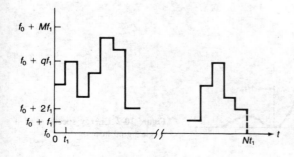

Figure 10-1 A frequency-hopped signal.

319

spectrum of the signal is simply the sum of the energy spectra of the individual hops. The energy spectrum of a single hop of the signal is obtained from the squared magnitude of the Fourier transform of Eq. (10-1).

This yields

$$|F_s(f)|^2 \simeq \frac{t_1^2}{2} \left\{ \left[\frac{\sin \pi t_1 (f - f_0 - q f_1)}{\pi t_1 (f - f_0 - q f_1)} \right]^2 + \left[\frac{\sin \pi t_1 (f + f_0 + q f_1)}{\pi t_1 (f + f_0 + q f_1)} \right]^2 \right\} \qquad (10\text{-}2)$$

and is shown in Fig. 10-2 for positive frequencies only. Frequently it is desirable to let $f_1 = 1/t_1$, but this is not a necessary requirement. When the hopping takes place over M frequencies separated by integer multiples of f_1, the bandwidth of the unmodulated frequency-hop signal is given approximately by

$$B_s \simeq M f_1 \qquad (10\text{-}3)$$

However, the actual shape of the spectrum within this bandwidth depends upon the type of message modulation employed and the relationship between the message bit rate and the hop rate. It is not possible, therefore, to write a general expression for the modulated signal spectral density as was done in the direct-sequence case.

The relation of the hop duration t_1 to the message bit duration t_m depends upon the number of hops per message bit. If this number is k, then

$$t_m = k t_1$$

and

$$B_m = \frac{1}{t_m} = \frac{1}{k t_1}$$

Thus the processing gain of the frequency-hop signal becomes

$$\text{PG} = \frac{B_s}{B_m} = M k f_1 t_1$$

when there is no frequency multiplication. If there is frequency multiplication before the final transmission, and if this multiplication is by a factor K, then the

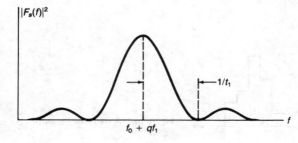

Figure 10-2 Energy spectrum of a single signal hop.

processing gain becomes

$$PG = MKkf_1t_1 \qquad (10\text{-}4)$$

For a given processing gain, the interference rejection and the jamming margin are not necessarily the same as that determined previously for direct-sequence systems. The evaluation of intercept energy also may be different, particularly if the observation time is small, because the signal energy during any one frequency hop may be much larger in a small bandwidth than it would be for a direct-sequence signal. Therefore, energy detection of single hops may be possible even under conditions in which a direct-sequence signal could not be detected.

10-2 INTERFERENCE REJECTION IN A FREQUENCY-HOPPING RECEIVER

Unlike the direct-sequence receiver, the frequency-hopping receiver cannot reduce narrow-band interference by averaging it over a message bit interval. Thus when interference exists near one of the hopping frequencies, its effect on the receiver during that hop is essentially the same as it would be for a typical unhopped FSK receiver. The reduction in interference in the receiver output comes about because the receiver hops near the interfering signal only a small fraction of the time.

Because the interference at the receiver output has an intermittent nature, the use of an output signal-to-noise ratio as a measure of receiver performance is no longer appropriate. Instead it is more desirable to consider the message bit error probability as a quality measure. To do this it is necessary to consider the FSK demodulation in a little more detail. This is done with the aid of Fig. 10-3, which illustrates one method of accomplishing noncoherent FSK demodulation.

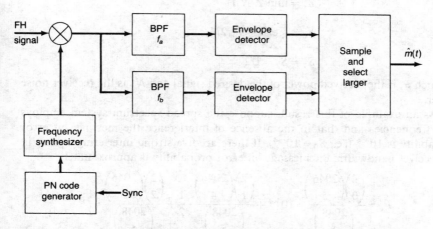

Figure 10-3 Noncoherent FSK demodulation in a frequency-hopping receiver.

Although the output of the mixer will contain only one desired frequency component during each hop, the demodulator must be able to respond to either of the two frequencies that might have occurred as a result of message modulation. Thus there are two frequencies that are capable of producing interference in the receiver output. If an interfering signal at either of these two frequencies is larger than the desired signal, the potential for destroying the message information during that hop exists.

An exact analysis of the message bit error probability for this situation is quite involved. A fairly simple result can be obtained, however, if several assumptions are made. These assumptions are

1. The number of frequencies over which the signal is hopped is M.
2. The number of narrow-band interfering signals is $N_j < M$ and the power of each interfering signal is much greater than that of the desired signal.
3. The interfering frequencies are independently and uniformly distributed over the M signal frequencies.
4. Message bit errors occur independently.

The resulting message bit error probability is given approximately by (ref. 3, p. 74)

$$P_e \simeq \sum_{n=0}^{n_1} \frac{\binom{2}{n}\binom{M-2}{N_j-n}}{\binom{M}{N_j}} S_n \qquad (10\text{-}5)$$

where

$$\binom{c}{r} = \text{the binomial coefficient}$$

$$n_1 = \min(2, N_j)$$

$$S_0 = \frac{1}{2} \exp(-P_r/2N_R)$$

$$S_1 = S_2 = 1/2$$

in which P_r is the received power of the desired signal and N_R is the receiver noise power.

As an example of this result, suppose the spread-spectrum system employs 2048 frequencies and that in the absence of interference the message bit error probability is 10^{-4} (i.e., $S_0 = 10^{-4}$). If there are five strong interfering signals in the receiver bandwidth, the message bit error probability is approximately

$$P_e \simeq \frac{\binom{2}{0}\binom{2046}{5}}{\binom{2048}{5}} \cdot 10^{-4} + \frac{\binom{2}{1}\binom{2046}{4}}{\binom{2048}{5}} \cdot \frac{1}{2} + \frac{\binom{2}{2}\binom{2046}{3}}{\binom{2048}{5}} \cdot \frac{1}{2}$$

$$= 3.46 \times 10^{-3}$$

It may be noted that the ability of a frequency-hopping spread-spectrum system to reject interference depends primarily on the ratio of the number of interfering signals to the number of hopping frequencies and is almost independent of the interference power so long as that power is greater than the desired signal power. Thus the processing gain of a frequency-hopping system is not a good measure of its ability to reject interference, except in the manner in which it depends upon the number of frequencies.

In Eq. (10-4) it is shown that the processing gain depends upon the product of the number of frequencies M and the frequency multiplication factor K. It is clear from the discussion here that obtaining a large processing gain by reducing the value of M and increasing K does not achieve the desired result in rejecting interference.

The approximate result in Eq. (10-5) can be extended to cases in which the interference power is not large compared with the desired signal power by appropriately defining the conditional probabilities S_1 and S_2. The results are quite complicated, however, and are not reproduced here.

10-3 THE TIME-HOPPED SIGNAL

A typical time-hopped signal is represented in Fig. 10-4. The duration of each frame is T_f seconds and in each frame there is one transmission burst that will contain k message bits. If there are M such burst periods in each frame, then the width of each burst is T_f/M and the frame duration must be

$$T_f = kt_m$$

Since these are transmitted in T_f/M seconds, the duration of each chip within a burst must be

$$t_1 = \frac{T_f}{kM} \quad \text{biphase modulation}$$

$$= \frac{2T_f}{kM} \quad \text{quadriphase modulation} \tag{10-6}$$

Note that this result depends upon whether biphase modulation or quadriphase modulation is used.

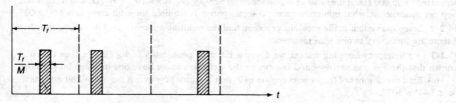

Figure 10-4 A time-hopping signal.

Since the bandwidth of the transmitted signal is equal to

$$B_s = \frac{2}{t_1}$$

this becomes

$$B_s = \frac{2kM}{T_f} \qquad \text{biphase}$$

$$= \frac{kM}{T_f} \qquad \text{quadriphase} \qquad (10\text{-}7)$$

Again this depends upon whether biphase or quadriphase modulation is used.

The processing gain of the time-hopped signal is the ratio of the signal bandwidth to the message bandwidth and becomes

$$PG = \frac{B_s}{B_m} = \frac{2kM/T_f}{1/t_m} = 2M \qquad \text{biphase}$$

$$= \frac{kM/T_f}{1/t_m} = M \qquad \text{quadriphase} \qquad (10\text{-}8)$$

For a given processing gain, the interference rejection and the jamming margin are not the same as for direct-sequence systems. Also the evaluation of intercept energy may be quite different, because the energy is concentrated in such a small time interval. Because of this small time interval, the transmitted power is large during this time and hence signals may be detected on a single-hop basis that could not be detected if they used the same energy as a direct-sequence system.

10-4 REFERENCES

1. Dixon, R. C.: *Spread Spectrum Systems*, John Wiley & Sons, Inc., New York, 1976.
2. Cahn, C. R.: *Spread Spectrum Applications and State of the Art Equipment*, AGARD-NATO Lecture Series no. 58, Paper no. 5, May–June 1973.
3. Torrieri, D. J.: *Principles of Military Communication Systems*, Artech House, Dedham, MA, 1981.

10-5 PROBLEMS

10.2-1 A frequency-hopping spread-spectrum communications system employs 1024 frequencies separated by 75,000 Hz. If the noise one-sided spectral density, N_0, is 5×10^{-18} W/Hz, and there is one hop per message bit, what minimum received signal power is required for a bit error rate of 0.0002?

10.2-2 Given the system in the previous problem, find the resulting approximate bit error probability if there are three very strong tone jammers.

10.2-3 In a frequency-hopping system, we desire a bit error probability of 1×10^{-4}. Given that it is known that there is one narrow-band jammer in the vicinity, and we have one message bit per hop:

(a) Find the number of frequencies (power of 2) that we must hop over if the overall bit error is to be $\leq 3.441 \times 10^{-4}$.

(b) Assuming the average received signal power is relatively constant over all hops at 3.349×10^{-11} watts, and $K = 1$, $f_1 t_1 = 1$, find the PN code generator clock rate required by the synthesizer.

10.3-1 A time-hopping spread spectrum system is to have the following parameters:

Message bit rate = 2400 b/s (after error correction coding)
Message bits per time frame = 16
Processing gain = ≥ 35 dB

(a) Find the smallest number of time slots per time frame if this number is to be a power of 2.

(b) Find the duration of each time frame.

(c) Find the duration of each time slot.

10.3-2 (a) Find the bandwidth of the signal in Problem 10.3-1 assuming quadriphase modulation is used.

(b) If the average transmitted power over all time is to be 10 watts, what must be the transmitted power during each burst?

ELEVEN

GENERATION OF SPREAD-SPECTRUM SIGNALS

This chapter considers methods of generating signals that might be useful in a spread-spectrum communication system. Both direct-sequence signals and frequency-hopping signals are discussed. Clearly there is insufficient space to go into detail about the specific hardware that might be used. Therefore, the discussion is limited to outlining the basic concepts and illustrating the principal methods with block diagrams. Direct-sequence signals are considered first.

11-1 SHIFT-REGISTER-SEQUENCE GENERATORS

It was noted in an earlier chapter that the most useful form of PN code is the m-sequence because these sequences can be generated by means of linear shift registers. The general form of such a shift register is shown in Fig. 11-1.[3] Note that this consists of m stages of shift register connected in cascade with the outputs of certain stages added together modulo-2 and fed back to the input. All stages of the shift register are clocked simultaneously, and each time a clock pulse occurs, a new binary digit appears at the output. For the resulting sequence of binary digits to be an m sequence, it is necessary that the feedback connections be made in a

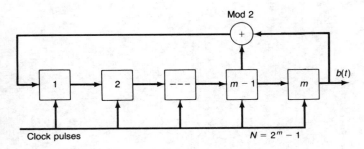

Figure 11-1 Generation of shift register sequences.

particular fashion. The manner of doing this is best illustrated by the polynomial representation that is considered next.

Consider a polynomial in x denoted by

$$f(x) = x^m + a_{m-1}x^{m-1} + \cdots + a_1x + 1 \tag{11-1}$$

in which the coefficients of the various powers of x are either zero or one. That is, each a_i where i ranges from 0 to m, is either a zero or a one. This is an mth-degree polynomial, since a_m must equal 1, and is said to be irreducible if it cannot be factored into polynomials of smaller degree whose coefficients are also only zero or one. Only a small fraction of all such polynomials is irreducible, but for every integer value of m, there is at least one irreducible polynomial, and there may be more than one. All of the usual arithmetic operations can be performed on such polynomials, provided that binary arithmetic is used when dealing with the coefficients. The condition that is of primary interest here is that in which $f(x) = 0$. This can be written as

$$x^m = a_{m-1}x^{m-1} \cdots + a_1x + 1 \tag{11-2}$$

Note that no minus sign occurs because -1 is not a permissible number in this form of binary arithmetic.

A shift register circuit that describes this polynomial is shown in Fig. 11-2, in which the outputs of the various stages are fed back to the input through weighting coefficients a_1 through a_{m-1}. Any a_i equal to 1 implies that this connection exists, whereas any a_i equal to zero implies that there is no connection at this particular point. This is best illustrated by means of a specific example. Consider the irreducible third-degree polynomial

$$f(x) = x^3 + x + 1$$

and write it as

$$x^3 = x + 1$$

Note that both a_0 and a_1 are 1, while a_2 is zero. Therefore, the shift register that represents this polynomial is as illustrated in Fig. 11-3, in which there are feedback connections from the third stage and the second stage, but none from the first stage.

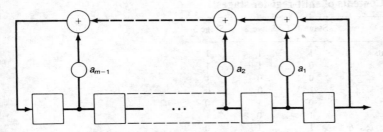

Figure 11-2 Implementation of a polynomial.

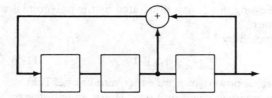

Figure 11-3 Shift register generator for $m = 3$.

When the shift registers are loaded with any set of binary digits except all zeroes and are clocked, the output will be an m sequence. To illustrate this consider the specific case in which the shift registers are initially loaded with all ones and then determine the state of the various shift registers that occurs for each succeeding clock pulse. This operation is illustrated in Table 11-1. Note that after seven clock pulses, the original state of the shift register is restored. Since the output of this shift register generator is taken from the third stage, the resulting sequence that is produced is given by the third column in Table 11-1.

This is

$$1 \quad 1 \quad 1 \quad 0 \quad 0 \quad 1 \quad 0$$

To convert this into the binary sequences that we have been discussing, it is necessary to identify the $+1$ with a $+1$ and the 0 with a -1. Hence the above sequence implies the sequence:

$$1 \quad 1 \quad 1 \quad -1 \quad -1 \quad 1 \quad -1$$

When this is converted to a time function, the resulting PN signal is as shown in Fig. 11-4. This time function repeats itself periodically every $7\ t_1$ seconds. It may also be noted from Table 11-1 that all combinations of three binary digits, except the all-zero sequence, occur in the process of clocking the shift registers seven times. Hence regardless of how the shift registers might have been loaded initially, the same periodic sequence will be produced, although with a different starting point.

It is possible, however, to obtain a different sequence if another irreducible third-degree polynomial exists. In fact it does, and this polynomial is

$$f(x) = x^3 + x^2 + 1$$

Table 11-1 Contents of shift-register stages

		Stage 1	Stage 2	Stage 3
Initial contents		1	1	1
Clock pulse	1	0	1	1
	2	0	0	1
	3	1	0	0
	4	0	1	0
	5	1	0	1
	6	1	1	0
New period	7	1	1	1

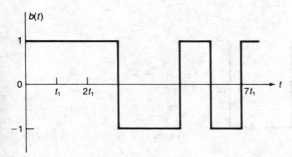

Figure 11-4 PN code for $m=3$.

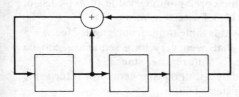

Figure 11-5 Alternative shift register generator for $m=3$.

which may be written as

$$x^3 = x^2 + 1$$

The shift register representation of this polynomial is presented in Fig. 11-5, in which it is shown that there is feedback from the third stage and the first stage, but not from the second stage.

The operation of this shift register generator is illustrated in Table 11-2, in which the initial contents of the shift registers are again taken to be all ones. As clock pulses are introduced, the output sequence is again the third column of this table. This sequence produces the PN signal shown in Fig. 11-6.

Note that this also has a period of $7 t_1$, as it must, but it is not the same sequence as that produced by the previous shift register generator nor will it be the same for any starting point. It is clear from the previous discussion that to construct a shift register generator for sequences of any permissible length, it is only necessary

Table 11-2 Contents of shift-register stages

		Stage 1	Stage 2	Stage 3
Initial contents		1	1	1
Clock pulse	1	0	1	1
	2	1	0	1
	3	0	1	0
	4	0	0	1
	5	1	0	0
	6	1	1	0
New period	7	1	1	1

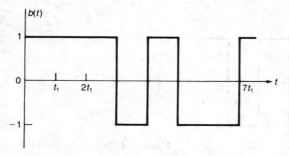

Figure 11-6 Alternative PN code for $m = 3$.

to know the irreducible polynomial for the corresponding value of m. A list of some such polynomials is shown in Table 11-3, for certain values of m.

Note that the m sequences described by this table range from length 7 for $m = 3$ to length 524,287 for $m = 19$. It is clear that even very long sequences can be generated with a relatively modest number of shift register stages.

The chip rate of the PN code generated by a shift register generator is the same as the clock rate f_c applied to the shift registers. That is,

$$f_c = \frac{1}{t_1}$$

Code rates cannot be made arbitrarily large. In general the maximum code rate is determined by the switching rate of the shift registers or by the delay in the feedback network. Different feedback connections can result in different maximum code rates.[1] This is illustrated in Fig. 11-7, in which two different ways of accomplishing the same feedback connections are shown. In the top drawing there are three summing points in cascade and, therefore, the delays of these three points add. In the bottom drawing, however, two summing points essentially operate in

**Table 11-3 Polynomials for
m-sequence generators**

m	$f(x)$
3	$x^3 + x + 1$
4	$x^4 + x + 1$
5	$x^5 + x^2 + 1$
6	$x^6 + x + 1$
7	$x^7 + x + 1$
8	$x^8 + x^6 + x^5 + x^4 + 1$
10	$x^{10} + x^3 + 1$
12	$x^{12} + x^{10} + x^2 + x + 1$
14	$x^{14} + x^{12} + x^2 + x + 1$
15	$x^{15} + x + 1$
17	$x^{17} + x^3 + 1$
19	$x^{19} + x^5 + x^2 + x + 1$

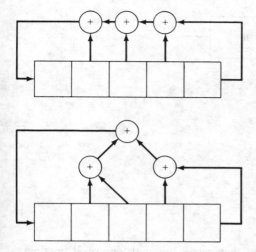

Figure 11-7 Changing feedback connections to increase speed.

parallel, and hence their delays do not add. Therefore, the corresponding code rate may be higher.

The maximum code rate also depends upon the length of the code since for more stages of shift register there is more total delay. For very long codes, rates of 50 to 100 Mb/s are obtainable with existing hardware, while for shorter codes rates of 300 to 500 Mb/s are obtainable. Although permissible code rates undoubtedly will increase as faster hardware becomes available, the existence of such hardware still places a limit on the maximum spreading that can be achieved in a direct-sequence spread-spectrum system.

As shown in the next section, shift register sequences also play an important role in the generation of frequency-hop signals. In this case, however, the amount of spreading is not limited by the code rate and, in general, code rates can be quite small compared with the value they would have in a direct-sequence system for the same bandwidth.

11-2 DISCRETE-FREQUENCY SYNTHESIZERS

Frequency-hopping signals are usually generated by means of a discrete-frequency synthesizer. There are two major classes of such synthesizers, usually referred to as the direct method and the indirect method. Both of these are discussed in the following.[1,2]

Direct Method (Add and Divide)

The direct frequency synthesizer consists of several stages, each of which has the form shown in Fig. 11-8. Note that this stage takes the input frequency, divides it

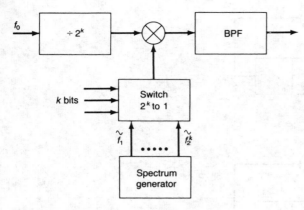

Figure 11-8 The direct frequency synthesizer.

by a number that is a power of 2, and then mixes this with one of 2^k possible frequencies that are produced by a spectrum generator. Which of these frequencies is actually used is selected by means of a set of k binary digits and these k binary digits usually come from a shift-register-sequence generator. The output of the mixer is then filtered and delivered to the next stage where the operation is repeated. If there are n stages, and 2^k frequencies per stage, then the total number of frequencies produced by the direct frequency synthesizer is

$$M = (2^k)^n = 2^{kn}$$

The operation of the direct frequency synthesizer is best illustrated by means of a specific example. For this purpose assume that $k = 2$, hence $2^k = 4$ frequencies per stage; that is, the spectrum generator must produce four frequencies. These same four frequencies will be produced in every stage of the direct frequency synthesizer. Next let us arbitrarily pick the initial frequency as 60 MHz and the spectrum generator frequencies as

$$f_0 = 60 \text{ MHz}$$

$$\tilde{f}_1 = 33 \text{ MHz} \quad \tilde{f}_3 = 49 \text{ MHz}$$

$$\tilde{f}_2 = 41 \text{ MHz} \quad \tilde{f}_4 = 57 \text{ MHz}$$

The reason for these choices will become apparent subsequently. Since the spectrum generator produces four frequencies, the output of the first stage can be any one of four frequencies. The input frequency, 60 MHz, is divided by 4 and then added to one of the spectrum generator frequencies. Thus the four output frequencies are

$$\frac{60}{4} + 33 = 48 \text{ MHz}$$

$$\frac{60}{4} + 41 = 56 \text{ MHz}$$

$$\frac{60}{4} + 49 = 64 \text{ MHz}$$

$$\frac{60}{4} + 57 = 72 \text{ MHz}$$

The output of the second stage will again be four frequencies for every one of the input frequencies. Since there are four possible input frequencies, there now will be 16 possible output frequencies. These are obtained by taking each of the input frequencies, dividing it by 4, and adding to it one of the spectrum generator frequencies. The resulting set of frequencies is illustrated in the following table.

	33	41	49	57
48	45	53	61	69
56	47	55	63	71
64	49	57	65	73
72	51	59	67	75

Thus there is a total of 16 frequencies spaced 2 MHz apart. The total bandwidth then is approximately given by

$$B_s = Mf_1 = 16 \times 2 = 32 \text{ MHz}$$

where f_1 is the separation of frequencies.

The above results can be generalized. Specifically the separation in the final set of frequencies can be expressed as

$$f_1 = \frac{\Delta F}{2^{k(n-1)}} \qquad (11\text{-}3)$$

in which ΔF is the separation of frequencies in the spectrum generator, which was 8 MHz in the foregoing example. The eventual bandwidth is given by

$$B = Mf_1 = 2^{kn} \cdot \frac{\Delta F}{2^{k(n-1)}} = 2^k \, \Delta F \qquad (11\text{-}4)$$

and this bandwidth is centered at the frequency f_0, which is the frequency into the first stage if the spectrum generator frequencies are centered at $f_0(1 - 1/2^k)$.

Note that the overall bandwidth is determined by the number of frequencies generated and by the separation between frequencies rather than by the rate at which these frequencies are generated. Thus the PN code generator that selects the k bits supplied to each stage of the direct frequency synthesizer can run at a rate that is much smaller than the overall bandwidth. In fact the maximum hop rate is usually determined by the switching speed of the switch shown in the direct frequency synthesizer rather than by the PN code rate. In a typical frequency synthesizer, switching times may be of the order of a few microseconds. The reason these switching times are so large is the need for reducing spurious output frequencies in the bandpass filters at the output of each stage. These bandpass filters are necessary to suppress other modulation products and need to have a relatively sharp cutoff in order to do an effective job. Therefore, there is a trade-off between switching speed and spurious outputs.

The switching speed may also be a function of the way in which the code command is applied to the frequency gating switches. In particular it is desirable to apply the code sequentially to the various stages rather than simultaneously. Finally, the period of a hop should be at least 10 times as long as the time required to set a frequency in order for the frequency to remain relatively stable during the frequency-hop time. Hence, if the switching time is of the order of 5 μs, the total hop duration will be of the order of 50 μs, and the hop rate, is limited to about 20,000 hops per second.

Finally, it should be noted that the phases of the frequency-hopped signal are incoherent from hop to hop. This is so because there is no specific phase relationship between the frequencies of the spectrum generator.

Indirect Method

The indirect method of generating frequency-hopped signals is illustrated in Fig. 11-9. Note that this is a closed-loop system in which the input to the low-pass filter is the difference between the reference oscillator frequency f_1 and the output frequency divided by an integer n_i. Hence the voltage control oscillator inside the loop is formed to be at a frequency that is $n_i \times f_1$. If the frequency divider can be programmed to change its division ratio, then the output frequency will change correspondingly. The frequency divider thus is programmable to divide by any integer from n_1 to $n_1 + 2^m - 1$. If this requires a large number of frequencies, several dividers may be necessary—connected in cascade. The output frequency at the ith hop is, therefore,

$$f_i = n_i f_1$$

To illustrate this method, assume that we again want 16 frequencies centered approximately at 60 MHz. Let

$$f_1 = 2 \text{ MHz}, \quad M = 16 = 2^4$$

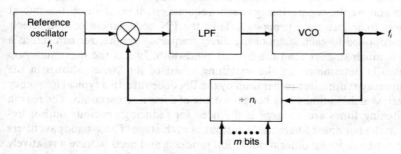

Figure 11-9 Indirect frequency synthesizer.

and use
$$n_1 = 23, \quad n_1 + 2^m - 1 = 23 + 16 - 1 = 38$$
Hence the lowest frequency becomes
$$\text{Lowest frequency} = 23 \times 2 = 46 \text{ MHz}$$
and the highest frequency is
$$\text{Highest frequency} = 38 \times 2 = 76 \text{ MHz}$$

Since the programmable frequency divider divides by all integers from 23 through 38, the resulting set of output frequencies is

46	54	62	70
48	56	64	72
50	58	66	74
52	60	68	76

In the indirect method of generating frequencies, the maximum hop rate is determined by the settling time of the frequency control loop. Usually rise times may be of the order of tens of microseconds. In this case the rise time must be balanced against the short-term stability of the frequency in each hop. Since indirect frequency synthesizers are inherently slower than direct frequency synthesizers, it is of interest to consider techniques for reducing switching time. Several such techniques have been proposed. These include the sampling loop filter, digital-to-analog frequency-command conversion, and the use of multiple loops. A brief discussion of each of these techniques follows.

In the sampling loop filter, a sample-and-hold circuit is used to reduce the jitter induced by the phase detector. This permits greater freedom in selecting the reference frequency and filter parameters and leads to an overall reduction in possible switching time.

In the frequency-command conversion technique, the coded frequency command applied to the programmable divider is converted to a dc control signal and used to set the VCO to a value that is close to the correct frequency. Lock time is reduced because the frequency uncertainty is less, and so requires less time for the feedback loop to pull the frequency to exactly the right value.

An even greater increase in hop rate can be achieved by the use of multiple loops. By using two frequency synthesizers producing alternate outputs, one loop can be adjusting its frequency to the proper value while the other loop is providing the frequency that is actually being used. The second loop is then used on the next hop and the first loop proceeds to adjust itself to its new correct value. Such a system, for example, can increase the hop rate by a factor of 10 when the rise time is 10 percent of the hop duration.

General Comments

It is appropriate to make some comparisons of the two different methods of generating frequency-hopped signals. In the direct frequency synthesizer, one of the k bits into one stage of the synthesizer is a message bit. All of the others come from the PN code generator that is producing the hopping sequence. In the indirect frequency synthesizer, one of the m bits controlling the programmable divider is the message bit and the others all come from a PN code generator.

Under normal circumstances the direct frequency synthesizer can produce signals at a higher hop rate than can the indirect frequency synthesizers. The penalty that must be paid for this, however, is usually an increased complexity and an increased power requirement. Therefore, in situations where a low hop rate, of the order of a few hundred hops per second, is acceptable; the indirect frequency synthesizer is preferred. When higher hop rates—of the order of a few thousand hops per second—are required, the direct frequency synthesizer is normally used.

The indirect frequency synthesizer can produce coherent phase from hop to hop. Whether or not this is an advantage depends upon the nature of the transmission medium. In a multipath situation, for example, the coherent phase is not retained at the receiver, and hence there is no advantage in insisting on phase coherence at the transmitter.

In both types of frequency synthesizers, the frequency control bits come from a PN sequence generator. When there is one hop per message bit ($t_1 = t_m$), the PN code rate is $(m - 1)/t_m$. This is usually much less than the code rate for direct sequence for the same frequency spread. As an illustration of this point, consider a case in which

$$m = 12, \quad M = 2^{12} = 4096$$

and pick the frequency f_1 as 5 kHz. Thus

$$B_s = Mf_1$$

$$= 4096 \times 5000$$

$$= 20.48 \text{ MHz}$$

The required PN code rate depends upon the message bit rate, and if this is selected to be

$$\text{Message bit rate} = 2400 \text{ b/s}.$$

then

$$\text{PN code rate} = (12 - 1)2400$$

$$= 26,400 \text{ b/s}$$

It may be noted that to obtain the same degree of spreading in a direct sequence system,

$$\text{PN code rate} = \frac{B_s}{2} = 10.24 \text{ Mb/s}$$

A further advantage of using frequency synthesizers is that they can be programmed to avoid specific portions of the spectrum. Therefore, if there are certain frequencies that must be avoided, frequencies on either side of this can be used to increase the total overall bandwidth, but still avoid interference with specific types of service within the band.

11-3 SAW DEVICE PN GENERATORS

Although the use of shift registers to generate PN codes is a very convenient technique, it is not the only method that can be used. An alternative method is to use a tapped delay line with appropriate weighting of the outputs of the taps. Although this may seem like an awkward and inefficient way of generating a PN code, there are certain situations in which it can be advantageous. For example, the chip rate can be much higher than can be generated with shift registers in some cases. Furthermore, the same tapped delay line may also be used in the receiver by employing techniques that are discussed in a subsequent chapter. Thus for short PN codes, for example, when the code length is equal to the message bit duration, this may be a useful technique. It is not useful, however, for very long codes. The general form of tapped-delay-line PN code generator is illustrated in Fig. 11-10.[2,5]

A specific example is useful in illustrating how this system can be used to generate a PN code. Assume that we wish to generate the code of length 7, as shown in Fig. 11-11. For this purpose the input to the tapped delay line is a rectangular pulse of length t_1 as shown in the upper part of Fig. 11-12. The tapped delay line itself is shown in the lower part of this figure. Note that the weighting

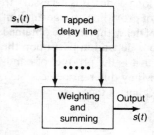

Figure 11-10 Tapped-delay-line PN code generator.

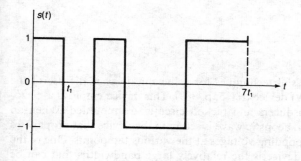

Figure 11-11 PN code for $n = 7$.

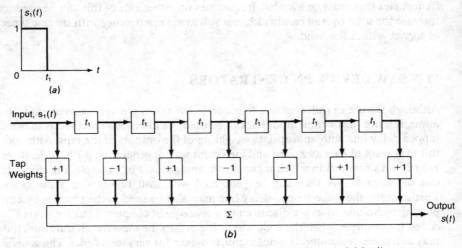

Figure 11-12 Generation of the $n=7$ code. (*a*) Input waveform; (*b*) the tapped delay line.

coefficients connected to the various taps are assigned values of $+1$ or -1 in accordance with the desired polarity of $s(t)$. Hence when the short rectangular pulse is applied to the input, delayed versions of it, either reversed in polarity or not, will be added up and appear at the output. The output $s(t)$ then will be of the form shown in Fig. 11-11.

It is clear from this discussion that to obtain very long codes, it is necessary to have a tapped delay line with many taps and capable of providing a large amount of delay. There are serious limitations on the amount of delay that can be obtained from a practical tapped delay line, and these limitations depend in part upon the bandwidth that is required in the system. In particular it is the product of bandwidth and delay that is normally the limiting factor. Many different devices have been employed for purposes of accomplishing delay. Some of these are[2]:

1. Lumped constant.
2. Acoustic.
3. Volume propagation.
4. Surface waves.
5. Microwave acoustic.
6. Charge-coupled devices.
7. Digital LSI.

One of the most successful techniques for achieving delay is the use of the surface acoustic wave (SAW) device (ref. 5, p. 389). This device employs a piezo-electric slab on which transducers and pickoff circuits are mounted. When an electric voltage is applied, an acoustic wave is set up in the device that propagates down it and induces corresponding voltages at the various tap points. One of the advantages of this type of device is the relatively large bandwidths that can be achieved, of the order of several hundred megahertz. A disadvantage, however, is that the maximum delay that can be achieved is relatively small, normally less than 100 μs.

The combinations of bandwidth and delay that are possible are illustrated in Fig. 11-13, and Fig. 11-14 illustrates the construction of a SAW device that might be used for generating a PN code of length 7.

Although the SAW tapped delay line may have a maximum delay of the order of 100 μs and a maximum bandwidth of the order of 500 MHz, these two quantities cannot be obtained simultaneously. Hence the available processing gain, which is the product of delay and bandwidth, is of the order of 2000, rather than the 50,000 that might be implied by the product of the two numbers above.

The reason that long delays cannot be achieved is that there is a substantial attenuation as the wave propagates down the device. This attenuation depends upon the loss parameters of the device and the center frequency as shown by

$$\text{Loss} = (VAC)f_0^2 + (AIR)f_0 \quad \text{dB}/\mu\text{s}$$

where

$$VAC = \text{loss in vacuum}$$

$$AIR = \text{air coupling loss}$$

$$f_0 = \text{frequency, GHz}$$

In a typical situation the values of VAC and AIR might be

$$VAC = 0.88$$

$$AIR = 0.19$$

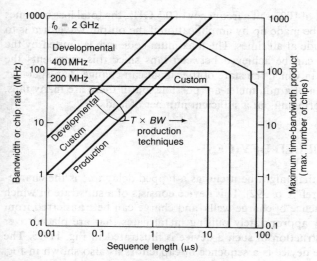

Figure 11-13 Practical limits on SAW devices.

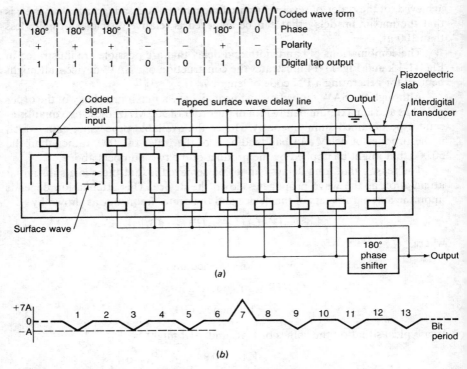

180°	180°	180°	0	0	180°	0	Phase
+	+	+	–	–	+	–	Polarity
1	1	1	0	0	1	0	Digital tap output

Figure 11-14 Construction of a SAW device for generating an $n = 7$ PN code.

Thus for a total duration of 100 μs at a frequency of 0.5 GHz, the total loss becomes 31.5 dB. This loss must be made up by amplification if the output waveform is to have a constant amplitude at all times. The maximum code rate is limited by the minimum spacing that can be achieved between taps since this represents the minimum time interval in which the state of the signal can change. Tap spacings of the order of a fraction of a millimeter are possible, but if there are many taps, there are difficulties in bringing out a sufficient number of leads.

11-4 CHARGE-COUPLED DEVICES

Another type of device that might be useful as a tapped delay line is the charge-coupled device (CCD) (ref. 5, p. 392). This device consists of a substrate in which there are formed a sequence of charge wells, and charge can be transferred from one well to the next by appropriately pulsing metal gates that are placed over them. The general construction of such a device is illustrated in Fig. 11-15. The equivalent circuit of the device is a sequence of capacitors as also shown in Fig. 11-15.

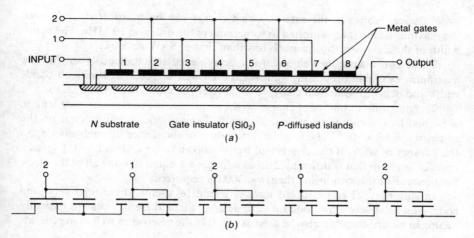

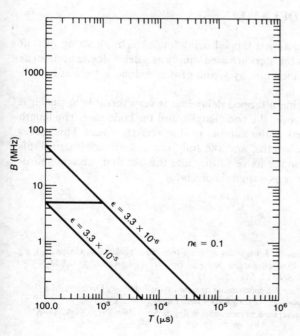

Figure 11-15 A charge-coupled device and its equivalent circuit.

Figure 11-16 Limits on practical charge-coupled devices.

The maximum code length is determined by the number of charge wells and this can be quite large, on the order of several thousand. The maximum delay depends upon the rate at which charges are transferred from one well to the next and maximum delays on the order of $10^4 \ \mu s$ are possible. Thus this device can achieve delays that are significantly greater than those that are obtainable in a

SAW device. However, the rate at which charge can be transferred, and hence the maximum chip rate, is limited to something of the order on 10 MHz. The bandwidth of these devices thus is much less than that of SAW devices.

These two maximum values cannot be achieved simultaneously so that the maximum available processing gain is on the order of 500. The limits on bandwidth and delay for practical charge-coupled devices are shown in Fig. 11-16.

The maximum number of charge wells, and hence the maximum code length, is limited by the charge transfer inefficiency. Typically this inefficiency is on the order of $\varepsilon = 3.3 \times 10^{-5}$. The frequency response of the device depends also upon the number of wells. If the number of wells is taken to be n, then $n\varepsilon < 0.1$ gives a transfer function that is down by 2 dB at a frequency equal to one half of the clock frequency. For this condition, then, $n < 3000$ is required.

In summary, charge-coupled devices would be useful where relatively long codes at a relatively slow code rate are desired. Since these devices are basically analog in nature, they can also be used as a part of the receiver as well as in generating the signal.

11-5 DIGITAL TAPPED DELAY LINES

An alternative way of constructing a tapped delay line is to use a string of shift registers.[2] In this case the shift registers are used simply as a delay device and not as a means of generating a specific code by virtue of the feedback. No feedback is employed.

This method of constructing a tapped delay line is very versatile in that it is relatively easy to change the code, the code length, and the code rate. The length of the code, however, is limited to the number of shift register stages. This places a serious constraint on code lengths, and the only reasons for considering this as a means of signal generation are its versatility and the fact that the same hardware can be used in the receiver as a digital correlator.

11-6 REFERENCES

1. Dixon, R. C.: *Spread Spectrum Systems*, John Wiley & Sons, Inc., New York, 1976, Chaps. 3, 4.
2. Cahn, C. R.: *Spread Spectrum Applications and State of the Art Equipments*, AGARD-NATO Lecture Series no. 58, Paper no. 5, Sec. 3.5, May–June 1973.
3. Torrieri, D. J.: *Principles of Military Communication Systems*, Artech House, Dedham, MA, 1981.
4. Golomb, S. W.: *Shift Register Sequences*, Aegean Park Press, Laguna Hills, CA, 1982.
5. Holmes, J.K.: *Coherent Spread Spectrum Systems*, John Wiley & Sons, Inc., New York, 1982.

11-7 PROBLEMS

11.1-1 A PN code generator has four stages with feedback from the fourth and third stages to the input. Find the sequence generated if the generator starts with a "1" at the output of every stage.

11.1-2 Given the polynomial in x as $f(x) = x^4 + x + 1$:

(a) Is this polynomial reducible?

(b) What would be the length of the shift register representation of this polynomial?

(c) Sketch the shift register representation of this polynomial.

(d) Show that the polynomial $f(x) = x^3 + 1$ is reducible.

11.1-3 For the polynomial in the first part of the previous problem:

(a) Find the sequence generated if all the shift registers are initially loaded with all ones.

(b) Show that the same sequence is generated (with a different starting point) if the shift registers are initially loaded with (0, 0, 0, 1).

11.1-4 Given the irreducible polynomial

$$f(x) = x^{24} + x^7 + x^2 + x + 1$$

(a) Find the length N of this sequence.

(b) If the clock rate is 0.5 Mb/s, find the period of this sequence, in minutes.

11.1-5 A direct-sequence spread-spectrum system is to have the following parameters:

PG ≥ 37 dB.

Analog message band-limited to 10,000/3 Hz quantized to 64 levels.

SN 74S74 for use as a shift register.

PN signal with period of 11.45 minutes.

(a) Neglecting the feedback delay, what is the absolute maximum clock frequency that we can use?

(b) How many stages of shift registers should we use?

(c) What is the approximate maximum RF bandwidth?

(d) What is the processing gain achieved with the above system assuming that we sample the analog message at the Nyquist rate?

11.2-1 We are interested in generating frequency-hopping signals by use of a single-stage direct frequency synthesizer.

(a) Show that if the frequencies to be generated are centered at f_0 and the separation between the frequencies produced by the spectrum generator is ΔF, the spectrum generator frequencies required can be found by solving the following equation for $\tilde{f}_1$:

$$\sum_{j=0}^{2^k-1} (\tilde{f}_1 + j\,\Delta f) = f_0 \left(1 - \frac{1}{2^k}\right) 2^k = f_0(2^k - 1)$$

Then

$$\tilde{f}_2 = \tilde{f}_1 + \Delta F; \quad \tilde{f}_3 = \tilde{f}_2 + \Delta F = \tilde{f}_1 + 2\Delta F, \ldots, \tilde{f}_{2^k} = \tilde{f}_1 + (2^k - 1)\Delta F.$$

(b) If it is desired that the center frequency be 288 MHz and the overall bandwidth is 80 MHz, find the necessary spectrum generator frequencies and the final output frequencies for this single-stage synthesizer, given that it should utilize a set of three binary digits.

11.2-2 A direct frequency synthesizer is to be designed with a bandwidth of 48 MHz. Assuming 2 bits are to be used to control the spectrum generator, and we desire 16 frequencies:

(a) Find the center frequency and the output frequencies if the spectrum generator only produces frequencies from 72 to 108 MHz.

(b) If the minimum switching time is 2.5 μs, find the maximum hop rate.

(c) If the K code bits are fed to the spectrum generator switches sequentially, what would be the maximum PN code generator rate into the second stage if there is one hop per message bit?

11.2-3 The *ideal* mixer shown in Fig. 11-8 will produce the sum of the two input frequencies, but, in addition, it will also output the difference of the two.

(a) For the previous problem, list the spurious or difference frequencies that will exist at the output of the second synthesizer stage mixer.

(b) Note that the difference frequencies as found above will not overlap the desired band of frequencies. To remove the difference frequencies, we can filter them out with a bandpass filter. Assuming we desire to use a fifth-order Butterworth bandpass filter whose power transfer function is as shown, find the required half-power bandwidth, $B_{1/2}$, so that the nearest difference frequency is 40 dB down referenced from the center of the filter f_0.

$$|H(f)|^2 = \frac{1}{1 + \left[\dfrac{f - f_0}{B_{1/2}/2}\right]^{2n}} \qquad n = \text{order}$$

11.2-4 An indirect frequency synthesizer is used to hop 128 frequencies around in a pseudorandom fashion to obtain an overall bandwidth of 25.6 MHz. Given that the lowest output frequency desired is 43.4 MHz and the message bit rate is 3180 b/s after error-correction coding:

(a) Find the reference oscillator frequency f_1.

(b) What is the range of divisors required in the programmable frequency divider? What is the center frequency?

(c) If there are three hops per message bit, find the required PN code generator rate.

11.2-5 A frequency-hopping spread-spectrum system is to be constructed using an indirect frequency synthesizer, with the following given parameters:

$M = 2048$ frequencies
$B_S = 61.44$ MHz
Center frequency $= 375.015$ MHz
Final multiplication $= K = 15$

(a) Find the reference oscillator frequency f_1.

(b) Calculate the divider values necessary.

(c) If one divider stage requires 4 bits and, therefore, will divide up to $2^4 = 16$, how many cascaded dividers will be required?

11.2-6 In an effort to reduce the number of required divisions, we can modify the frequency synthesizer used in the previous problem, to the one shown here.

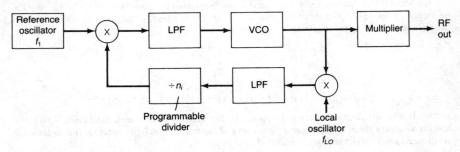

Assuming we desire the lowest divisor to be 27:

(a) Find the highest divider.

(b) Find the required local oscillator frequency, f_{LO}.

(c) Find the minimum required cutoff frequency for the low-pass filter in the divider path.

TWELVE
DETECTION OF SPREAD-SPECTRUM SIGNALS

In a spread-spectrum communication system, the PN codes used in a direct-sequence system or the hopping patterns used in a frequency-hopping system may have periods that are many times longer than the message bit duration. To receive such signals successfully, it is necessary that the receiver have a reference signal that is a replica of the transmitted signal and is in time synchronization with it. The establishment of the communication link under these circumstances involves two phases. The first phase is the acquisition phase, in which the receiver adjusts its time so that its own reference signal is within one time chip of being in step with the incoming signal. Having accomplished acquisition, the second phase, synchronization, involves a finer adjustment and maintains the local reference signal in step with the desired received signal. This chapter discusses both operations for both direct-sequence systems and frequency-hopped systems. The problem of synchronization is considered first.

12-1 COHERENT DIRECT-SEQUENCE RECEIVERS

A block diagram of a typical coherent receiver for direct-sequence signals is shown in Fig. 12-1.[2] This block diagram involves three major portions. First is the carrier tracking loop in the upper right-hand corner of the diagram. Next is the code tracking loop, which is the major portion of the bottom of the diagram. Finally there is provision for accommodating Doppler shift in the received signal, which is done by a combination of mixers and the carrier tracking loop. To illustrate the performance of this receiver, consider the incoming signal to have a frequency of f_0, the nominal center frequency, plus some unknown Doppler shift f_d. This signal is first mixed with a local oscillator whose frequency is $f_0 + f_1$ and the difference signal, at a frequency of $f_1 - f_d$, goes into the carrier tracking loop. Because of the feedback in the carrier tracking loop, the voltage-controlled oscillator (VCO) in that loop also establishes a frequency of $f_1 - f_d$. The incoming signal also goes into two other mixers, each of which has a local oscillator at a frequency of $f_0 + f_1$. The outputs of these mixers, at a frequency of $f_1 - f_d$, go into bandpass filters, centered at a frequency of f_1, but having a bandwidth great enough to accommodate

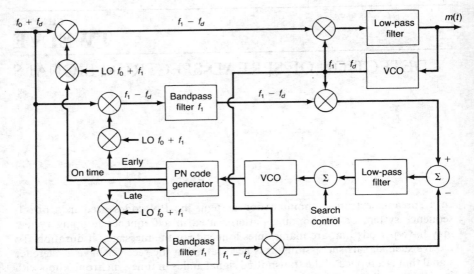

Figure 12-1 Coherent direct sequence receiver.

the Doppler shift. The output of the bandpass filters is then mixed with the output of the carrier tracking loop.

The two local oscillators are further modulated by outputs from the PN code generator. One output is producing a code that is early in time and one output is producing a code that is late in time. If the receiver PN code generator is exactly at the time of the received signal, the two signals coming out of the bandpass filters are identical and the result is a zero error signal into the low-pass filter that is controlling the voltage-controlled oscillator that provides the clock function for the PN code generator. However, if the PN code generator is not at quite the proper time, there will be an error signal produced and the polarity of this error signal depends upon whether the local code generator is early or late. The nature of this error signal is illustrated in Fig. 12-2.

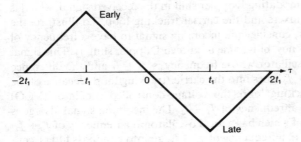

Figure 12-2 The delay-lock-loop error signal.

The operation described assumes that the local code generator is within one time chip of the received signal; that is, acquisition has already taken place. During the acquisition phase, which is discussed in more detail later, coherent detection is not possible and it is necessary to insert envelope detectors following the bandpass filters.

Once acquisition has been accomplished, these envelope detectors may be eliminated and the circuit becomes as shown in Fig. 12-1. This portion of the circuit is referred to as a delay-lock loop, and is one of the ways in which a receiver is made to track an incoming direct-sequence signal.

An analysis of the delay-lock loop is made in a subsequent section. However, before doing that, it is worth considering other methods for accomplishing carrier tracking.

12-2 OTHER METHODS OF CARRIER TRACKING

The carrier tracking loop shown in Fig. 12-1 is only one method of obtaining a signal that is locked to the incoming carrier. Another method of doing this is the so-called squaring loop, which is shown in Fig. 12-3 (ref. 3, p. 121). The squaring loop is useful when the carrier is modulated with biphase phase-shift keying, because the square of such a signal has no modulation at all. To see this, consider the output of the squaring circuit in Fig. 12-3. If the input is modeled as a sinusoid with phase modulation, the output is

$$y(t) = x^2(t) = A^2 \sin^2[\omega_s t + \psi(t)] + 2An(t)\sin[\omega_s t + \psi(t)] + n^2(t)$$

$$= \frac{A^2}{2} - \frac{A^2}{2} \cos[2\omega_s t + 2\psi(t)] + \text{(noise terms)} \tag{12-1}$$

Note in the second term of this equation that the phase has been multiplied by 2, thus if the phase is either zero or π, twice that value is either zero or 2π, which is exactly the same as zero. The phase modulation has been removed by the squaring operation. The net result is that a normal carrier tracking loop can be established involving a low-pass filter and a voltage-controlled oscillator at twice the carrier frequency without having to contend with the presence of phase modulation. If

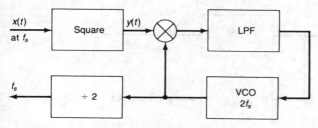

Figure 12-3 The squaring loop for carrier tracking.

the output of the VCO is then divided by 2, it produces an unmodulated carrier frequency. When there is noise present at the input to the squaring loop, there is a resulting random phase modulation of the output signal. The variance of this phase variation can be shown to be

$$\sigma_\phi^2 = \frac{N_0 B_L \mu}{A^2/2} \tag{12-2}$$

in which

$$\mu = 1 + \frac{N_0 B_s}{A^2}$$

$$B_L = \text{loop bandwidth} \ll B_s$$

Note that the variance of the phase becomes small as the input SNR becomes large. One of the disadvantages of the squaring loop as compared with the carrier tracking loop shown in Fig. 12-1 is that it does not produce a demodulated message output. This disadvantage can be circumvented by means of the Costas loop.

The Costas loop, which is sometimes also called a quadrature loop, is shown in Fig. 12-4 (ref. 3, p. 129). In this loop the incoming signal is mixed with the output of the VCO, both before its phase is shifted and after its phase is shifted by 90 degrees. These two outputs are then filtered, multiplied together, filtered again,

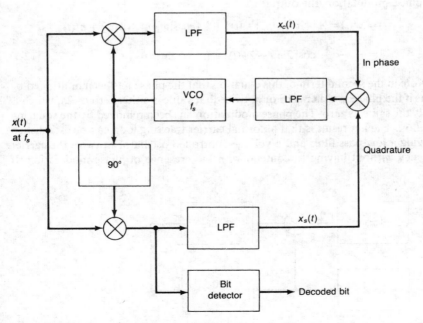

Figure 12-4 The Costas loop for carrier tracking.

and used to control the frequency of the voltage-controlled oscillator. The decoded message bit can be taken from one of the mixer outputs.

To see how this circuit works, consider the outputs from the two low-pass filters. These are

$$x_c(t) = A \cos \psi(t) + n_c(t) \tag{12-3a}$$

$$x_s(t) = A \sin \psi(t) + n_s(t) \tag{12-3b}$$

When these two outputs are multiplied together, the product is

$$x_s(t)x_c(t) = \frac{A^2}{2} \sin 2\psi(t) + n_{eq}(t) \tag{12-4}$$

Note that this product term involves the sine of twice the phase angle. Hence as the phase changes from 0 to π, the double angle changes from 0 to 2π, and so does not contribute a variation to the voltage-controlled oscillator.

The noise into the VCO does produce a random phase variation and the variance of this phase can be shown to be the same as it is for the squaring loop. The advantage of the Costas loop over that of the carrier tracking loop shown in Fig. 12-1 is that the message output is taken before filtering and, therefore, is not distorted by the low-pass filter in the feedback portion of the loop.

12-3 DELAY-LOCK LOOP ANALYSIS

In this section we consider a simplified analysis of the performance of the delay-lock loop. To do this we look first at a simplified model for this loop as shown in Fig. 12-5.[2] In this model τ_d is the amount by which the early and late codes from the code generator are delayed and advanced relative to the on-time code. The remainder of the loop is simply a low-pass feedback circuit in which the error

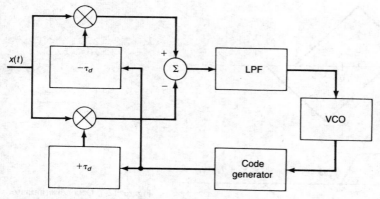

Figure 12-5 Simplified model of the delay-lock loop.

signal is filtered with a low-pass filter, and then used to control the frequency of a VCO, which in turn serves as a clock for the code generator.

For purposes of analysis, it is convenient to represent the loop of Fig. 12-5 in equivalent mathematical form as shown in Fig. 12-6. Note that the input, the error signal, and the feedback signal in this loop are shown as times rather than as voltages. This is so because of the linear relationship between delay and voltage that occurs in the error signal as shown in Fig. 12-2. Thus it is possible to represent an error voltage in terms of the error in time. The quantity A in Fig. 12-6 is the gain of the delay discriminator and the quantity K is the gain of the low-pass filter.

The voltage-controlled oscillator has a transfer characteristic of $1/j\omega$ since it acts as an ideal integrator, and $F(\omega)$ is the transfer function that corresponds to the low-pass filter. The shape of the delay discriminator characteristic depends upon the magnitude of τ_d compared with the chip duration t_1. This is illustrated in Fig. 12-7.

This figure is drawn for a situation in which τ_d is smaller than t_1 and hence the discriminator characteristic has a flat top. However, the slope of the discriminator characteristic as it passes through 0 is independent of the value of τ_d and this slope

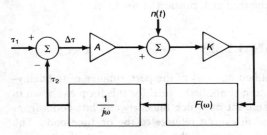

Figure **12-6** Equivalent linear mathematical model of the delay-lock loop.

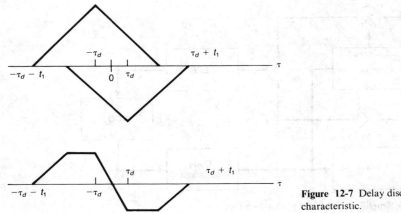

Figure **12-7** Delay discriminator characteristic.

is simply $2/t_1$. Therefore, the parameter A is[2]

$$A = \frac{2\sqrt{P_r}}{t_1} \tag{12-5}$$

The transfer function that relates the error in delay to the delay of the input signal is obtained from normal feedback circuit analysis as

$$H_e(\omega) = \frac{j\omega}{j\omega + AKF(\omega)} \tag{12-6}$$

In a similar way, the transfer function that relates the noise in the error signal to the input noise is

$$H_n(\omega) = \frac{KF(\omega)}{j\omega + AKF(\omega)} \tag{12-7}$$

The equivalent noise bandwidth of the loop is related to the area under the square of this transfer function. Thus this bandwidth becomes

$$B_L = \frac{A^2}{2\pi} \int_0^\infty |H_n(\omega)|^2 \, d\omega \tag{12-8}$$

The random error in delay due to the noise now has a variance given by

$$\sigma_{\Delta\tau}^2 = \frac{1}{2\pi} \int_{-\infty}^\infty N_0 |H_n(\omega)|^2 \, d\omega$$

$$= \frac{2N_0 B_L}{A^2} = \frac{N_0 B_L t_1^2}{2P_r} \tag{12-9}$$

It is convenient to relate the standard deviation of this error to the magnitude of the chip duration t_1 as

$$\frac{\sigma_{\Delta\tau}}{t_1} = \sqrt{\frac{N_0 B_L}{2P_r}} \tag{12-10}$$

The delay-lock loop will not lose lock as long as the error in delay remains smaller than t_1. However, if the error in delay becomes larger than t_1, then lock may be lost and it may be necessary to reacquire the signal. Therefore, it is of interest to evaluate the probability of this occurrence. This is simply the probability that $\Delta\tau$ is greater than t_1 and is given by

$$Pr[\Delta\tau > t_1] = 2Q\left[\frac{t_1}{\sigma_{\Delta\tau}}\right]$$

$$= 2Q\left[\sqrt{\frac{2P_r}{N_0 B_L}}\right] \tag{12-11}$$

Note that the quantity involved in the radical in the Q function is simply the SNR in the equivalent noise bandwidth of the loop. It is desirable, therefore, to make the loop bandwidth as small as possible. However, when the loop bandwidth is made small and the loop becomes slow in responding to changes in input time—if, for example, the receiver is moving with respect to the transmitter so that the delay is changing—it may not be possible for the loop to follow such a change. Thus the actual choice of equivalent noise bandwidth for the loop is a compromise between the desire to minimize the probability of losing lock and the desire to allow the loop to follow changes in input delay.

12-4 TAU-DITHER LOOP

The delay-lock loop discussed in the preceding section is a fairly complicated device that involves the generation of PN codes with three different times, several mixers, and several filters. A somewhat simpler method of performing the same tasks can be accomplished by the tau-dither loop as shown in Fig. 12-8 (ref. 3, p. 494).

In this case the voltage-controlled oscillator that serves as a clock for the PN code generator is phase modulated by a square wave at some low frequency f_τ. As a result of this phase modulation, the relative time of the PN code generator shifts back and forth between two values. As a result of this shift, the output of the correlator is amplitude-modulated in accordance with the square-wave generator. When the two values that the PN code generator switches between are equally spaced around the desired time, the two outputs have the same magnitude as shown in Fig. 12-9.

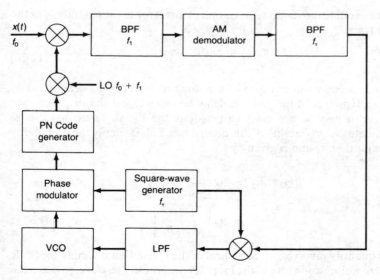

Figure 12-8 The tau-dither loop for delay locking.

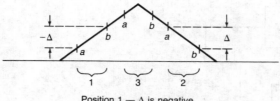

Position 1 — Δ is negative
Position 2 — Δ is positive
Position 3 — Δ is zero

Figure 12-9 Relative outputs of the tau-dither loop.

However, when the two outputs are both on one side of the desired time, then the difference will be either positive or negative, as also shown in Fig. 12-9. Thus by determining the polarity of the change in amplitude, it is possible to determine the direction of the error and to use this to produce an error signal that can drive the VCO to the proper position and maintain it there.

The amount by which the PN code generator is modulated is normally quite small—on the order of one tenth to one fifth of a chip duration. The tracking error of a tau-dither loop is somewhat larger than that of a delay-lock loop, normally by something on the order of 1 to 2 dB. Also, when the loop is properly tracking, the correlation is never at its peak value and the loss in correlation will be on the order of $\Delta\tau/2t_1$. As an example, suppose that $\Delta\tau/t_1$ is 0.1; in this case the loss is on the order of 0.5 dB.

12-5 COHERENT CARRIER TRACKING

A third and still simpler method of accomplishing code tracking is to make the carrier frequency a multiple of the code rate. One method of doing this is illustrated in Fig. 12-10.[1] Note that the reference frequency that is driving the code generator is multiplied by a factor n to become the carrier frequency, and this is then multiplied by the output of the code generator to produce the modulated PN code.

The receiver for such a signal is shown in Fig. 12-11. Note that there is also a frequency reference in the receiver that may be a voltage-controlled oscillator and the output of this frequency reference controls both the code generator and is multiplied by a factor n to produce a coherent carrier reference. The product of the

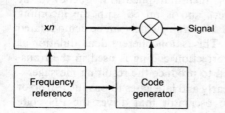

Figure 12-10 Transmitter for coherent carrier tracking.

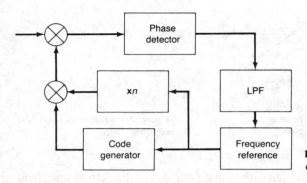

Figure 12-11 Receiver for coherent carrier tracking.

locally generated signal and the incoming signal is then detected in a phase detector to produce the error signal.

There are several problems that may arise in this type of carrier tracking. First, there may be a phase ambiguity that results from locking to a frequency that is related to the code rate, but is different from it. This normally can be resolved by means of a + or − 1-bit correlation search, but it does involve additional hardware. The second problem is that coherence between the clock rate and the carrier may be destroyed by multipath. Under these conditions the receiver may lose lock.

Although this method of receiving signals will track Doppler shifts, coherence between the code and the carrier may be lost if the Doppler is excessive. Because of the difficulties mentioned, this method of tracking carrier and code may be useful only in special circumstances.

12-6 NONCOHERENT FREQUENCY-HOP RECEIVER

As noted previously frequency-hop signals usually must be received in a noncoherent fashion even if they are transmitted coherently. This is true because the characteristics of the channel may destroy any phase coherence that existed among the various transmitted frequencies.

A diagram of a typical noncoherent frequency-hopping receiver is shown in Fig. 12-12.[2] Note that this receiver contains a digital frequency synthesizer that is driven by $m - 1$ bits; that is, one less bit than the number of bits that was used to generate the signal in the first place. This is then multiplied in frequency (if the transmitter employs a frequency multiplier), and is mixed with the incoming signal. The output of the mixer is a normal FSK signal, which can be demodulated to recover the message in the usual fashion. This is noncoherent demodulation as discussed in a previous chapter. If error-correction coding is used in the transmitter, the corrections must now be applied to produce the resulting message.

The FSK signal can also be used with early and late gates to produce an error signal that controls the voltage-controlled oscillator that drives the PN code

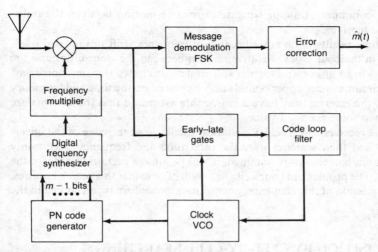

Figure 12-12 Noncoherent frequency-hopping receiver.

generator. The operation of the early and late gates relies upon the fact that if the code generator is not in step with the received signal, the correct frequency will occur at the beginning of a message bit (if the error is one direction), and at the end of a message bit (if the error is in the other direction). This information can be used to generate the error signal.

A detailed analysis of the operation of the noncoherent frequency-hopping receiver is very difficult and will not be considered here.

A further question that arises in connection with this receiver is whether it can follow Doppler shifts. The means for doing this is not shown in Fig. 12-12, but it can be accomplished by adding a circuit that is analogous to the carrier tracking loop at the point where the FSK signal arises.

12-7 ACQUISITION OF SPREAD-SPECTRUM SIGNALS

Now that methods of maintaining synchronization between the locally generated signal and the received signal have been discussed, it is appropriate to consider ways in which the synchronization is accomplished in the first place. This is known as the acquisition process. To accomplish acquisition the receiver must have some knowledge concerning the nature of the received signal. It is assumed here that the receiver knows exactly the PN sequence that is to be received or the frequency-hopping pattern of the signal. In addition the receiver must have an approximate knowledge of the time associated with the code clock of the received signal and of the frequency associated with the received signal. Both the time and frequency may be in error, either because there has been some appreciable elapsed time since the last communication, or because of relative motion between the transmitter and the receiver.

Elapsed time results in an error because of frequency drift in the local oscillator or drift in the code clock. Relative motion between the transmitter and the receiver results in an unknown Doppler shift and an unknown transmission delay. Under most circumstances upper bounds may be placed on the time and frequency errors so that the receiver may have a reasonable assurance that these errors are not greater than some specified values.

Thus if the receiver sets its clock and local oscillator at the edge of the uncertainty region and then searches over the total time and frequency uncertainty until a maximum correlation is obtained, it will be almost certain to acquire the desired signal. The problem, of course, is the length of time that this search requires. A detailed discussion of this elementary acquisition procedure is considered in the next section.

12-8 ACQUISITION BY CELL-BY-CELL SEARCHING

It is convenient to represent the time–frequency uncertainty region as shown in Fig. 12-13.[2] The maximum time uncertainty is denoted as plus or minus ΔT and the maximum frequency uncertainty as plus or minus ΔF. These maximum values can be specified in terms of the parameters of the system. The maximum time uncertainty may be represented as[4]

$$\Delta T = \Delta t + \frac{\Delta R}{c} + F_c T + \frac{1}{2} F_c' T^2 + n(T) \qquad (12\text{-}12)$$

where

$$\Delta t = \text{initial time offset}$$

$$\Delta R = \text{range uncertainty}$$

$$c = 3 \times 10^8 \text{ m/s}$$

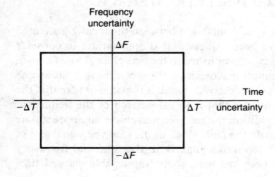

Figure 12-13 The time–frequency uncertainty region.

$$F_c = \frac{\Delta f_c}{f_c}, \text{ fractional clock offset}$$

$T = $ elapsed time since last synchronization

$$F'_c = \frac{\Delta \alpha_c}{f_c}, \text{ fractional clock drift}$$

$n(T) = $ random component of clock time

This representation is appropriate only over elapsed time intervals in which the fractional clock drift F'_c may be considered constant. Over long time intervals, this fractional clock drift is also a random quantity and its contribution is included in $n(T)$. The point here is that ΔT does not in fact become unbounded as T becomes large. However, for the values of elapsed time that are normally of interest in a communication system, this representation is adequate. In a similar manner, the maximum frequency uncertainty can be represented as

$$\Delta F = \Delta f + \frac{\Delta v f_0}{c} + F'_0 f_0 T + m(T) \tag{12-13}$$

where

$\Delta f = $ initial frequency offset

$\Delta v = $ range-rate uncertainty

$f_0 = $ nominal center frequency

$$F'_0 = \frac{\Delta \alpha_0}{f_0}, \text{ fractional frequency drift}$$

$m(T) = $ random component of frequency

To illustrate the use of these equations, consider a system in which the various parameters are assumed to have the following values.

$$\Delta t = 6 \ \mu s$$

$$\Delta R = 70 \text{ km}$$

$$F_c = 10^{-5}$$

$$F'_c = 10^{-10}$$

$$\Delta f = 1000 \text{ Hz}$$

$$\Delta v = 300 \text{ m/s}$$

$$f_0 = 500 \text{ MHz}$$

$$F'_0 = 10^{-10}$$

These parameters might arise in a ground-to-air radio system in which relatively poor crystal clocks are used to control the frequencies. Both $n(T)$ and $m(T)$, the random components, are neglected. The maximum time uncertainty and the maximum frequency uncertainty both are computed by inserting these parameters into the equations given above. Elapsed times of 1 hour and 24 hours are assumed and the results are as follows:

T (hours)	ΔT (seconds)	ΔF (Hz)
1	0.037	1680
24	1.25	5820

Note that when the elapsed time is 1 hour, which may well be a reasonable value, the maximum time uncertainty is 37 ms. While this may seem to be a very small value, in a direct-sequence system running at 10 Mb/s, it corresponds to 370,000 time chips. Obviously searching over this many time chips in order to establish the correct location for the local code could be a very time-consuming operation indeed. It is probably not realistic to use the above equations for an elapsed time of 24 hours, but this value is included simply to demonstrate that the time uncertainty increases drastically with the elapsed time.

The time required to search over the time–frequency uncertainty depends upon the area of this region that can be examined at any given time. This area is conveniently specified in terms of the signal ambiguity function. The signal ambiguity function is defined as

$$A(\tau, \beta) = \frac{1}{Ti} \int_0^{Ti - \tau} s(t)s(t + \tau)e^{j2\pi\beta t} \, dt \tag{12-14}$$

and represents the response of the correlator to a signal that is delayed by τ seconds from the time reference in the correlator and shifted in frequency by β hertz from the reference signal in the correlator.

During the acquisition phase, neither the phase nor the frequency of the received signal is known and, therefore, noncoherent detection is essential. Under these circumstances it is the envelope of the signal ambiguity function that is significant. This envelope, designated as $\text{Env}[A(\tau, \beta)]$, represents the output from a noncoherent receiver in which the reference signal differs from the incoming signal in both time and frequency by τ and β respectively.

The nature of the signal ambiguity function is illustrated in Fig. 12-14 as a

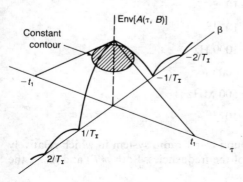

Figure 12-14 Ambiguity function for spread-spectrum signals.

surface over the (τ, β) plane. Since the integration in Eq. (12-14) takes place over one time chip only, the ambiguity function is zero outside of this range. However, it may extend to plus and minus infinity in the β direction. Note that along the β axis the ambiguity function envelope goes to zero at frequencies that are multiples of the reciprocal of the integration time. This integration time will be equal to the message bit duration in a direct-sequence system, and equal to one frequency-hop duration in a frequency-hopping system.

Under the assumption of a constant envelope, Eq. (12-14) is readily evaluated for PN signals to yield an envelope of[2]

$$\text{Env}[A(\tau, \beta)] = \left| \frac{\sin \pi \beta T_I \sin \pi \beta(t_1 - |\tau|)}{\pi \beta T_I \sin \pi \beta t_1} \right| \qquad |\tau| \leq t_1$$

$$= 0 \qquad |\tau| > t_1 \qquad (12\text{-}15)$$

in which T_I is the integrator time. In a similar manner, when a frequency-hopping signal is employed, the envelope of the ambiguity function is given by[2]

$$\text{Env}[A(\tau, \beta)] = \left| \frac{\sin \pi \beta(t_1 - |\tau|)}{\sin \pi \beta t_1} \right| \qquad |\tau| \leq t_1$$

$$= 0 \qquad |\tau| > t_1 \qquad (12\text{-}16)$$

It may be recalled that the envelope of the ambiguity function represents the output of a correlator for a noncoherent receiver. For this output to be detected and thereby establish the existence of a signal, it is necessary that it be sufficiently large that it not be confused with noise. Therefore, the size of the search cell in the uncertainty region is determined by the area of a contour of constant amplitude on the envelope of the ambiguity function picked at some level that is reasonably close to the maximum value. For purposes of the analysis here, this level will be assumed to be at a point 0.707 times the peak value of the ambiguity function. This contour of constant amplitude has dimensions in the τ direction and the β direction that can be determined readily from the equations previously given. For example, for PN signals the value in the τ direction is given by

$$\text{Env}[A(\tau_1, 0)] = 1 - \frac{|\tau_1|}{t_1} = 0.707$$

which leads to a value of τ_1 of

$$\tau_1 = 0.293 t_1 \qquad (12\text{-}17a)$$

Similarly, in the β direction, the envelope is

$$\text{Env}[A(0, \beta_1)] = \frac{\sin \pi \beta T_I}{\pi \beta_1 T_I} = 0.707$$

which leads to a value of β_1 of

$$\beta_1 = \frac{0.443}{T_I} \qquad (12\text{-}17b)$$

In an analogous manner, the equation for the frequency-hop signal can be solved to yield

$$\tau_1 = 0.293 t_1 \tag{12-18a}$$

$$\beta_1 = \frac{0.443}{t_1} \tag{12-18b}$$

We consider first the acquisition search in a direct-sequence system. A simplified block diagram of this procedure is shown in Fig. 12-15.[2]

The output of the mixer and bandpass filter is passed through an envelope detector and then an integrate-and-dump circuit, in which the integration time is equal to one message bit or less. The output of the integrate-and-dump circuit is compared with a threshold, and if the threshold is not exceeded, the code is stepped to the next position. If the threshold is exceeded, the search procedure is terminated and the receiver is synchronized by means of the delay-lock loop, as discussed previously. To quantify the acquisition time, define a search rate as r chips per second; therefore, the time required to search over a total time uncertainty of $2\,\Delta T$ is

$$T_t = \frac{2\,\Delta T}{r t_1}$$

The amount by which the code position is changed from one step to the next is normally one half of a chip. This ensures that the maximum will not be missed by taking too large a step. Thus if there are two steps per chip, the search rate will be $1/2 T_I$ where T_I is the integration time of the integrate-and-dump circuit. The total time required to search for one frequency is, therefore

$$T_t = \frac{4 T_I \, \Delta T}{t_1} \tag{12-19}$$

The search may have to be carried out at several frequencies, however. The number of such frequencies is determined by

$$N_f = \left[\frac{2\,\Delta F}{2\beta_1} \right] + 1 = \left[\frac{\Delta F T_I}{0.443} \right] + 1 \tag{12-20}$$

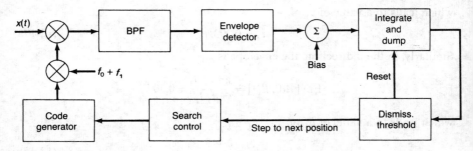

Figure 12-15 Receiver model for direct-sequence acquisition search.

where the square brackets imply the integer part of the quantity inside the brackets. This is required because the number of frequencies searched must be an integer.

The total maximum acquisition time is simply the product of the time required to search at one frequency and the number of frequencies to be searched. It thus becomes

$$T_A = N_f T_t = \frac{4 T_I \, \Delta T}{t_1} \left\{ \left[\frac{\Delta F T_I}{0.443} \right] + 1 \right\} \tag{12-21}$$

The required integration time T_I depends upon the desired probability of acquisition. If T_I is too small, there is always a chance that the peak will be missed or misinterpreted as noise. If T_I is too large, the acquisition time becomes large. A typical rule of thumb for determining the appropriate acquisition time is[2]

$$\frac{P_r T_I}{N_0} \geq 10 \tag{12-22}$$

To evaluate the acquisition time in a typical situation, consider a direct-sequence system in which the chip rate is 10^7 chips per second and the message bit rate is 20,000 b/s. If it is desired to achieve an output SNR of 14 dB, then

$$(\text{SNR})_0 = 14 \, \text{dB} = 25$$

$$= \frac{2 P_r t_m}{N_0}$$

from which the required ratio of received signal power to noise spectral density can be obtained as

$$\frac{P_r}{N_0} = \frac{(\text{SNR})_0}{2 t_m} = \frac{25 \times 20000}{2} = 2.5 \times 10^5$$

Using this value in Eq. (12-22) leads to an integration time of

$$T_I \geq \frac{10}{P_R/N_0} = \frac{10}{2.5 \times 10^5} = 4 \times 10^{-5}$$

If the previously determined values of ΔT and ΔF are assumed, the resulting acquisition time is as follows:

T (hours)	ΔT (seconds)	ΔF (Hz)	N_f	T_A (seconds)
1	0.037	1680	1	59
24	1.25	5820	1	2000

It is clear from this result that allowing the clock to go as long as 24 hours without being updated is not acceptable, and even permitting 1 hour between updates leads to an acquisition time that is almost a minute in length. However, acquisition times of the order of 100 seconds or more are not unusual in direct-sequence systems.

Next consider the acquisition search procedure for a frequency-hop system.

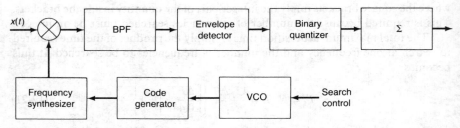

Figure 12-16 Receiver model for frequency-hopping acquisition search.

A block diagram of the receiver that might be used for this is shown in Fig. 12-16.[2] Again the output of the mixer and bandpass filter is passed through an envelope detector. The output of the envelope detector is compared with a threshold and the number of times the threshold is exceeded is summed in a counter. The value of this count is used to decide whether a signal is present, and the search should be terminated, or a signal is not present and the code generator should be stepped to the next position.

Since frequency hopping is an avoidance system, the SNR during each hop is relatively high. However, errors in counts can arise if there are interference frequencies that occur at the same place as one of the hop frequencies. These interference frequencies may add in phase and thus not cause a problem, or they may add out of phase and thus reduce the signal level to the point where a count does not occur. This must be taken into account in evaluating the performance of the system.

If the number of hops that are summed over in the counter is designated as C, then the search rate is

$$r = \frac{1}{Ct_1} \quad \text{hops per second}$$

The required value of C depends upon the desired probability of acquisition. This probability can be expressed as

$$P_A = \sum_{i=C_t}^{C} \binom{C}{i}(1-\alpha)^i\alpha^{C-i} \tag{12-23}$$

where

$$\binom{C}{i} = \text{the binomial coefficient}$$

$$C_t = \text{threshold to decide signal present}$$

$$\alpha = \text{fraction of the frequencies occupied by interference}$$

The threshold value appearing in Eq. (12-23) depends upon the desired probability of false alarm. This probability is given by

$$P_F = \sum_{i=C_t}^{c} \binom{C}{i} \alpha^i (1 - \alpha)^{C-i} \tag{12-24}$$

Since there is no desired signal present, all of the interfering signals will produce outputs.

In a typical situation, the values of C and C_t may be

$$\alpha = 0.1, \quad P_A = 0.984, \quad P_F = 0.001, \quad C = 6, \quad C_t = 4$$

Curves of the probability of acquisition and the probability of false alarm as a function of the parameter α are displayed in Fig. 12-17. Note that α must be fairly small in order for the system to yield satisfactory performance.

The acquisition time for a frequency-hop system can be computed in a manner similar to that for the direct-sequence system. In particular the time required to search the total time uncertainty of $2\,\Delta T$ at one frequency is

$$T_t = 2C(\Delta T) \tag{12-25}$$

Correspondingly, the number of frequencies that must be searched over is

$$N_f = \left[\frac{2\,\Delta F}{2\beta_1} \right] + 1 = \left[\frac{\Delta F t_1}{0.443} \right] + 1 \tag{12-26}$$

Hence the total maximum acquisition time is

$$T_A = N_f T_t = 2C(\Delta T) \left\{ \left[\frac{\Delta F t_1}{0.443} \right] + 1 \right\} \tag{12-27}$$

As an example consider a system in which the message bit rate is 20,000 b/s and there is one frequency hop per bit. Therefore, $t_1 = 1/20,000$. If $C = 6$ as described, and if ΔT and ΔF have the same values as previously computed, the resulting

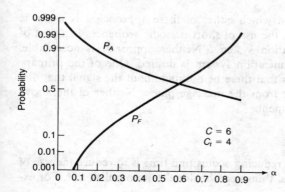

Figure 12-17 Probabilities of acquisition and false alarm.

acquisition time becomes

T (hours)	ΔT (seconds)	ΔF (Hz)	N_f	T_A (seconds)
1	0.037	1680	1	0.444
24	1.25	5820	1	15

Note that the acquisition time has been drastically reduced as compared with that for the direct-sequence system. This is one of the major advantages of frequency hopping.

12-9 REDUCTION OF ACQUISITION TIME

It was noted in the previous section that the cell-by-cell search technique yields extremely large acquisition times in direct-sequence systems, and relatively large acquisition times in frequency-hop systems. Often it is desirable to make the acquisition time much less than any of the values quoted previously. This is particularly true in a system in which "push-to-talk" operation is required. Therefore, it is desirable to consider techniques that might be used to reduce the acquisition time. Several such techniques are possible. In general these may be classified as either acquisition-aiding waveforms or acquisition-aiding techniques. Both are discussed here.[2]

Acquisition-Aiding Waveforms

The philosophy behind the use of acquisition-aiding waveforms is to modify the transmitted signal during the acquisition phase to a form that makes it easier to acquire it. There are at least two ways in which this can be done.

First, the signal can be changed so that the ambiguity function is wider. That is, the size of the cell is increased.

The second way is to use periodic short sequences so that the initial search only has to take place over a short period associated with each sequence. These sequences can then be modulated to give information as to the exact time of the code.

The number of situations in which either of these approaches is useful is somewhat limited. For example, the use of short periodic sequences is a part of certain spread-spectrum navigation systems. Neither approach is acceptable, however, when a secure communication system is desired. One of the primary requirements of such a system is that there be nothing about the signal that distinguishes the acquisition phase from the message phase. Neither of the above approaches satisfies that requirement.

Acquisition-Aiding Techniques

The most obvious technique for reducing acquisition time is to reduce the size of the region of uncertainty. Several things can be done to accomplish this. For example, the time uncertainty ΔT can be made smaller by reducing either Δt or ΔR when T is small. These quantities contribute the most to the time uncertainty in this situation. However, if T is large, then F_c and F_c' are the most important contributions and the technique would then be to improve the characteristics of the clock.

When T is small, Δf and Δv are the most important contributions to the frequency uncertainty ΔF. These can be reduced by providing additional information as to the relative velocity of the transmitters and receivers or by having more accurate frequency control of the transmitter frequency. On the other hand, when T is large, the most important contribution to ΔF is F_0'. Reducing this quantity requires an oscillator with a lower drift rate.

A second technique for reducing acquisition time involves the use of multiple correlators. If, for example, we search in N_f different frequency paths simultaneously and over N_t time segments simultaneously, then the total time–frequency uncertainty region is divided into cells of size $2\,\Delta F/N_f$ by $2\,\Delta T/N_t$ as shown in Fig. 12-18. The total number of correlators required to achieve this is

$$N = N_f N_t$$

This results in reducing the acquisition time by a factor of approximately N. The price that must be paid for this, however, is a greatly increased complexity in the receiver because of the need for implementing N correlators that operate simultaneously. A somewhat simpler implementation is achieved if N_t is made equal to one, so that only one code generator is required, but the search is carried out over several frequencies simultaneously. This requires several local oscillators and mixers, but does not unduly increase the number of code generators.

The third technique that might be used to decrease acquisition time is the use of matched filters. Because this is a very important technique, it is discussed separately in a subsequent section.

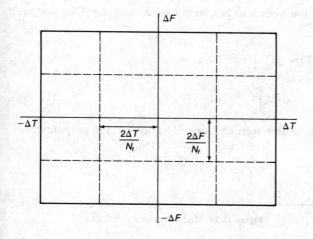

Figure 12-18 Partitioning the time–frequency uncertainty region with multiple correltors.

Finally there is the possibility of using frequency-estimation techniques to reduce the need for searching in frequency. In this case an estimate is made of the center frequency of the spectrum by performing a Fourier transform and, having identified the center frequency, searching only once in time.

Likewise, in a frequency-hopping system, it is possible to estimate the frequency of N successive hops for an N-stage code generator and then compute the position in the sequence. The difficulty with this is that the signal-to-interference ratio must be relatively high; perhaps greater than -15 dB. This might be achievable in a system operating in the presence of noise only, but it probably is not achievable in a system that must operate in the presence of other systems or intentional jamming.

12-10 ACQUISITION WITH MATCHED FILTERS

As noted in a previous chapter, the matched filter is a device that maximizes the SNR at a particular time. For purposes of the analysis here, the model that will be used for the matched filter is as shown in Fig. 12-19.[2] The noise at the input to the matched filter is assumed to be white with a one-sided spectral density of N_0 and the impulse response of the matched filter is

$$h(t) = s(T-t) \qquad 0 \le t \le T$$

The response of the matched filter to the signal is given by

$$m_0(t) = \sqrt{P_r} \int_0^T h(u)s(t-u)\, du$$

$$= \sqrt{P_r} \int_0^T s(T-u)s(t-u)\, du \qquad (12\text{-}28)$$

At the time at which the output SNR is to be maximized, namely, $t = T$, the output becomes

$$m_0(T) = \sqrt{P_r} \int_0^T s^2(T-u)\, du$$

$$= \sqrt{P_r} \int_0^T s^2(t)\, dt = \sqrt{P_r}\, T \qquad (12\text{-}29)$$

since $s(t)$ is normalized to have unit average power. The noise at the output of the

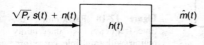

$$\sqrt{P_r}\, s(t) + n(t) \qquad \boxed{h(t)} \qquad \hat{m}(t)$$

Figure 12-19 Model for the matched filter.

matched filter is easily found to be

$$\sigma_n^2 = \frac{N_0}{2} \int_0^T h^2(u) \, du$$

$$= \frac{N_0}{2} \int_0^T s^2(T - u) \, du = \frac{N_0 T}{2} \tag{12-30}$$

Hence the maximum output SNR is given by

$$(\text{SNR})_0 = \frac{m_0^2(T)}{\sigma_n^2} = \frac{P_r T^2}{N_0 T/2} = \frac{2P_r T}{N_0} \tag{12-31}$$

Note that this is exactly the same value as obtained with an active correlator. The matched filter and the active correlator thus perform in exactly the same way in the presence of white noise.

The advantage of the matched filter is that there is no need to search in time. The matched filter will respond whenever the signal occurs and so time synchronization is not required. However, the receiver may have to search in frequency or use parallel filters if the frequency uncertainty is in excess of the frequency width of the ambiguity function. To determine whether or not a signal is present, the output of the matched filter is compared with the threshold. This operation is shown in Fig. 12-20. If the threshold γ is exceeded, it is assumed that a signal is present and the synchronization mode of operation is initiated. If the threshold is not exceeded, the system is allowed to continue to look for a signal or to search in frequency, if that is necessary.

The value of the threshold is a compromise between the desired probability of detection and probability of false alarm. These probabilities depend upon the SNR as shown in Fig. 12-21. It is clear, therefore, that to achieve a given probability

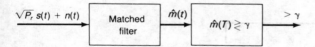

Figure 12-20 Threshold detection with a matched filter.

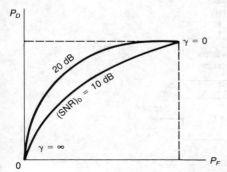

Figure 12-21 Receiver operating characteristic.

of detection and probability of false alarm, the threshold must be adjustable, since it is a function of the noise and interference power. One way of achieving this is to use a device known as a self-referencing matched filter detector.[2] The block diagram of this device is shown in Fig. 12-22. Note that the matched filter and envelope detector are paralleled by another filter whose impulse response is $g(t)$, and another envelope detector. The requirement on $g(t)$ is that it be orthogonal to $h(t)$ over the time interval from zero to T; that is,

$$\int_0^T g(t)h(t)\, dt = 0 \tag{12-32a}$$

If this is true, then any response out of $g(t)$ at time T must be due to interference rather than to desired signal. Hence this provides an independent measurement of the interference and noise and this, when multiplied by an appropriate gain G, can be compared with the output of the matched filter to decide whether a signal is present or not. It is also desirable that $g(t)$ satisfy the requirement

$$\int_0^T g^2(t)\, dt = 1 \tag{12-32b}$$

so that its gain be the same order of magnitude as that of the impulse response of the matched filter. The gain that follows the envelope detector can then be used to control the probability of false alarm.

While the use of the matched filter for acquisition is very desirable from the standpoint of reducing acquisition time, its application is actually quite limited. This is so because matched filters can be built only for waveforms that are of relatively short duration. It is possible to build matched filters for short PN sequences or for short frequency-hopped sequences, and both of these are con-

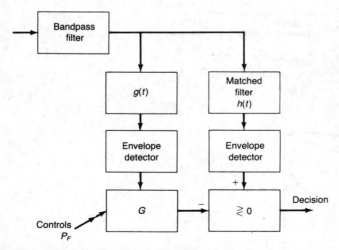

Figure 12-22 Self-referencing matched filter detection.

sidered in subsequent sections. However, the time–bandwidth product, and hence, the processing gain, that can be achieved with such filters is quite limited, and this limits their application.

12-11 MATCHED FILTERS FOR PN SEQUENCES

The matched filter for a PN sequence is most easily implemented as a tapped delay line, having essentially the same form as the tapped delay lines that were discussed previously as a means of generating PN sequences. The general form of such a filter is illustrated in Fig. 12-23 (ref. 3, p. 457).

As an illustration of this type of filter, consider again the specific case for $N = 7$. The waveform for this case is shown in Fig. 12-24. The impulse response of the matched filter must be

$$h(t) = s(7t_1 - t) \qquad 0 \le t \le 7t_1 \tag{12-33}$$

The form of the tapped-delay-line matched filter for this particular signal is shown in Fig. 12-25. Note that the tap weights in this filter are in reverse order from the input signal states. This is so because the part of the signal that arrives first must be delayed by the greatest amount, while the part of the signal that arrives last must be delayed by the least amount in order for all time chips to add together at the output.

Tapped delay lines for several hundred time chips, or even one or two thousand time chips, are feasible; however, for much longer signals it becomes very difficult to

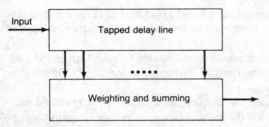

Figure 12-23 Matched filter for PN signals.

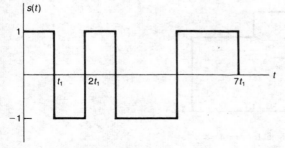

Figure 12-24 PN code for $N = 7$.

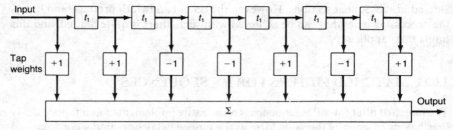

Figure 12-25 Tapped-delay-line matched filter for $N = 7$ PN code.

build a filter. It has been assumed up to now that the matched filter is designed for a specific center frequency f_0. If the signal is received in the presence of Doppler shift, then the received frequency is shifted by an amount f_d. The peak output of the matched filter is then reduced by the factor

$$\left| \frac{\sin \pi f_d T}{\pi f_d T} \right|$$

This factor becomes zero when the Doppler shift is equal to $1/T$ and is 0.637, or -3.92 dB, when the Doppler shift is $1/2T$. Thus when T is large and f_d is unknown, a search in frequency may still be required.

12-12 MATCHED FILTERS FOR FREQUENCY-HOPPED SIGNALS

It is also possible to use a tapped delay line to build a matched filter for a frequency-hopped signal.[5] As an illustration of this, consider the specific frequency-hopped signal shown in Fig. 12-26.

The tapped delay line has a bandpass filter connected to each tap; each bandpass filter being associated with a particular frequency component. This is illustrated in Fig. 12-27.

Note that the sequence in which the frequencies appear in the matched filter is opposite to the sequence in which the frequencies arise in the incoming signal.

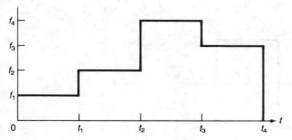

Figure 12-26 A frequency-hopped signal for $M = 4$.

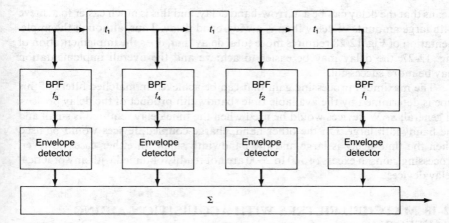

Figure 12-27 Tapped-delay-line matched filter for the $M = 4$ frequency-hopped signal.

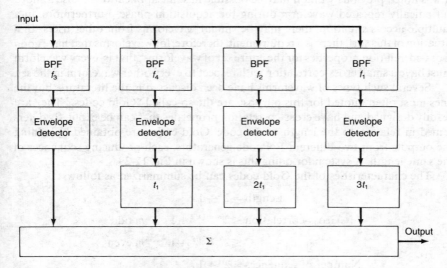

Figure 12-28 Alternative form for tapped delay line matched filter.

This is so for exactly the same reason as in the direct-sequence matched filter. The first portion of the signal must be delayed the most.

One of the difficulties with the matched filter shown in Fig. 12-27 is that the tapped delay line must have both a large bandwidth and a large time delay, and, therefore, it may not be possible to build a matched filter using this particular configuration. An alternative configuration simplifies this problem somewhat.[6] This alternative is shown in Fig. 12-28. Note that in this case the bandpass filters are introduced first and the delay is placed after the envelope detectors. This means that the delay can be a narrow-band delay, and this is much easier to achieve with large amounts of delay than is a wide-band delay. Thus, although the implementation of Fig. 12-28 requires more total delay than does the implementation of Fig. 12-27, the delay may be easier to achieve and the overall implementation may be more successful.

The maximum processing gain that can be achieved in matched filters of this type is determined by the available time–bandwidth product of the delay devices. In general, SAW devices would be used when the time delay required is small and the bandwidth large. On the other hand, charge-coupled devices would be used when the time delay is large and the bandwidth small. In either case, however, processing gains in excess of 500 to 2000 are not readily obtained with any practical delay devices.

12-13 MATCHED FILTERS WITH ACQUISITION-AIDING WAVEFORMS

It was noted previously that it may be possible in some applications to use a short, periodically repeated waveform during the acquisition phase. Furthermore, in a multiple-access situation, there may be similar waveforms from other users. In a situation of this sort, there is a requirement, therefore, for waveforms that have good cross-correlation properties for the entire set of waveforms; that is, every waveform must have a small cross-correlation with respect to every other waveform in the set.

Several such types of waveforms have been discussed in the literature, but the ones most often quoted for this purpose are the so-called "Gold codes." They are readily designed and have cross-correlation properties that can be explicitly determined in relation to the length of the code. Gold codes are obtained by adding the outputs from two different PN code generators, each producing sequences of the same length. A system for doing this is shown in Fig. 12-29.

The characteristics of the Gold codes can be summarized as follows:[7]

$$\text{Length} = 2^m - 1$$

$$\text{Cross-correlation} \le 2^{(m+1)/2} + 1 \qquad m \text{ odd}$$

$$\le 2^{(m+2)/2} + 1 \qquad m \text{ even}$$

$$\text{Number of sequences} = 2^m + 1$$

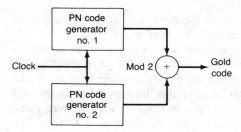

Figure 12-29 Implementation of Gold codes.

Note that the number of codes available may be quite large if m is large. Thus this is a useful technique for creating a large number of signals for a multiple-access spread-spectrum communication system. As an example of the cross-correlation properties of such codes, consider a simple case in which

$$m = 7, \quad N = 2^7 - 1 = 127$$

The maximum cross-correlation then becomes

$$\text{Cross-correlation} \leq 2^{(7+1)/2} + 1$$

$$\leq 17$$

This maximum cross-correlation may be compared with the peak value of auto-correlation of 127. The number of sequences obtained with this code is $2^7 + 1$ or 129.

Gold codes are widely used in multiple-access spread-spectrum communication systems, but the implementation of such codes is limited to fairly short sequences on the order of 1000 or so. However, they do provide the desired result of achieving a short acquisition time and providing a large number of signals guaranteed to have a minimal amount of mutual interference.

12-14 REFERENCES

1. Dixon, R. C.: *Spread Spectrum Systems*, John Wiley & Sons, Inc., New York, 1976, Chaps. 5, 6.
2. Cahn, C. R.: *Spread Spectrum Applications and State of the Art Equipments*, AGARD-NATO Lecture Series no. 58, Paper no. 5, Sec. 2.5, May–June 1973.
3. Holmes, J. K.: *Coherent Spread Spectrum Systems*, John Wiley & Sons, Inc., New York, 1982.
4. Barnes, J. A.: Atomic Timekeeping and the Statistics of Precision Signal Generators, *Proc. IEEE*, vol. 54, no. 2, pp. 207–220, February 1966.
5. Cooper, G. R., and R. W. Nettleton; A Spread Spectrum Technique for High Capacity Mobile Communications, *IEEE Trans. Vehic. Tech.*, vol. VT-27, pp. 264–275, November 1978.
6. Cooper, G. R., and D. P. Grybos: A Receiver Feasibility Study for the Spread Spectrum High Capacity Mobile Radio System, *Conf. Rec., 28th IEEE Vehicular Technology Conf.*, pp. 98–104, 1978.
7. Gold, R.: Optimal Binary Sequences for Spread Spectrum Multiplexing, *IEEE Trans. Inf. Theory*, pp. 619–621, October 1967.

12-15 PROBLEMS

12.3-1 A delay-lock loop system is used as a spread-spectrum receiver. Assuming the received signal power is 2×10^{-13} watts and the noise spectral density, N_0, is 5×10^{-17} W/Hz:

(a) Show that the transfer function that relates the noise in the error signal to the input noise is

$$H(j\omega) = \frac{KF(\omega)}{j\omega + AKF(j\omega)}$$

(b) If the low-pass filter as shown in Fig. 12-6 is just a constant—that is $F(j\omega) = 1$—find the value of the low-pass filter gain, K, so that the probability of loss of lock is 0.002. The PN code rate used is 50 Mb/s.

12.3-2 Assuming the received signal power and spectral density of the noise are the same as in the previous problem, as well as the PN code rate:

(a) Show that if the filter used in the previous problem is changed to a lead–lag network

$$F(s) \doteq \frac{cs+1}{bs} \qquad \begin{array}{l} c = 8.95 \times 10^{-4} \\ b = 1.8 \times 10^{-4} \end{array}$$

the transfer function reduces to the ratio of polynomials in s as

$$H_n(s) = \frac{K(sc+1)}{s^2 b + cAKs + AK} = \frac{(K/b)(sc+1)}{s^2 + (c/b)AKs + (AK/b)}$$

(b) Find the necessary gain of this filter if the probability of loss of lock is the same as in the previous problem (0.002).

12.8-1 A direct-sequence communication system uses a PN code running at 40 Mb/s on a carrier frequency of 1600 MHz. The transmission is from a geostationary satellite to an airborne receiver whose maximum velocity relative to the satellite is 1800 mi/h, either approaching or receding, and this relative velocity is unknown to the receiver. If serial search in both time and frequency is employed, the maximum search rate in time is 400 chips per second, and the initial time uncertainty is ± 100 μs, what is the maximum acquisition time?

12.8-2 A direct-sequence spread-spectrum system is to be used as a point-to-point data transmitter with the following parameters:

$\Delta t = 5 \times 10^{-6}$ seconds
$\Delta f_c = 1000$
$\Delta f = 2$ kHz
$f_o = 120$ MHz
Data bit rate $= 500$ b/s
PN code rate $= 10$ Mb/s
$F_c' = 1 \times 10^{-7}$
$F_0' = 1 \times 10^{-8}$
$(SNR)_t = -19$ dB

(a) Find the maximum time and frequency uncertainty if the elapsed time since the last transmission was 15 minutes. Repeat for 1 hour.

(b) Find the acquisition time if the elapsed time since the last transmission was that of part (a).

12.8-3 In the previous problem, the overall acquisition times were found to be unacceptable. Assuming the clocks were redesigned by using specially designed crystal ovens and the oscillators were made more stable with the result that the new parameters are as shown below, repeat the previous problem.

$F_c' = 1 \times 10^{-11}$
$F_0' = 1 \times 10^{-10}$
$\Delta f_c = 200$

12.8-4 In a particular frequency-hopping system, it is desired that the probability of false alarm be $P_F = 0.00865$. Assuming that eight hops are to be summed over in the counter:

(a) Find the threshold, C_t, if 20 percent of the desired frequencies are occupied by interference.

(b) Find the probability of acquisition, P_A.

12.8-5 For the frequency-hopping system described in the previous problem, find the acquisition time T_A, for which the following parameters apply. Compare with your answer for the direct-sequence case.

$\Delta t = 5 \times 10^{-6}$
$F_c = 2 \times 10^{-5}$
$F_c' = 1 \times 10^{-11}$
$T =$ elapsed time $= 15$ minutes
$\Delta f = 2000$ Hz
$F_0' = 1 \times 10^{-10}$
$f_0 = 120$ MHz
Data rate $= 500$ b/s with one hop per message bit

12.8-6 A frequency-hopping system has a bandwidth of 102.4 MHz centered at 600 MHz. A direct frequency synthesizer (with no frequency multiplier at its output) creates 4096 frequencies. The system is modulated with a message at 2500 b/s and there are 10 hops per message bit. If the maximum time uncertainty is 0.01 second, the maximum frequency uncertainty is 12,000 Hz, and the receiver detector output counter sums over 12 hops, what is the maximum acquisition time?

12.9-1 In the system of Problem 12.8-1, assume that serial search in time and parallel search in frequency are employed.

    (a) Find the number of filters in the parallel filter bank.

    (b) Find the maximum acquisition time.

12.9-2 A direct-sequence communication system uses a PN code running at 48 Mb/s on a carrier at 1600 MHz. The message bit rate is 2400 b/s and binary phase modulation is used for both the message and the code. Communication is between a fixed ground station and aircraft whose maximum velocity relative to the ground station is 300 m/s and the actual relative velocity is unknown. The maximum distance of the aircraft from the ground station is 500 km and this distance is known to within ± 20 km. Initial time offset, initial frequency offset, and errors due to clock frequency and local oscillator frequency may be neglected.

    During acquisition two correlators, each covering a different range of frequencies, are used in parallel and serial search in both time and frequency is employed in each correlator. If the search rate is 200 chips per second, what is the maximum acquisition time?

12.13-1 Two m sequences corresponding to $m = 3$ are shown here. Construct a Gold code from these two m sequences. What is the upper bound on the cross-correlation?

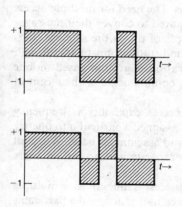

THIRTEEN

APPLICATION OF SPREAD SPECTRUM TO COMMUNICATIONS

13-1 GENERAL CAPABILITIES OF SPREAD SPECTRUM

Now that we have discussed methods for generating and detecting spread-spectrum signals, it is appropriate to consider some of the applications. There are many possible applications and only a few are presented here. One of the most important applications is usually referred to as multiple access. The need for multiple access arises when a number of independent users are required to convey their messages through a common facility. An outstanding example of this is the satellite communication system in which all messages from ground stations must pass through a common satellite repeater. Another such application is ground-based mobile communications in which mobile vehicles must communicate with a central base station.

The classical method of providing multiple-access capability is frequency division (FDMA). In frequency division each user is assigned a particular frequency channel, either on a permanent basis or on a demand assignment basis. When all channels are occupied, the system has reached its capacity and no further users may be accommodated.

A more recent technique for providing multiple access is time-division multiple access (TDMA). In time-division multiple access, each user is assigned a particular time slot within a time frame, and during this time slot, transmits a portion of a message by any standard digital technique. Time-division multiple access is an important application for satellite communications as well as ground-based digital communication systems. Again, however, when all time slots are occupied, the system is operating at capacity and no additional users can be accommodated.

The third general class of multiple-access communication systems is usually referred to as code-division multiple access (CDMA). Code-division multiple access is always accomplished by means of spread spectrum. In this system each user is assigned a particular code, that is, either a particular PN sequence or a particular frequency-hopping pattern. It is the fact that each user has a unique code that enables the messages to be separated at the receiving point. In contrast to the previously mentioned methods of multiple access, code-division multiple access does not have any sharply defined system capacity. As the number of users increases, the signal-to-interference ratio becomes smaller and there is a gradual degradation in performance until the SNR falls below threshold. Thus the system can tolerate significant amounts of overload if the users are willing to tolerate poorer performance. An additional advantage of CDMA is that the messages intended for one user are not readily decodable by other users, because they may not know the proper codes or have the equipment for generating the appropriate reference signals. Thus there is a privacy feature that is not available in other multiple-access techniques. An important consideration in CDMA is the number of users that can be accommodated simultaneously.[1] This problem is addressed in a subsequent section. A more complete discussion of spread-spectrum multiple access is found elsewhere.[2]

A second application of spread spectrum is usually referred to as selective calling. In a sense selective calling is the inverse of multiple access, that is, there is a central station that must communicate with a number of different receiving points and it wishes to do so on a selective basis. The message is intended for one receiving point and should not be received at other receiving points. This is accomplished by having distinct codes assigned to each receiving point. An example of such a system might be a ground-based radio system in which there is a central station that is communicating with a number of mobile receivers.

Another application of spread spectrum arises from its ability to resist the effects of intentional jamming. This antijam capability was, in fact, the primary reason for early consideration of spread-spectrum communications by the military. Although a discussion of this capability has already been presented, it is reconsidered in the light of some of the difficulties that may arise in the practical system.

An application of spread spectrum that is also of interest to the military is that of covert communication or low probability of intercept. In this case it is desired to maintain a signal spectral density that is sufficiently small that its presence is not readily detected, or that it cannot be detected on a chip-by-chip basis.

A characteristic of spread spectrum that is of particular interest in mobile communications is the ability of a wide-band signal to resist the effects of multipath fading. A property of multipath fading is that frequencies separated by only a few hundred kilohertz may fade essentially independently. Thus at any given time when the signal has a large bandwidth, only a small portion of the bandwidth will be in a fade. The average received signal power thus can be made more nearly constant than it would be for a narrow-band signal. This resistance to fading is an important consideration in the potential application of spread spectrum to mobile communication situations.

A sixth reason for using spread spectrum lies in its ability to accommodate a secure communication system. If codes are made sufficiently complex, it becomes very difficult to break them, and for unauthorized listeners to determine what the message actually is. This secrecy capability is important not only to the military, but in many commercial applications. The development of spread-spectrum codes intended for secure communications is closely related to the problems of cryptography that have appeared recently in the literature.

A final application of spread spectrum lies in its ability to yield accurate distance information. It is well known that a broadband signal can be resolved in time much more precisely than a narrow-band signal. Thus by transmitting a signal with a large bandwidth, it is possible to measure delay times much more accurately and obtain more accurate range information. This is of importance not only in radar systems, but also in navigation systems. Thus there are many spread-spectrum radars that are intended for high-resolution target detection, and there are several electronic navigation systems incorporating spread-spectrum signals.

13-2 MULTIPLE-ACCESS CONSIDERATIONS

A model for the receiver in a multiple-access system is shown in Fig. 13-1. In this figure the signals are designated as $s_1(t)$ through $s_U(t)$ for a total of U active users and each received signal power is designated as P_1 through P_U. Thus it follows that the average power of $s_i(t)$ is 1 for all i, that is,

$$\frac{1}{t_m} \int_0^{t_m} s_i^2(t)\, dt = 1 \qquad i = 1, 2, \ldots, U$$

The receiver model also includes noise $n(t)$, which is assumed to be white with a one-sided spectral density of N_0, and a reference $v_j(t)$. This reference signal is equal to the signal $s_j(t)$ if the receiver is intended to receive that particular signal and if the reference is properly synchronized. Finally there is an integrator that averages the product of the reference and the incoming signals over the duration of one message bit. The output of this integrator at a time equal to the end of the integration time can be represented as

$$\hat{m}(t_m) = m_{jj} + \sum_{\substack{i=1 \\ i \neq j}}^{U} m_{ij} + n_j \tag{13-1}$$

In this equation m_{jj} is the response of the receiver to the desired signal, the various m_{ij}'s are the responses of the receiver to all of the other signals, and n_j is the response of the receiver to the noise.

When the receiver is properly synchronized with the desired signal $s_j(t)$, then

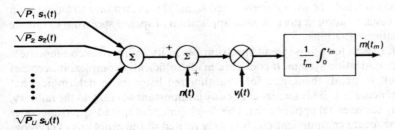

Figure 13-1 Model of a multiple-access system.

$v_j(t)$ is equal to $s_j(t)$ and the quantity m_{jj} becomes

$$m_{jj} = \frac{1}{t_m} \int_0^{t_m} \sqrt{P_j} s_j(t) v_j(t) \, dt = \sqrt{P_j}$$

In a similar way, the m_{ij} may be represented as

$$m_{ij} = \frac{1}{t_m} \int_0^{t_m} \sqrt{P_i} s_i(t) v_j(t) \, dt = \sqrt{P_i} \rho_{ij} \tag{13-2}$$

in which ρ_{ij} is the normalized correlation coefficient between signal $s_i(t)$ and signal $s_j(t)$. In a similar fashion, the noise term becomes

$$n_j = \frac{1}{t_m} \int_0^{t_m} n(t) v_j(t) \, dt \tag{13-3}$$

The output signal-to-noise plus interference ratio for the jth signal may now be written as

$$(\text{SNR})_j = \frac{m_{jj}^2}{E\left\{ \left[\sum_i m_{ij} + n_j \right]^2 \right\}}$$

which may be further expressed as

$$(\text{SNR})_j = \frac{P_j}{E\left\{ \left[\sum_i m_{ij} \right]^2 \right\} + 2E\left\{ \sum_i m_{ij} n_j \right\} + E\{n_j^2\}} \tag{13-4}$$

It is now necessary to evaluate the various expected values in the denominator of Eq. (13-4). The noise term is easily disposed of under the assumption that the noise is white; hence

$$E[n_j^2] = \frac{1}{t_m^2} \int_0^{t_m} \int_0^{t_m} E[n(t)n(u)] v_j(t) v_j(u) \, dt \, du$$

$$= \frac{N_0}{2t_m^2} \int_0^{t_m} \int_0^{t_m} \delta(t - u) v_j(t) v_j(u) \, dt \, du$$

$$= \frac{N_0}{2t_m^2} \int_0^{t_m} v_j^2(t) \, dt = \frac{N_0}{2t_m} \tag{13-5}$$

Similarly, the term involving the expected value of noise and signal is

$$E\left[\sum_i m_{ij} n_j \right] = 0$$

since the noise has zero mean and is independent of all of the signals. Finally the expected value of the m_{ij}'s is given by

$$E\left\{ \left[\sum_i m_{ij} \right]^2 \right\} = \sum_{\substack{i=1 \\ i \neq j}}^{U} \sum_{\substack{k=1 \\ k \neq j}}^{U} E|[m_{ij} m_{kj}]|$$

Without going through the details, let us simply state that this result can be shown to become

$$E\left\{\left[\sum_i m_{ij}\right]^2\right\} = \sum_{\substack{i=1 \\ i \neq j}}^{U} P_i E[\rho_{ij}^2] \tag{13-6}$$

and that

$$E[\rho_{ij}^2] \simeq \frac{2}{t_m} \int_0^{t_m} R_s^2(\tau)\, d\tau \tag{13-7}$$

where $R_s(\tau)$ is the autocorrelation function of any one signal. In this analysis all signals have been assumed to have the same autocorrelation function. This would certainly be true, for example, if they were PN codes at the same chip rate.

It is desirable to relate the autocorrelation function to the energy spectrum of the signal. This can be done by defining an effective bandwidth as

$$
\begin{aligned}
B_{\text{eff}} &= \frac{\left[\int_0^\infty S_s(f)\, df\right]^2}{\int_0^\infty S_s^2(f)\, df} \\
&= \frac{1}{2} \frac{\left[\int_{-\infty}^\infty S_s(f)\, df\right]^2}{\int_{-\infty}^\infty S_s^2(f)\, df}
\end{aligned} \tag{13-8}
$$

where $S_s(f)$ is the energy density spectrum of the signal. As an example of this, consider the ideal rectangular energy spectrum as shown in Fig. 13-2. The effective bandwidth for this spectrum is readily shown to be

$$B_{\text{eff}} = \frac{[S_1(f_2 - f_1)]^2}{[S_1^2(f_2 - f_1)]} = f_2 - f_1$$

Note also that in this particular case

$$B_{\text{eff}} = B_{1/2}$$

where $B_{1/2}$ is half-power bandwidth of the spectrum. This will certainly not be true in all cases, but there are many situations in which the effective bandwidth is of the same order of magnitude as the half-power bandwidth.

By applying Parseval's theorem to the numerator and denominator of Eq.

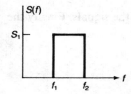

Figure 13-2 An ideal spectral density.

(13-8), the effective bandwidth can also be expressed as

$$B_{\text{eff}} = \frac{R_s^2(0)}{2 \displaystyle\int_{-\infty}^{\infty} R_s^2(\tau)\, d\tau} = \frac{1}{4 \displaystyle\int_{0}^{\infty} R_s^2(\tau)\, d\tau} \tag{13-9}$$

In a spread-spectrum system, it is always true that the duration of the message bit is much greater than the width of $R_s(\tau)$. This is illustrated in Fig. 13-3. This implies, therefore, that the integral of the square of the autocorrelation function from 0 to t_m in Eq. (13-7) can be replaced by an integral from 0 to infinity. Thus

$$\int_0^{t_m} R_s^2(\tau)\, d\tau \simeq \int_0^{\infty} R_s^2(\tau)\, d\tau$$

It follows, therefore, that

$$\int_0^{t_m} R_s^2(\tau)\, d(\tau) \simeq \frac{1}{4B_{\text{eff}}}$$

Using this result in Eq. (13-7) leads to

$$E[\rho_{ij}^2] = \frac{2}{t_m} \cdot \frac{1}{4B_{\text{eff}}} = \frac{1}{2t_m B_{\text{eff}}} \tag{13-10}$$

Finally the signal-to-noise-plus-interference ratio becomes

$$(\text{SNR})_j = \frac{P_j}{\displaystyle\sum_{\substack{i=1 \\ i \neq j}}^{U} P_i\left(\frac{1}{2t_m B_{\text{eff}}}\right) + \frac{N_0}{2t_m}} \tag{13-11}$$

which may be written in more convenient form as

$$(\text{SNR})_j = \frac{2t_m B_{\text{eff}} P_j}{N_0 B_{\text{eff}} + \displaystyle\sum_{\substack{i=1 \\ i \neq j}}^{U} P_i} \tag{13-12}$$

Note that the factors $2t_m B_{\text{eff}}$ represent the processing gain in this situation. Thus we have defined the processing gain in terms of a more precise definition of bandwidth.

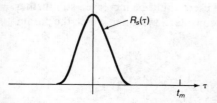

Figure 13-3 Comparison of message bit duration with the width of the signal autocorrelation function.

Number of Active Users—Equal Powers

In this section we wish to find the number of users that can be active simultaneously when the received powers from all users are the same. To do this let us first define the desired output SNR to be

$$(SNR)_0 = (SNR)_j$$

Furthermore let

$$P_j = P_r \qquad \text{all } j$$

The desired output SNR thus can be expressed as

$$(SNR)_0 = \frac{2t_m B_{\text{eff}} P_r}{N_0 B_{\text{eff}} + (U-1)P_r} \tag{13-13}$$

This equation may be solved for U, the number of users, to yield

$$U = 1 + \left[2t_m B_{\text{eff}} \left\{ \frac{1}{(SNR)_0} - \frac{1}{(SNR)_1} \right\} \right] \tag{13-14}$$

where

$$(SNR)_1 = \frac{2t_m P_r}{N_0} > (SNR)_0$$

and

$$[x] = \text{integer part of } x$$

Note that $(SNR)_1$ is the output SNR that would be obtained when there is only *one* user. Note also that

$$t_m P_r = E_b \qquad \text{energy per bit}$$

Therefore, this SNR can be expressed as

$$(SNR)_1 = \frac{2E_b}{N_0} \tag{13-15}$$

The variation of the number of active users U with the parameter E_b/N_0 is illustrated in Fig. 13-4. As E_b/N_0 becomes large, the number of users asymptotically approaches an upper bound that depends upon the processing gain and the desired output signal-to-noise ratio. It should be remembered, however, that this maximum number of users is a maximum only in the sense that more users would result in a decreased value of output signal-to-noise ratio. If one were willing to tolerate a smaller value of output SNR, the number of users could be increased still further.

As an example of the computation of U, consider a system in which the parameters are

$$(SNR)_0 = 14$$

$$R = 30{,}000 = \frac{1}{t_m}$$

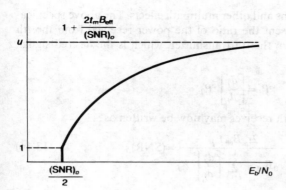

Figure 13-4 Number of active users in a multiple-access system.

$$\frac{E_b}{N_0} = 12.5$$

$$B_{\text{eff}} = 20 \text{ MHz}$$

The maximum number of users, therefore, becomes

$$U = 1 + \left[\frac{2}{30,000} \times 20 \times 10^6 \left(\frac{1}{14} - \frac{1}{25} \right) \right] = 42$$

Doubling the number of users from 42 to 84, say, would roughly have the effect of reducing the output SNR by 3 dB.

Number of Active Users—Unequal Powers

Since there are many ways in which the received powers can be unequal, it is necessary to specify a particular situation in order to compute specific results. For this purpose assume that all users transmit with equal powers, but are different distances from the jth receiver. Then the received power from the ith transmitter may be represented as

$$P_i = \frac{P_0}{d_i^{\alpha}}$$

where

$$P_0 = \text{received power at unit distance}$$

$$d_i = \text{distance from the } i\text{th transmitter to the } j\text{th receiver}$$

$$\alpha = \text{propagation law}$$

The parameter α is the propagation law and depends upon the medium in which the transmission takes place. In free space, for example, α would be equal to 2. At ultrahigh frequency over an ideal earth, α tends to range between 3 and 4. This is true because of ground reflections and other multipath effects. The above specification makes it possible to represent the ratio of the power received from the ith transmitter to that received from the jth transmitter, which is the desired signal. This is shown by

$$P_i = \left[\frac{d_j}{d_i}\right]^\alpha P_j$$

The SNR at the output of the jth receiver may now be written as

$$(\text{SNR})_{0j} = \frac{2t_m B_{\text{eff}} P_j}{N_0 B_{\text{eff}} + \sum\limits_{\substack{i=1 \\ i \neq j}}^{U} \left[\frac{d_j}{d_i}\right]^\alpha P_j} = (\text{SNR})_0 \tag{13-16}$$

This equation can be solved for the quantity that is related to the distances, to obtain

$$\sum_{\substack{i=1 \\ i \neq j}}^{U} \left[\frac{d_j}{d_i}\right]^\alpha \leq 2t_m B_{\text{eff}} \left[\frac{1}{(\text{SNR})_0} - \frac{1}{(\text{SNR})_1}\right] \tag{13-17}$$

Note that when d_i is less than d_j, this particular term in the summation is larger and contributes more than its share to the total sum. Thus fewer terms can be summed until the sum becomes equal to the right-hand side, and this results in a smaller number of active users. This is the so-called "near–far" problem.

To illustrate the near–far problem, assume that all transmitters are at the same distance from the receiver except for one, that is,

$$d_i = d_j \qquad i \neq 1, j$$

$$d_1 = \frac{d_j}{2.5}$$

$$\alpha = 3.68$$

Each term in the summation is now 1 except for the $i = 1$ term and this one is 2.5 raised to the 3.68 power. The summation becomes

$$\sum_{\substack{i=1 \\ i \neq j}}^{U} \left[\frac{d_j}{d_i}\right]^\alpha = (2.5)^{3.68} + U - 2$$

Use of this summation in Eq. (13-17) leads to a value of U of

$$U = 1 + \left[1 - (2.5)^{3.68} + 2t_m B_{\text{eff}} \left\{\frac{1}{(\text{SNR})_0} - \frac{1}{(\text{SNR})_1}\right\}\right]$$

Evaluating this for the same parameters as considered in the equal-power case leads to

$$U = 1 + \left[1 - (2.5)^{3.68} + \frac{2 \times 20 \times 10^6}{30,000}\left(\frac{1}{14} - \frac{1}{25}\right)\right]$$

$$= 14$$

Note that the number of users has been reduced by a factor of 3 simply by virtue of one of the transmitters being 2.5 times closer than all of the others. It might also be noted that if

$$d_1 < \frac{d_j}{2.78}$$

this system would fail as a multiple-access system since only one user could be supported and none of the others would be able to be received with the desired output signal-to-noise ratio.

Bandwidth-Limited Channels

There are many situations in which the bandwidth of a channel is limited either by virtue of physical considerations in the channel or by regulatory considerations. It is of interest, therefore, to see how the number of active users is constrained when the bandwidth of the channel is the limiting factor rather than the received signal power. The number of active users was shown to be

$$U = 1 + \left[2t_m B_{\text{eff}} \left\{ \frac{1}{(\text{SNR})_0} - \frac{1}{(\text{SNR})_1} \right\} \right] \tag{13-18}$$

If the SNR for a single user is now allowed to become arbitrarily large, this result becomes

$$\lim_{(\text{SNR})_1 \to \infty} U = 1 + \left[\frac{2B_{\text{eff}}}{R(\text{SNR})_0} \right] \tag{13-19}$$

where R is the message bit rate. Note also that if more than one user is to be supported by the channel, the maximum SNR is limited by

$$\text{Max}(\text{SNR})_0 = \frac{2B_{\text{eff}}}{R} \tag{13-20}$$

This may not be a very serious limitation, unless the constraint on the bandwidth is quite severe or the message bit rate quite large.

Power-Limited Channels

There are other situations in which the power that can be transmitted in the channel is limited. This is true, for example, in many satellite communication channels because of the finite power resources in the satellite repeater. If one holds the power in the channel fixed and allows the bandwidth to become arbitrarily large, then Eq. (13-18) becomes

$$\lim_{B_{\text{eff}} \to \infty} U = \infty$$

provided that

$$(\text{SNR})_1 > (\text{SNR})_0$$

Note that this suggests that the number of users can be made arbitrarily large if the bandwidth is arbitrarily large even when there is a limitation on power per user. This is not inconsistent with previous results because the total power would also become arbitrarily large. However, a more meaningful measure of the number of users might be the number of users per unit bandwidth. In this case

$$\lim_{B_{\text{eff}} \to \infty} \frac{U}{B_{\text{eff}}} = \frac{2}{R} \left[\frac{1}{(\text{SNR})_0} - \frac{1}{(\text{SNR})_1} \right]$$

It may be noted that the number of users per unit bandwidth remains finite even as the bandwidth becomes infinite.

Hard-Limited Channels

In early satellite communication systems, it was conventional to introduce a hard limiter prior to the satellite power amplifier as shown in Fig. 13-5. The purpose of this was to enable the power amplifier to operate more nearly at its point of maximum efficiency. When such a hard-limited satellite repeater is used with frequency-division multiple access, there results an appreciable amount of intermodulation that reduces the number of channels that can be handled.

In a time-division multiple-access system, there is only one signal at a time in the repeater, and hence the presence of the hard limiter does not produce intermodulation and does not reduce the number of channels available. The question remains, however, as to what happens in a code-division multiple-access system in which there may be a number of pseudorandom sequences in the hard limiter at the same time. For the case of the linear channel, it was shown that

$$(\text{SNR})_j = \frac{2t_m B_{\text{eff}} P_j}{N_0 B_{\text{eff}} + \sum\limits_{\substack{i=1 \\ i \neq j}}^{U} P_i} \tag{13-21}$$

When all of the powers are the same and the same as the desired signal power, it can be shown that the hard limiter reduces the effective received power by a factor of

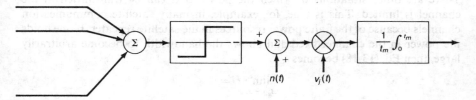

Figure 13-5 Multiple-access system with a hard limiter.

$\pi/4$. Thus SNR now becomes

$$(\text{SNR})_j = \frac{\left(\dfrac{\pi}{2}\right) t_m B_{\text{eff}} P_j}{N_0 B_{\text{eff}} + \displaystyle\sum_{\substack{i=1 \\ i \neq j}}^{U} P_i} \tag{13-22}$$

This only represents a degradation in SNR of about 1 dB and in itself is not particularly serious. However, the effect on the number of active users that can be supported may be considerably more severe. Considering the equal-power case again, the number of active users becomes

$$U = 1 + \left[\frac{\pi}{2} t_m B_{\text{eff}} \left(\frac{1}{(\text{SNR})_0} - \frac{4}{\pi (\text{SNR})_1} \right) \right] \tag{13-23}$$

Using the same parameters that were previously considered for the linear channel, the number of active users is now evaluated as

$$U = 1 + \left[\frac{\pi \times 20 \times 10^6}{2 \times 30,000} \left(\frac{1}{14} - \frac{4}{\pi \times 25} \right) \right]$$
$$= 22$$

This value must be compared with the value 42 for the linear channel and it is seen that the reduction in the number of active users is considerably greater than might have been anticipated from the degradation in signal-to-noise ratio.

A somewhat different situation arises if one signal is much larger than all of the others. It can be shown in this case that the small equal-power signals are reduced by a factor of 4. Thus the P_j in Eq. (13-22) would be replaced by $P_j/4$.

13-3 ENERGY AND BANDWIDTH EFFICIENCY IN MULTIPLE ACCESS

The degree to which a communication system can utilize the available signal energy and the available bandwidth was discussed in considerable detail in a previous chapter. These same concepts are now applied to the case of a multiple-access spread-spectrum communication system.

As shown previously the output SNR for the jth channel in such a system is

$$(\text{SNR})_j = \frac{2 t_m B_{\text{eff}} P_j}{N_0 B_{\text{eff}} + \displaystyle\sum_{\substack{i=1 \\ i \neq j}}^{U} P_i} \tag{13-24}$$

We now wish to consider a special case in which the sum of all the powers is a constant, that is,

$$\sum_{i=1}^{U} P_i = P_T$$

and assume that P_T is fixed. An example of this might be a satellite repeater in which P_T is the available power output from the power amplifier in the satellite. We also assume that all of the received signal powers are equal so that

$$P_i = \frac{P_T}{U} \qquad \text{all } i$$

Furthermore the total message bit rate in the system is taken to be

$$R = U R_1$$

where R_1 is the bit rate for each user. For a general M-ary system, the energy per bit is

$$E_b = \frac{P_i}{R_1} = \frac{P_T}{U} \cdot \frac{T}{\log_2 M}$$

Thus the sum of the powers can be expressed as

$$\sum_{\substack{i=1 \\ i \neq j}}^{U} P_i = (U-1) \frac{P_T}{U} = \frac{U-1}{U} \cdot \frac{E_b U \log_2 M}{T}$$

$$= (U-1) \frac{E_b \log_2 M}{T}$$

Furthermore we note that

$$t_m P_j = E_b$$

The output SNR for the jth channel thus becomes

$$(\text{SNR})_j = \frac{2 B_{\text{eff}} E_b}{N_0 B_{\text{eff}} + (U-1) \frac{E_b \log_2 M}{T}} \qquad (13\text{-}25)$$

$$= \frac{2(E_b/N_0)}{1 + \frac{(U-1)\log_2 M}{B_{\text{eff}} T} (E_b/N_0)}$$

If we wish all channels to have the same output SNR, that is,

$$(\text{SNR})_j = (\text{SNR})_0 \qquad \text{all } j$$

and that the output SNR is fixed at some desired quality level, the above result can be solved to obtain the required E_b/N_0 in the channel. This yields

$$\frac{E_b}{N_0} = \frac{B_{\text{eff}} T (\text{SNR})_0}{2 B_{\text{eff}} T - (U-1)\log_2 M (\text{SNR})_0} \qquad (13\text{-}26)$$

For an M-ary system, the duration of a signal is

$$T = \frac{\log_2 M}{R_1} = \frac{U \log_2 M}{R}$$

Thus

$$\frac{E_b}{N_0} = \frac{\dfrac{B_{eff}}{R}(SNR)_0}{\dfrac{2B_{eff}}{R} - \dfrac{U-1}{U}(SNR)_0} \tag{13-27}$$

The above result has been expressed in terms of the effective bandwidth. However, the effective bandwidth can be related to the equivalent noise bandwidth of the channel by

$$B_{eff} = \eta B$$

where B is the equivalent noise bandwidth of the channel and the parameter η depends upon the type of spread-spectrum signal and the channel filters that are employed. In a typical situation, η is approximately one, but it may differ from one by as much as 50 percent in some cases.

In using these relations, Eq. (13-27) becomes

$$\frac{E_b}{N_0} = \frac{\eta\left(\dfrac{B}{R}\right)(SNR)_0}{2\eta\left(\dfrac{B}{R}\right) - \dfrac{(U-1)}{U}(SNR)_0} \tag{13-28}$$

This equation represents the trade-off between energy utilization efficiency and bandwidth utilization efficiency. Such a trade-off is displayed in Fig. 13-6. Note that there is a minimum value of bandwidth utilization, which is given by

$$(B/R)_{min} = \frac{U-1}{U}\frac{(SNR)_0}{2\eta}$$

There is also a minimum value of energy utilization given by

$$\left(\frac{E_b}{N_0}\right)_{min} = \frac{1}{2}(SNR)_0$$

Finally, the total signal power and the noise in the channel specifies a channel

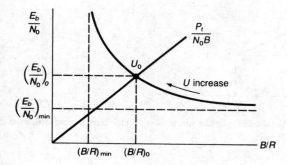

Figure 13-6 Energy and bandwidth utilization in a spread-spectrum multiple-access system with fixed power.

SNR that defines a straight line in the plane of Fig. 13-6. The intersection of this straight line with the trade-off curve of Eq. (13-28) specifies an operating point. At this operating point, the value of bandwidth utilization is

$$(B/R)_0 = \frac{\eta(B/R_1)(SNR)_0\left[1 + \dfrac{P_T}{N_0 B}\right]}{\dfrac{P_T}{N_0 B}\left[2\dfrac{\eta B}{R_1} + (SNR)_0\right]} \tag{13-29}$$

Likewise the value of energy utilization is

$$\left(\frac{E_b}{N_0}\right)_0 = \frac{\eta(B/R_1)(SNR)_0\left[1 + \dfrac{P_T}{N_0 B}\right]}{2\eta(B/R_1) + (SNR)_0} \tag{13-30}$$

The number of active users that can be supported at this particular point becomes

$$U_0 = \frac{(P_T/N_0 B)[2\eta(B/R_1) + (SNR)_0]}{(SNR)_0\left[1 + \dfrac{P_T}{N_0 B}\right]} \tag{13-31}$$

The above result assumes that the total power is fixed. An alternative situation might be to assume that the total bandwidth is fixed. In this case we can eliminate U from the previous result and obtain

$$\frac{E_b}{N_0} = \frac{\eta(B/R)(SNR)_0}{\eta(B/R)\left[2 + \dfrac{(SNR)_0}{\eta(B/R_1)}\right] - (SNR)_0} \tag{13-32}$$

This is a different trade-off between energy utilization and bandwidth utilization and is displayed in Fig. 13-7.

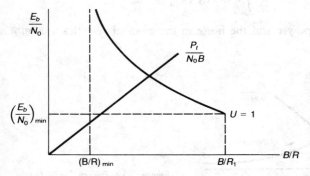

Figure 13-7 Energy and bandwidth utilization in a spread-spectrum multiple-access system with fixed bandwidth.

There is again a minimum bandwidth utilization given by

$$(B/R)_{min} = \frac{(SNR)_0}{2\eta + \dfrac{(SNR)_0}{B/R_1}}$$

and a minimum energy utilization given by

$$\left(\frac{E_b}{N_0}\right)_{min} = \frac{1}{2}(SNR)_0$$

The maximum number of users that can be supported in this case is

$$U_{max} = \frac{2\eta(B/R_1) + (SNR)_0}{(SNR)_0} \tag{13-33}$$

As an illustration of the performance of a system in which there is a total power limitation, assume that our system has the following parameters

$$R_1 = 30,000$$

$$\eta = 1$$

$$\frac{P_T}{N_0} = 90 \text{ dB} = 10^9$$

$$B = 100 \text{ MHz}$$

$$(SNR)_0 = 14$$

The energy utilization becomes

$$\left(\frac{E_b}{N_0}\right)_0 = \frac{(1)\left(\dfrac{10^8}{30,000}\right)(14)\left[1 + \dfrac{10^9}{10^8}\right]}{2(1)\left(\dfrac{10^8}{30,000}\right) + 14} = 76.84$$

and the number of users that can be supported at the operating point is

$$U_0 = \frac{10\left[2(1)\dfrac{10^8}{30,000} + 14\right]}{14[1 + 10]} = 433$$

The maximum number of users that can be supported by this system is

$$U_{max} = \frac{2(1)\left(\dfrac{10^8}{30,000}\right) + 14}{14} = 477$$

13-4 SELECTIVE CALLING AND IDENTIFICATION

As noted previously the need for selective calling arises when a central station needs to be able to address distinct receivers individually without having messages intended for that receiver be received by other receivers also. This implies that each user is assigned a distinct address. To minimize interference between users, it is necessary that all pairs of signals have a small cross-correlation for all relative time shifts. To explore this requirement in more detail, consider the receiver model shown in Fig. 13-8.

When the receiver contains a reference signal for the jth signal and receives the ith signal, the receiver output is given by

$$\hat{m}_i(t_m) = \frac{1}{t_m} \int_0^{t_m} \sqrt{P_i} s_i(t - \tau) v_j(t) \, dt \tag{13-34}$$

When

$$v_j(t) = s_j(t)$$

the output may be expressed as

$$\hat{m}_i(t_m) = \frac{\sqrt{P_i}}{t_m} \int_0^{t_m} s_i(t - \tau) s_j(t) \, dt \tag{13-35}$$

Since both signals have unit average power, the integral in Eq. (13-35) defines a normalized cross-correlation function as

$$R_{ij}(\tau) = \frac{1}{t_m} \int_0^{t_m} s_i(t - \tau) s_j(t) \, dt \tag{13-36}$$

Note that for $i = j$, which is autocorrelation,

$$R_{ii}(0) = 1$$

Thus the response of a receiver matched to $s_j(t)$ to the signal $s_i(t)$ is given by

$$\hat{m}_i(t_m) = \sqrt{P_i} R_{ij}(\tau) \tag{13-37}$$

This response is really a random variable depending upon the value of τ and the statistics of $R_{ij}(\tau)$. We would like to have the rms and peak values of this correlation as small as possible. Since $R_{ij}(\tau)$ is a bandpass function, it may be more convenient to talk about its envelope. The mean square value of the envelope of $R_{ij}(\tau)$ is given by

$$|R_{ij}|^2 = \frac{1}{t_m} \int_{-t_m}^{t_m} R_{ij}^2(\tau) \, d\tau \tag{13-38}$$

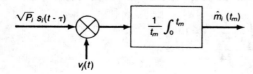

Figure 13-8 Receiver model for selective calling.

From Parseval's theorem this can be related to the energy spectra of the signals by

$$|R_{ij}|^2 = \frac{1}{t_m} \int_{-\infty}^{\infty} S_{ii}(f) S_{jj}(f)\, df \tag{13-39}$$

where $S_{ii}(f)$ is the energy spectrum of $s_i(t)$ and $S_{jj}(f)$ is the energy spectrum of $s_j(t)$. When these two energy spectra are the same, the mean square value of the envelope becomes

$$|R_{ij}|^2 = \frac{1}{t_m} \int_{-\infty}^{\infty} S_{jj}^2(f)\, df$$

$$= \frac{1}{2 t_m B_{\text{eff}}} \tag{13-40}$$

Thus the rms value of the envelope for aperiodic cross-correlation becomes

$$|R_{ij}| = \frac{1}{\sqrt{2 t_m B_{\text{eff}}}} \tag{13-41}$$

for any signal pair.

This result may also be extended to the case in which signals repeat themselves periodically. It can be shown that in this case

$$|R_{ij}| = \frac{1}{\sqrt{t_m B_{\text{eff}}}} \tag{13-42}$$

There is no analytic procedure for bounding the peak value of the cross-correlation, and in some cases this peak value can become quite large unless the signals are carefully designed.

Another aspect of selective calling that is of considerable importance is the number of addresses that can be obtained. If we consider direct sequence first, we know that if we have signals that have a length of

$$N = 2^m - 1$$

the number of such sequences is given by

$$\frac{\phi(2^m - 1)}{m}$$

where $\phi(k)$ is the Euler-Totient function, which is the number of positive integers less than k and relatively prime to it. As an example of this, consider a set of signals having the parameters

$$m = 13$$

$$N = 2^{13} - 1 = 8191 \text{ (prime)}$$

$$\frac{\phi(2^m - 1)}{m} = \frac{8190}{13} = 630 \text{ sequences}$$

Thus there are 630 different addresses that could be obtained with sequences that have a length of 8191. Not all of these sequences have good cross-correlation properties, and in fact there is no general analysis of the cross-correlation properties of m sequences. For short sequences the cross-correlation properties can be computed and general results indicate that the peak cross-correlation may be significantly higher than those obtained from the Gold codes discussed previously.[3] If small cross-correlation is desirable, it is important to use Gold codes. Fortunately the number of Gold codes is quite large. Hence, in the previous example there would be 8192 Gold codes whose cross-correlation is guaranteed to be no greater than 129 compared with a maximum autocorrelation of 8191.

We consider next the situation for frequency-hopping signals. If these signals are generated from m sequences, then there will be as many frequency-hopping signals as there are corresponding m sequences—but very little is known about the cross-correlation properties of such signals. There is one class of frequency-hopping signal, however, for which the cross-correlations can be well controlled.[4] Although this class of signals is not useful for cases in which the sequences are very long, it is useful in situations where short sequences are acceptable. The good cross-correlation characteristics of such signals result from a so-called "one-coincidence" property. To see how this arises, assume that the frequency sequence is defined by

$$f_i = f_0 + a_i f_1$$

where

$$a_i = 1, 2, \ldots, m$$

$$f_1 = \text{minimum frequency separation}$$

The parameter m is selected to be $p-1$, where p is any prime number. If the primitive root of this prime number is b, then it is possible to define a permutation sequence that has the form

$$P = [b^0, b^1, b^2, \ldots, b^{m-1}]$$

This is normally referred to as a power residue sequence because the successive powers of b are computed modulo-p.

Since b is a primitive root of p, the resulting sequence will contain all integers from 1 to m once and only once. One can then start the process by defining any initial frequency-hopping sequence in terms of a sequence of a_i values. Let this be

$$S_1 = (a_1, a_2, \ldots, a_m)$$

A second sequence is then obtained by permuting the elements of S_1 in accordance with the permutation sequence P. That is,

- Element in position b^0 moves to position b^1.
- Element in position b^1 moves to position b^2.
-
-
-
- Element in position b^{m-1} moves to position b^0.

This is repeated until the original sequence is generated again. The net result is that we now have m sequences of length m. Furthermore, when any sequence is shifted in time with respect to any other sequence, there will be at most one time chip in which the frequencies are coincident. This is the "one-coincidence" property.

Since the cross-correlation between any pair of sequences depends upon the differences between frequencies, coincident frequencies contribute the most. By limiting these coincidences to only one, the peak value of the cross-correlation is thereby limited to relatively modest values. To illustrate the design procedure, consider a specific case for which

$$p = 5, \quad m = 4, \quad b = 2$$

$$P = [1, 2, 4, 3]$$

$$S_1 = [3, 1, 4, 2]$$

$$S_2 = [4, 3, 2, 1]$$

$$S_3 = [2, 4, 1, 3]$$

$$S_4 = [1, 2, 3, 4]$$

It has been demonstrated by computation that the peak value of the cross-correlation is seldom more than two or three times the rms value, even when the time–bandwidth products of the sequences are very large. Thus these frequency-hopping sequences form the best sequences known to date. They are most useful for short sequences and matched-filter receivers and some applications of these have been proposed in the mobile communication area.[5]

The number of sequences can be further increased by partitioning each of the m sequences generated by the procedure outlined above into shorter sequences of length n. These shorter sequences will also have the one coincidence property, and if m is sufficiently large, can also have very large time–bandwidth products. The net result is that millions of frequency-hop signals can be defined that will have controlled cross-correlation properties and are useful in the selective calling application. The number of discrete addresses that can be obtained with this type of frequency-hopping signal is far in excess of the number of signals available from PN sequences having the same time–bandwidth product.

13-5 ANTIJAM CONSIDERATIONS

As noted previously, one of the earliest applications of spread spectrum was that of reducing the effects of intentional jamming. This application is based on the ability of a spread-spectrum signal to suppress the effects of interference by an amount that is roughly proportional to the processing gain. The purpose of this section is to discuss this application in somewhat more detail. A more complete discussion is available elsewhere.[6]

Many different types of jamming signals have been proposed or used in an effort to minimize the effectiveness of a communication system. Five of the most common types are

1. The pulse jammer,
2. The tone jammer.
3. The partial band jammer.
4. The barrage or wide-band jammer.
5. The intelligent jammer.

The pulse jammer transmits short, high-intensity pulses at either a regular or irregular rate. These pulses are of sufficient strength to destroy completely that portion of the desired signal that occurs at the same time. The tone jammer, on the other hand, transmits one or more narrow-band sinusoidal signals. These are spread across the desired bandwidth and are present all of the time but do not occupy all of the frequency spectrum. When the tones are concentrated in only a portion of the spread-spectrum signal bandwidth, the situation is frequently referred to as partial band jamming. Spread-spectrum techniques are effective in combating all three types of jamming, as discussed subsequently.

The barrage jammer frequently consists of wide-band noise transmitted at a high-power level. If the bandwidth of this noise is as great as the bandwidth of the spread spectrum, then the application of spread spectrum does little to counteract the effects of the jamming. However, it should be noted that the barrage jammer must utilize large amounts of power since the jammer must provide signal energy at all times and at all frequencies within the band. Intelligent jamming usually refers to radar jamming in which attempts are made to produce signals that simulate the actual signals but contain false information. This is difficult to accomplish in a spread-spectrum communication system, but it is not impossible. The present discussion of antijam characteristics, however, will be limited to pulse and tone jamming.

Jamming Direct-Sequence Systems

When a direct-sequence spread-spectrum system is to be jammed, the optimum strategy for the jammer is to place a single tone at or near the carrier frequency of the direct-sequence signal. Under ideal conditions the jamming power of this single tone is suppressed by an amount equal to the processing gain. Therefore, the SNR, as given in previous chapters, becomes

$$(\text{SNR})_o = \frac{P_r}{\dfrac{N_0 B_m}{2} + \dfrac{J}{\text{PG}}} \tag{13-43}$$

where J is the jamming power. The jamming margin was defined in a previous section as

$$M_j(\text{dB}) = (\text{PG})(\text{dB}) - L(\text{dB}) - (\text{SNR})_o(\text{dB}) \tag{13-44}$$

where L is the system loss.

Calculations made previously indicated that very effective suppression of jammer power can be accomplished in a direct-sequence system. However, in an actual system, this suppression may not be nearly as good as the theoretical value described by Eq. (13-43). There are several reasons for this. In the first place, in an ideal system, there is no discrete carrier term, but in an actual system, the carrier in the reference signal may not be adequately suppressed in the balanced modulator. At the receiver the single jamming tone may mix with the residual carrier and produce a large component at the difference frequency. Since this difference frequency is in the same range of frequencies as the message frequencies, severe interference can result.

The second reason that jamming suppression may not be ideal is that the PN code may not be sufficiently balanced in some message bit periods. When discussing PN sequences, it was pointed out that in the entire period of such a sequence, the number of positive chips differs from the number of negative chips by only one. However, when the total period of the sequence is much greater than the duration of a message bits, as it will be in most situations, then in some message bits, the number of positive chips and the number of negative chips may not be close. This results in a bias during that particular message bit that can seriously disturb the performance of the receiver. The net result of both of these factors is that the jamming margin in a practical direct-sequence system may be far less than that predicted by Eq. (13-44), unless extreme care is taken in balancing out the carrier and ensuring an adequate balance in message bit periods.

The above statements can be made more quantitative by considering the direct-sequence receiver when a tone jammer is operating at or near the carrier frequency. The model for this situation is illustrated in Fig. 13-9. The tone jammer is modeled as

$$j(t) = \sqrt{2J} \, \cos(\omega_j t + \theta_j) \tag{13-45}$$

and the reference signal, with a residual carrier component, as

$$v(t) = \sqrt{2} b(t - \tau)\cos(\omega_0 t + \theta) + \sqrt{2C_R} \, \cos(\omega_0 t + \theta) \tag{13-46}$$

where C_R is the average power of the residual carrier and $b(t - \tau)$ is the PN reference sequence.

The receiver output due to the jammer is

$$j_0(t_m) = \frac{1}{t_m} \int_0^{t_m} j(t)v(t) \, dt$$

The mean square value of this output is obtained by assuming that $b(t - \tau)$, θ, and

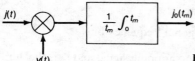

Figure 13-9 Direct-sequence receiver with a tone jammer.

θ_j are random variables. The result is

$$\overline{j_0^2} = J \left[\frac{t_1}{2t_m} + C_R \left(\frac{\sin \pi(f_j - f_0)t_m}{\pi(f_j - f_0)t_m} \right)^2 \right] \tag{13-47}$$

from which it is clear that the maximum degradation occurs when $f_j = f_0$.

It is of interest to define a degradation factor (DF) that indicates the amount that the jammer output power increases as a result of the residual carrier. This is

$$\text{DF} = \frac{\overline{j_0^2}(C_R)}{\overline{j_0^2}(C_R = 0)} = 1 + \frac{2t_m}{t_1} C_R = 1 + (\text{PG})C_R \tag{13-48}$$

when $f_j = f_0$. It is clear that the degradation is greater when the processing gain is larger. As an illustration of this, assume that the direct-sequence system has a processing gain of 40 dB and the residual carrier is 20 dB below the reference signal level. The degradation factor becomes

$$\text{DF} = 1 + 10^4 \times 10^{-2} = 101$$

Therefore, instead of obtaining a jammer suppression of 40 dB as the basic theory suggests, only 20-dB suppression would be achieved.

Jamming Frequency-Hopping Systems

It was noted previously that a frequency-hopping system is an avoidance system. Therefore, the optimum strategy is to space a series of tones across the band so that many of these tones will coincide with frequencies whenever they occur. The net result will be that many of the hops will have interfering frequencies present, and thus the information contained in those hops will be degraded. To accomplish this it is only necessary that the received power from each jammer tone slightly exceed the received signal power in each hop. When this type of jamming is used, the previous computation of jamming margin is no longer appropriate. Instead it is necessary to discuss the probability of error directly.

It can be shown that the conditional probability of error, given that there is a hit (i.e., that a jamming frequency coincides with a particular hop frequency) is[7]

$$P(\varepsilon | \text{hit}) = \frac{M - 1}{M} \tag{13-49}$$

where M is the size of the signaling alphabet, that is, an M-ary modulation method is being employed. Note that for $M = 2$, the binary case, this immediately becomes one half. Let

$$N_j = \text{number of jamming tones}$$

$$N_f = B_s t_1 = \text{number of orthogonal frequency bins}$$

where B_s is the total bandwidth of the frequency-hopping signal and t_1 is the duration of each hop. The probability of a hop decision error, therefore, is given by

$$P'_e = P(\varepsilon|\text{hit})P(\text{hit})$$

$$= \frac{M-1}{M} \cdot \frac{N_j}{N_f/M} = (M-1)\frac{N_j}{N_f} \qquad (13\text{-}50)$$

The jammer power required to produce a given degradation of performance, therefore, depends upon the error probability and may be much less than that predicted by the processing gain. To illustrate this point, consider a specific example in which P'_e is to be less than 10^{-2} and consider a system with 8192 frequencies. For such a system, the processing gain is 39.13 dB. If there are no system losses, and if the desired output SNR is 14 dB, then the jamming margin according to Eq. (13-44) becomes

$$M_j(\text{dB}) = 39.13 - 14 = 25.13 \text{ dB}$$

This result implies that the jamming power could be 25.13 dB greater than the desired signal power and still achieve acceptable receiver performance. Let us assume next that the jammer produces 100 tones, with the power of each tone equal to the received signal power from the frequency-hop signal. For a binary system, the probability of error becomes

$$P'_e = (2-1)\frac{100}{8192} = 0.012 > 0.01$$

Note that the total jamming power is only 20 dB above the received signal power and the probability of error is already greater than the minimum acceptable value. This situation would become much worse if the required probability of error were less. For example, if P'_e were to be less than 10^{-3}, it would require only nine jammer tones to degrade the system below its acceptable value. The degradation due to tone jamming can be alleviated somewhat by the use of error-correction coding as discussed in the next section.

It is of interest to note that there is a very convenient way of generating uniformly spread tone signals by using a PN code generator. Suppose, for example, that a frequency-hop system is being modulated by a message bit rate of 30,000 b/s. This implies that the frequency separation between hop frequencies is also 30,000 Hz. A PN code generator run at a rate that is of the same order of magnitude as the bandwidth of the frequency-hopping system and having a period of 1/30,000 second will produce harmonics separated by 30,000 Hz. Because of the high chip rate, these harmonics will have very nearly the same amplitude, and thus every frequency in the frequency-hop signal will have a sinusoidal tone associated with it. This could have a disastrous effect on the performance of the spread-spectrum system if sufficient jamming power were used. This technique is even more useful in partial band jamming because the PN code generator can run at a lower rate.

Pulse Jamming

A direct-sequence system that does not employ error-correction coding may be very vulnerable to pulses that are longer than a message bit. Such a pulse will completely destroy the message information contained in a given bit, and if these pulses occur with sufficient frequency, the resulting error probability will be lower than an acceptable value. Yet the average power of a pulse may be much less than that predicted by the expression involving processing gain when the degradation is specified in terms of the probability of error. The net result of pulse jamming is that the probability of error increases and, furthermore, errors tend to cluster together, particularly if the jamming pulses are several message bits in length. The probability of error due to pulse jamming is

$$P_e' \simeq \frac{1}{2}\eta$$

where

$$\eta = \text{pulse duty factor}$$

under the assumption that the pulse is longer than the message bit. The average jamming power required to produce a given probability of error, relative to the signal power, is

$$\frac{J_{av}}{P_r} = 2P_e' 10^{M_j/10} \tag{13-51}$$

To illustrate this consider the direct-sequence system in Chap. 9 for which the jamming margin was computed to be 37 dB. For the jammer to force the probability of error to be greater than 0.01 requires that

$$\frac{J_{av}}{P_r} > 2 \times 0.01 \times 10^{3.7} = 100$$

Note that in this case the effective jamming margin is only 20 dB, instead of 37 dB. The situation would be much worse if the probability of error that is required were considerably less. For example, if a probability of error of 10^{-3} were required, the effective jamming margin would only be 10 dB.

It was noted that pulse jamming causes the errors to be clustered together. Independence among bit errors can be restored by interleaving the message data. This is usually accomplished in a pseudorandom fashion. The effects of pulse jamming can also be alleviated by introducing error-correction coding.

13-6 ERROR-CORRECTION CODING

The previous discussion of jamming indicated that some improvement in spread-spectrum system performance can be achieved by the use of error-correction coding. It is appropriate, therefore, to present a very brief summary of error-correction techniques and to indicate something about the degree of improvement that can be achieved. For the interested reader, there are several excellent texts that treat this subject in detail.[8-10]

Many types of error-correction codes have been proposed, but most of these can be classified as one of three types. The first type is the *linear block code*. The second is a special class of linear block code, the *cyclic code*, and the third is an entirely different type of code known as the *convolutional code*. Simple examples of all three types will be considered.

An important parameter of error-correction codes is the code rate. This is most easily defined by reference to the block code. In a block code, let

n = number of digits in each block

k = number of message digits in each block

Then the rate of the code is

$$r = k/n = \text{code rate}$$

Thus the rate of the code is simply the fraction of the total number of digits transmitted that represents message information. In a code that contains k message digits in each block, there will be 2^k code words, since this is the number of different ways in which k binary digits can be arranged. Such codes are designated as (n, k) codes.

A block code is one in which the message digits are broken up into blocks of length k and to each block are added additional digits to make longer blocks of length n. The resulting code words are, therefore, n digits long. To be a linear block code, it is necessary that the sum of any two code words also be a code word. As a simple illustration of a linear block code, consider the (5, 2) code. In a (5, 2) code, the four different combinations of two message digits are each encoded into code words of length 5. As an example of this, consider the following case.

$$
\begin{array}{ccccccc}
0 & 0 & \rightarrow & 0 & 0 & 0 & 0 & 0 \\
0 & 1 & \rightarrow & 0 & 1 & 0 & 1 & 1 \\
1 & 0 & \rightarrow & 1 & 0 & 1 & 0 & 1 \\
1 & 1 & \rightarrow & 1 & 1 & 1 & 1 & 0 \\
\end{array}
$$

Note that this is a rate 2/5 code. It is also a single-error correcting code. This code will also correct two double errors, but which particular double errors are corrected depends upon the decoding algorithm selected.

A cyclic code is a linear block code for which an additional constraint has been imposed, namely, that every code word is a cyclic shift of some other code word. This is in addition to the requirement that the sum of any two code words also be a code word. As a simple illustration of a cyclic code, consider the 6, 2 code shown here.

$$
\begin{array}{ccccccccc}
0 & 0 & \rightarrow & 0 & 0 & 0 & 0 & 0 & 0 \\
0 & 1 & \rightarrow & 0 & 1 & 0 & 1 & 0 & 1 \\
1 & 0 & \rightarrow & 1 & 0 & 1 & 0 & 1 & 0 \\
1 & 1 & \rightarrow & 1 & 1 & 1 & 1 & 1 & 1 \\
\end{array}
$$

This is a rate 1/3 code and is also a single-error correcting code. Cyclic codes are of particular interest becuase the code words can be generated and can be decoded with shift registers. Thus it is much easier to implement cyclic codes than the more general linear block code.

The convolutional code is one in which the message digits are not broken up into blocks, but instead are convolved with a fixed pattern set into the encoder. A simple rate 1/2 convolutional coder is displayed in Fig. 13-10. For any three message digits stored in the three shift register stages, the first transmitted digit will be the sum of all three, and then the next transmitted digit will be the sum of the first and the last. After these two digits have been transmitted, a new message digit will be shifted into the shift register and the third digit will be dropped. The process will then be repeated and two new digits transmitted. It is clear that it is possible to implement convolutional codes in a very simple fashion, although the decoding may not be quite as easy. However, special algorithms have been developed for decoding that are quite efficient and result in extremely good error-correction capability.

As an indication of the improvement that can be obtained with appropriate error-correction techniques, several different coding techniques are compared in Table 13-1. The first line of this table indicates that without coding, an E_b/N_0 of 11.7 is required to achieve a channel error probability of 10^{-3}. For the other techniques, the probability of error after corrections have been made is still 10^{-3}, but the table lists the required channel probability of error, which is now considerably larger, and the required E_b/N_0, which is now smaller. Note that the convolutional codes are extremely efficient in this regard, although the complexity of the decoders may become quite significant.

The significance of this table for the antijam capability of spread-spectrum systems is revealed in the second column, which indicates the probability of error that can be permitted in the channel and still achieve a message bit error probability of 10^{-3}. For example, the rate 1/3 convolutional code can produce the desired message bit error probability when the channel probability of error is 0.082. Thus the frequency-hopping system described in the previous section could achieve a message bit error probability of 10^{-3} with 671 jamming tones present instead of only the eight tones that would be permissible without error-correction coding.

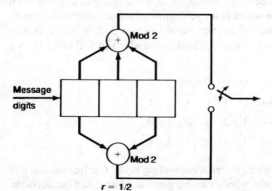

Message digits

Mod 2

Mod 2

$r = 1/2$

Figure 13-10 A rate 1/2 convolutional encoder.

Table 13-1 Comparison of coding techniques[11]

Technique	Required channel P_e	Required E_b/N_0	Complexity (no. of ICs)
No coding	10^{-3}	11.7	—
$r = 1/2$, conv.	0.0266	10.2	10
Golay (24, 12)	0.0266	10.2	20
Block (48, 24)	0.029	10.0	60
$r = 1/3$, conv.	0.082	10.7	(Viterbi) 80
$r = 1/2$, conv.	0.064	9.4	(Seq.) 500

Although the use of error correction can produce impressive improvement under some circumstances, it is not a universal cure for the effects of jamming. In particular, if the channel probability of error becomes too large, error-correction coding actually may make the situation worse. Although the point at which this occurs depends upon the particular code, the simple (6, 2) cyclic code described in the previous section serves to illustrate the point. When the channel probability of error is 0.012, this code will produce a message bit error probability of 10^{-3}. However, if the channel error probability is as large as 0.21, the message bit error probability is also 0.21 and no improvement is obtained. Thus under conditions of severe jamming, the use of error correction is not advisable.

13-7 INTERCEPT CONSIDERATIONS (AI)

Another objective of spread-spectrum techniques is to make it more difficult for unauthorized receivers to receive the message, or even to know that a message is being transmitted. Direct-sequence methods are generally more efficient for this purpose because the spectral density of the transmitted signal is more uniformly spread over the entire bandwidth. However, frequency-hopping methods, with appropriate regard for the separation of frequencies, are also useful.

One strategy that the transmitter can employ to reduce the likelihood of the message being intercepted is to utilize minimum power; that is, to transmit no more power than is absolutely essential to reach the receiver of interest. To do this it may be necessary to estimate the propagation loss or to minimize the data rate. Either of these techniques can result in minimizing the average power of the transmitted signal.

Another general technique that can be employed by the transmitter is to make the signals more noiselike. In the case of a direct-sequence system, one way of accomplishing this is to filter the transmitted signal before it is transmitted. If the filter has a bandwidth that is sufficiently narrow, the resulting random sequence of binary digits will result in amplitude changes that make the signal look very much like random noise. With an appropriate filter at the receiver to restore the proper shape, the signals can still be recovered, although there is some penalty in signal-to-noise ratio.

An item of considerable concern regarding the probability of intercepting spread-spectrum signals is the detectability. A parameter known as the detectability index has been defined and this parameter is a measure of how easy it is for a noncoherent receiver to detect the presence of the signal. To define this parameter, let

N = total information bits in the message

M = size of signal set (i.e., M-ary signaling)

$N_t = N/\log_2 M$ = total number of transmitted symbols

T_M = length of an M-ary symbol

P_I = power received by interceptor

B = intercept bandwidth

T = intercept integration time

d^2 = detectability index

In terms of these parameters then, the detectability index is[12]

$$d^2 = \frac{(N_t T_M)^2}{TB} \left[\frac{P_I}{N_0} \right]^2 \qquad (13\text{-}52)$$

To minimize unauthorized detection, the detectability index should be as small as possible. Note that this implies that N_t, the total number of transmitted symbols, should be as small as possible; that T should be as large as possible, which can be accomplished by spreading the signal in time, for example, by time hopping); and that B should be as large as possible (which can be accomplished by spreading the signal in frequency). It is also clear that the power received by the interceptor should be as small as possible. An extensive discussion of interception is given by Torrieri.[6]

13-8 MISCELLANEOUS CONSIDERATIONS

This section briefly discusses several techniques that may be useful in connection with antijam and anti-intercept.

Automatic Data-Rate Control

From the equations that have been presented thus far, it is clear that both the antijam and anti-intercept capabilities of the spread-spectrum system will improve as the data rate is made smaller, thus making it possible to reduce the data rate under conditions where jamming is present or where interception is likely will improve the overall performance of the system. However, this adjustment in data rate should be made with no distinguishable change in the signal. That is, the jammer or the intercept receiver should not be able to tell that the data rate has been changed. If the data rate can be made arbitrarily small, then jammers cannot deny communications, but can only minimize the rate. Furthermore, they cannot know how well they are doing and, therefore, cannot know whether they need to increase power or to change their strategy. Similarly, interceptors may find that the ratio of P_I/N_0 will be below their threshold so that they no longer can detect the signal.

Code Breaking

In some cases an interceptor may be interested only in knowing when a signal is being transmitted so that jamming can be initiated. In other cases, however, the interceptor may wish actually to attempt to decode the signal in order to determine the message. To decode the signal, the interceptor must be able to find out the particular PN sequence that is being generated. This may be relatively easy. If the intercept receiver has sufficient signal strength to detect the sequence on a chip-by-chip basis, and if that sequence is generated by an m-stage linear shift register, then the interceptor only needs to observe $2m$ successive chips to be able to compute the entire sequence and regenerate it. For this reason linear shift register codes are seldom used when security is an issue.

The difficulty of breaking a code can be significantly increased by adding nonlinear logic to the shift register sequence. One way in which this can be done is illustrated in Fig. 13-11. To illustrate the degree to which code-breaking difficulty can be increased, let m be the number of stages in the linear shift register and n be the maximum number of inputs to any single nonlinear logic unit at the output of the shift register. Then define the quantity[7]

$$_mN_n = \sum_{i=0}^{n} (i^m) \tag{13-53}$$

It now can be shown that the interceptor must receive 2_mN_n successive errorfree chips in order to break the code. Two examples of this illustrate the point. In

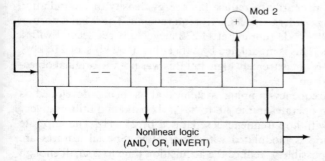

Figure 13-11 Generation of nonlinear shift register codes.

the first case, let

$$m = 17, \quad n = 4; \quad {}_{17}N_4 = 3214$$

For the second example,

$$m = 17, \quad n = 17; \quad {}_{17}N_{17} = 131{,}072$$

Note that a relatively modest increase in the number of inputs to a nonlinear unit has drastically increased the number of errorfree chips that must be received. Normally it would be very difficult to receive this many chips errorfree. Thus for practical purposes the code could be considered almost unbreakable.

13-9 EXAMPLES OF SPREAD—SPECTRUM SYSTEMS

The purpose of this section is to provide a very brief description of certain spread-spectrum systems that are either in operation or in the planning phase at the present time. More details concerning these systems are available in both the classified and unclassified literature.

One example of a direct-sequence system is the satellite navigation system known as the Global Positioning System (GPS). This system employs spread spectrum for two purposes: One purpose is to accommodate the multiple-access capability, since each user of the system must receive signals from four different satellites in order to make proper measurements of the user's position. These signals are arriving constantly, so it is necessary that the user be able to distinguish among them. Therefore, distinct codes are assigned to the various satellites so that users can tell which signals they are receiving. The second reason is to increase the resolution capability so that the accuracy of the system is improved. For this purpose there is a PN sequence running at a chip rate of 10.23 megachips per second and modulated with data at a rate of 50 b/s. This leads to a processing gain of about 56 dB and enables the system to operate with input signal-to-noise ratios of the order of -34 dB.

Another direct-sequence system is the Telecommunication Data Relay Satellite System (TDRSS), which relays data to and from the space shuttle. In this case spread spectrum is used primarily to reduce the energy density of the radiation at the surface of the earth in order to meet CCIR requirements. The S-band portion of this system utilizes a PN code running at 11.232 megachips per second with a code period of 2047 chips. This is modulated by data either at 32 kb/s. or 216 kb/s. Thus the processing gain is not extremely high, but this was not the original objective of using spread spectrum.

An example of a frequency-hopping system that is being developed is SINCGARS. This system operates in the 30- to 88-MHz band and utilizes a very long hopping sequence with hop frequencies separated by 25 kHz. The hop rate is quite low and each hop is modulated with binary FM. Special features are incorporated to make it possible to reduce the acquisition time to a small enough value that push-to-talk operation is possible.

An example of a hybrid system is the Joint Tactical Information Distribution System (JTIDS). This system is basically a time-division multiple-access system in which each pulse is coded with a PN code of 32 bits and is then further frequency-hopped over a frequency range from 960 to 1215 MHz. There is also error-correction capability employing a Reed-Solomon code. This code was not discussed here, but it is a very efficient coding method that enables a great deal of error correction.

There are numerous other spread-spectrum systems that might also be mentioned; however, the ones noted illustrate a number of the aspects of spread spectrum that were discussed in previous chapters.

13-10 REFERENCES

1. Weber, C. L., G. K. Huth, and B. H. Batson: Performance Consideration of Code Division Multiple-Access Systems, *IEEE Trans. Vehic. Tech.*, vol. VT-30, no. 1, pp. 3–10, February 1981.
2. Pursley, M. B.: Spread-Spectrum Multiple-Access Communications, in *Multi-User Communication Systems*, G. Longo, Ed., Springer-Verlag, Berlin, pp. 139–199, 1981.
3. Sarwate, D. V., and M. B. Pursley: Crosscorrelation Properties of Pseudorandom and Related Sequences, *Proc. IEEE*, vol. 68, no. 5, pp. 593–619, May 1980.
4. Cooper, G. R., and K. V. Cai: Design and Performance of Time-Frequency Coded Signal Sets for Spread-Spectrum Systems, *Record of MIDCON* 79, Chicago, November 1979.
5. Cooper, G. R., and R. W. Nettleton: A Spread-Spectrum Technique for High Capacity Mobile Communications, *IEEE Trans. Vehic. Tech.*, vol. VT-27, pp. 264–275, November 1978.
6. Torrieri, D. J.: *Principles of Military Communication Systems*, Artech House, Dedham, MA, 1981.
7. Ristenbatt, M. P., and J. L. Daws, Jr.: Performance Criteria for Spread Spectrum Communications, *IEEE Trans. Com.*, vol. COM-25, no. 8, pp. 756–763, August 1977.
8. McEliece, R. J.: The Theory of Information and Coding, in *Encyclopedia of Mathematics*, vol. 3, Addison-Wesley, Reading, Mass., 1977.
9. Clark, G. C., and J. B. Cain: *Error-Correction Coding for Digital Communications*, Plenum Press, New York, 1981.
10. Berlekamp, E. R.: *Algebraic Coding Theory*, McGraw-Hill Book Company, New York, 1968.
11. Jacobs, I. M.: Practical Applications of Coding, *IEEE Trans. Inf. Theory*, vol. IT-20, no. 3, pp. 305–310, May 1974.
12. Ristenbatt, M.: Estimating Effectiveness of Covert Communications, *Proc.* 1973 *Symposium on Spread Spectrum Communications*, San Diego, pp. 113–118, March 1973.

13-11 PROBLEMS

13.2-1 Show that

$$E\left\{\left[\sum_i m_{ij}\right]^2\right\} = \sum_{\substack{i=1 \\ i \neq j}}^{U} \sum_{\substack{k=1 \\ k \neq j}}^{U} E[m_{ij}m_{kj}]$$

$$= \sum_{\substack{i=1 \\ i \neq j}}^{U} P_i E[\rho_{ij}^2]$$

13.2-2 A code-division multiple-access receiver is designed to receive the signal $s_j(t)$ by use of a correlator and reference signal $v_j(t)$. Assuming the receiver receives the signal $s_i(t)$, find the output of the correlator if $s_i(t)$ and $v_j(t)$ are specified as

$$v_j(t) = b_j(t)\sqrt{2}\sin\omega_0 t$$

$$s_i(t) = b_i(t)\sqrt{2}\sin\omega_0 t$$

where $\omega_0 t$ is defined here as $40\pi t/t_1$

$$\left(f_0 = \frac{20}{t_1} = 20x \text{ PN code rate}\right)$$

and the received signal power of $s_i(t)$ is 49×10^{-8} watts.

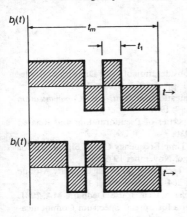

13.2-3 Given that the autocorrelation function $R_S(\tau)$ is of the form

$$R_S(\tau) = R_b(\tau)R_m(\tau) = \left[1 - \frac{|\tau|}{mt_1}\right]\left[1 - \frac{|\tau|}{t_1}\right]$$

where $b(t)$ and $m(t)$ are independent processes and m is an integer, show that

$$E[\rho_{ij}^2] = \frac{1}{PG} \simeq \frac{2t_1}{3t_m}$$

13.2-4 Suppose that in the previous problem there are four users with received signal powers as given, all with the same autocorrelation function.

$$P_1 = 3 \times 10^{-10}$$
$$P_2 = 2 \times 10^{-10}$$
$$P_3 = 6 \times 10^{-11}$$
$$P_4 = 9 \times 10^{-12}$$

and

$$N_0 = 4 \times 10^{-18} \text{ W/Hz}$$
$$1/t_m = 30{,}000 \text{ b/s}$$
$$1/t_1 = 45 \text{ Mb/s}$$

(a) Find the $(\text{SNR})_2$; that is, the SNR when the system is synchronized to user 2.
(b) Find the processing gain.

13.2-5 From the parameters of the previous problem, find the number of active users, U, when the received powers from all users (equal power) equals $P_r = 2 \times 10^{-10}$ watts, and we desire $(SNR)_o = 20$ dB.

If the value of E_b/N_o becomes arbitrarily large, what is the maximum number of users we can obtain?

13.2-6 In a certain multiple-access system, assume that all transmitters are at the same distance, except two of them, and these are

$$d_i = d_j; \quad i \neq 2, 4, j \quad d_2 = \frac{d_j}{1.2}; \quad d_4 = \frac{d_j}{1.5}$$

If the effective bandwidth is 25 MHz and the message bit rate is 20 kb/s, $\alpha = 3.73$, $E_b/N_0 = 15$, $(SNR)_0 = 10$.

(a) How many active users can the above system support?

(b) If one particular user elects to move two times closer (other than user 2, 4), how many overall users will the above now support?

13.2-7 From the previous problem:

(a) Is it possible to regain the original number of users by increasing the signal-to-noise ratio $(SNR)_1$? If so, how much more signal power would be required?

(b) If the nth user decided not to move two times closer, but instead many times closer, how close would that user have to come to cause the system to fail as a multiple-access system, assuming the original value of $(SNR)_1$?

(c) If we now desire to have the original number of users, how much lower $(SNR)_o$ would we have to tolerate?

13.2-8 For the parameters as given, find the number of users per unit bandwidth. Note that these are the same parameters as used in the previous two problems.

$$(SNR)_o = 10; \quad E_b/N_0 = 15; \quad R = 20,000$$

13.2-9 If a hard-limited channel is used with the following parameters

$B_{eff} = 33.75$ MHz

$1/t_m = 30,000$

$P_r = 2 \times 10^{-10}$ watts

$(SNR)_o = 20$ dB

$N_o = 4 \times 10^{-18}$ W/Hz

(a) Find the number of active users.

(b) If one of the users increases its power tenfold, will this system fail as a multiple-access system? How much lower $(SNR)_o$ must now be tolerated to still achieve the original number of users?

13.3-1 Show that for an M-ary system with equal output SNR for each channel, the minimum value of bandwidth utilization is

$$(B/R)_{min} = \frac{U-1}{U} \frac{(SNR)_o}{2\eta}$$

and that the minimum value of energy utilization is $(E_b/N_0)_{min} = \frac{1}{2}(SNR)_o$.

13.3-2 If the total channel SNR is 15 and

$R_1 = 10$ kb/s

$(SNR)_o = 12$

$\eta = 1.15$

$B_{eff} = 103.5$ MHz

(a) Find the bandwidth and energy utilization operating point, as well as the number of active users U_o, that can be supported at this particular point.

(b) What must the total channel SNR be in order to minimize the values of $(E_b/N_0)_0$ and $(B/R)_0$?

(c) For this point find the value of U_0.

13.3-3 A multiple-access communication system is constrained in bandwidth, and has the parameters of the previous problem. Find:
   (a) The minimum bandwidth utilization, $(B/R)_{\min}$.
   (b) The maximum number of users that can be supported.

13.4-1 Given the aperiodic cross-correlation function $R_{ij}(\tau)$ as

$$R_{ij}(\tau) = AR(\tau)\sin \omega_0\tau$$

where $R(\tau)$ is as defined in the sketch. Find the effective bandwidth B_{eff} if

$$\beta = 2.7735$$
$$R = 30,000$$
$$A = 10^{-2}$$

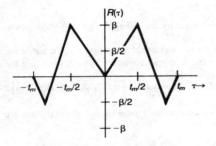

13.5-1 A pulse jammer sends a very short cosine pulse of power P_J for a duration of τ seconds. Given that the frequency of the pulse is f_0, find the energy density; that is,

$$|\mathscr{F}\{j(t)\}|^2$$

where

$$j(t) = \sqrt{2P_J}\cos 2\pi f_0 t \qquad 0 \leq t \leq \tau \quad \text{and} \quad \frac{1}{f_0} \ll \tau$$

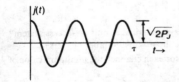

If the pulse jammer sends these pulses at a repetition period of 4τ, what is the energy spectrum?

13.5-2 A signal has the form

$$s(t) = \sqrt{2P}\cos[\omega_0(t - t_0) + \theta] \qquad |t - t_0| \leq t_1$$
$$= 0 \qquad |t - t_0| > t_1$$

Find $|F_s(\omega)|^2$, where $F_s(\omega)$ is the Fourier transform of $s(t)$.

13.5-3 A frequency-hopping system has a bandwidth of 102.4 MHz centered at 600 MHz. A direct frequency synthesizer (with no frequency multiplier at its output) creates 4096 frequencies. The system is modulated with a message at 2500 b/s and there are 10 hops per message bit. If this system is jammed with 256 tone jammers uniformly distributed across the band, what is the probability that one or more hops in a given message bit will receive a jamming tone?

13.5-4 A typical balanced modulator for use in 180-degree biphase modulation, for example, is shown below (also called a doubly balanced diode ring mixer).

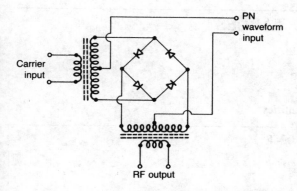

If care is taken so that the dynamic resistance of the diodes are matched and the taps on the transformers are as close to the electrical center of the windings as possible, the result will be an output with little residual carrier.

Consider the case in which the above modulator is used in a direct-sequence system with the following parameters.

$$\text{PN code rate} = 56 \text{ Mb/s}$$
$$\text{Message bit rate} = 2.5 \text{ kb/s}$$
$$(\text{SNR})_0 = 25$$
$$N_0 = 2 \times 10^{-17} \text{ W/Hz}$$
$$\text{Residual carrier suppression} = -35 \text{ dB}$$
$$\text{System losses} = 4 \text{ dB}$$

If the jammer is exactly on the desired carrier frequency, find the minimum amount of received jamming power that will degrade the above-established $(\text{SNR})_o$.

MATHEMATICAL TABLES

Table A-1 Trigonometric identities

$$\sin(A \pm B) = \sin A \cos B \pm \cos A \sin B$$

$$\cos(A \pm B) = \cos A \cos B \mp \sin A \sin B$$

$$\cos A \cos B = \tfrac{1}{2}[\cos(A + B) + \cos(A - B)]$$

$$\sin A \sin B = \tfrac{1}{2}[\cos(A - B) - \cos(A + B)]$$

$$\sin A \cos B = \tfrac{1}{2}[\sin(A + B) + \sin(A - B)]$$

$$\sin A + \sin B = 2 \sin \tfrac{1}{2}(A + B)\cos \tfrac{1}{2}(A - B)$$

$$\sin A - \sin B = 2 \sin \tfrac{1}{2}(A - B)\cos \tfrac{1}{2}(A + B)$$

$$\cos A + \cos B = 2 \cos \tfrac{1}{2}(A + B)\cos \tfrac{1}{2}(A - B)$$

$$\cos A - \cos B = -2 \sin \tfrac{1}{2}(A + B)\sin \tfrac{1}{2}(A - B)$$

$$\sin 2A = 2 \sin A \cos A$$

$$\cos 2A = 2 \cos^2 A - 1 = 1 - 2 \sin^2 A = \cos^2 A - \sin^2 A$$

$$\sin \tfrac{1}{2}A = \sqrt{\tfrac{1}{2}(1 - \cos A)} \quad \cos \tfrac{1}{2}A = \sqrt{\tfrac{1}{2}(1 + \cos A)}$$

$$\sin^2 A = \tfrac{1}{2}(1 - \cos 2A) \quad \cos^2 A = \tfrac{1}{2}(1 + \cos 2A)$$

$$\sin x = \frac{e^{jx} - e^{-jx}}{2j} \quad \cos x = \frac{e^{jx} + e^{-jx}}{2} \quad e^{jx} = \cos x + j \sin x$$

$$A \cos(\omega t + \phi_1) + B \cos(\omega t + \phi_2) = C \cos(\omega t + \phi_3)$$

where

$$C = \sqrt{A^2 + B^2 - 2AB \cos(\phi_2 - \phi_1)}$$

$$\phi_3 = \tan^{-1}\left[\frac{A \sin \phi_1 + B \sin \phi_2}{A \cos \phi_1 + B \cos \phi_2}\right]$$

$$\sin(\omega t + \phi) = \cos(\omega t + \phi - 90°)$$

Table A-2 Bessel function identities

$$J_n(x) = \sum_{m=0}^{\infty} \frac{(-1)^m \left(\dfrac{x}{2}\right)^{2m+n}}{m!(m+n)!}$$

$$J_n(-x) = (-1)^n J_n(x)$$

$$J_{-n}(x) = (-1)^n J_n(x)$$

$$e^{jx\sin\theta} = \sum_{n=-\infty}^{\infty} J_n(x)e^{jn\theta}$$

$$e^{j\omega t + jx\sin\theta} = \sum_{n=-\infty}^{\infty} J_n(x)e^{j(\omega t + n\theta)}$$

$$\cos(\omega t + x\sin\theta) = \sum_{n=-\infty}^{\infty} J_n(x)\cos(\omega t + n\theta)$$

$$\sin(\omega t + x\sin\theta) = \sum_{n=-\infty}^{\infty} J_n(x)\sin(\omega t + n\theta)$$

$$\cos(x\sin\theta) = J_0(x) + 2\sum_{n=1}^{\infty} J_{2n}(x)\cos 2n\theta$$

$$\sin(x\sin\theta) = 2\sum_{n=0}^{\infty} J_{2n+1}(x)\sin(2n+1)\theta$$

$$\cos(x\cos\theta) = J_0 + 2\sum_{n=1}^{\infty} (-1)^n J_{2n}(x)\cos 2n\theta$$

$$\sin(x\cos\theta) = 2\sum_{n=1}^{\infty} (-1)^n J_{2n+1}(x)\cos(2n+1)\theta$$

$$e^{jx\cos\theta} = \sum_{n=-\infty}^{\infty} j^n J_n(x)e^{jn\theta}$$

$$I_n(x) = j^{-n} J_n(jx) = \sum_{n=-\infty}^{\infty} \frac{(x/2)^{2m+n}}{m!(m+n)!}$$

$$I_n(-x) = (-1)^n I_n(x)$$

$$I_{-n}(x) = I_n(x)$$

$$e^{x\cos\theta} = \sum_{n=-\infty}^{\infty} I_n(x)e^{jn\theta} = I_0(x) + 2\sum_{n=1}^{\infty} I_n(x)\cos n\theta$$

(continued)

Table A-2 Bessel function identities (continued)

For large x:

$$J_n(x) \sim \left(\frac{2}{\pi x}\right)^{1/2} \left\{ \left[1 - \frac{(4n^2 - 1)(4n^2 - 3^2)}{2!(8x)^2} + \cdots \right] \cos\left[x - \frac{(2n+1)\pi}{4} \right] \right.$$

$$\left. - \frac{4n^2 - 1}{8x} \left[1 - \frac{(4n^2 - 3^2)(4n^2 - 5^2)}{3!(8x)^2} + \cdots \right] \sin\left[x - \frac{(2n+1)\pi}{4} \right] \right\}$$

$$I_n(x) \sim \frac{e^x}{(2\pi x)^{1/2}} \left[1 - \frac{4n^2 - 1}{8x} + \frac{(4n^2 - 1)(4n^2 - 3^2)}{2!(8x)^2} - \cdots \right]$$

Bessell function coefficients $J_n(\beta)$ representing 98 percent of the power in a tone-modulated FM carrier

β	\multicolumn{12}{c}{n}
	0	1	2	3	4	5	6	7	8	9	10	11
0.1	0.997											
0.2	0.990											
0.5	0.938	0.242										
1.0	0.765	0.440	0.115									
2.0	0.224	0.577	0.353	0.129								
5.0	−0.178	−0.328	0.047	0.365	0.391	0.261	0.131					
10.0	−0.246	0.043	0.255	0.058	−0.220	−0.234	−0.014	0.217	0.318	0.292	0.207	0.123

Table A-3 Hilbert transforms

$s(t)$	$\hat{s}(t)$
$s(t)$	$\dfrac{1}{\pi} \displaystyle\int_{-\infty}^{\infty} \dfrac{s(\lambda)}{t - \lambda}\, d\lambda$
$\hat{s}(t)$	$-s(t)$
$s(at)$	$\hat{s}(at)$
$\mathscr{F}^{-1}\{S(f)\}$	$\mathscr{F}^{-1}\{-j \operatorname{sgn} f\, S(f)\}$
$\sin \omega_0 t$	$-\cos \omega_0 t$
$\cos \omega_0 t$	$\sin \omega_0 t$
$\delta(t)$	$\dfrac{1}{\pi t}$
$\dfrac{\sin t}{t}$	$\dfrac{1 - \cos t}{t}$
$\dfrac{1}{1 + t^2}$	$\dfrac{t}{1 + t^2}$

Table A-4 Approximations

Taylor's series

$$f(x) = f(a) + \dot{f}(a)\frac{(x-a)}{1!} + \ddot{f}(a)\frac{(x-a)^2}{2!} + \cdots$$

Maclaurin's series

$$f(0) = f(0) + \dot{f}(0)\frac{x}{1!} + \ddot{f}(0)\frac{x^2}{2!} + \cdots$$

For small values of x

$$\frac{1}{1+x} \doteq 1 - x$$

$$(1+x)^n \doteq 1 + nx$$

$$e^x \doteq 1 + x$$

$$\ln(1+x) \doteq x$$

$$\sin x \doteq x$$

$$\cos x \doteq 1 - \frac{x^2}{2}$$

$$\tan x \doteq x$$

Table A-5 Derivatives

$$\frac{d}{dx}\left(\frac{u}{v}\right) = \frac{v\dfrac{du}{dx} - u\dfrac{dv}{dx}}{v^2}$$

$$\frac{d}{dx}u^n = nu^{n-1}\frac{du}{dx}$$

$$\frac{d}{dx}\log u = \frac{\log e}{u}\frac{du}{dx}$$

$$\frac{d}{dx}\ln x = \frac{1}{u}\frac{du}{dx}$$

$$\frac{d}{dx}a^u = a^u \ln a \frac{du}{dx}$$

$$\frac{d}{dx}e^u = e^u\frac{du}{dx}$$

$$\frac{d}{dx}u^v = vu^{v-1}\frac{du}{dx} + u^v \ln u \frac{dv}{dx}$$

$$\frac{d}{dx}\sin u = \cos u \frac{du}{dx}$$

$$\frac{d}{dx}\cos u = -\sin u \frac{du}{dx}$$

$$\frac{d}{dx}\tan u = \frac{1}{\cos^2 u}\frac{du}{dx}$$

Table A-6 Indefinite integrals

$$\int \sin ax \, dx = -1/a \cos ax \qquad \int \cos ax \, dx = 1/a(\sin ax)$$

$$\int \sin^2 ax \, dx = \frac{x}{2} - \frac{\sin 2ax}{4a}$$

$$\int x \sin ax \, dx = 1/a^2(\sin ax - ax \cos ax)$$

$$\int x^2 \sin ax \, dx = 1/a^3(2ax \sin ax + 2 \cos ax - a^2x^2 \cos ax)$$

$$\int \cos^2 ax \, dx = \frac{x}{2} + \frac{\sin 2ax}{4a}$$

$$\int x \cos ax \, dx = 1/a^2(\cos ax + ax \sin ax)$$

$$\int x^2 \cos ax \, dx = 1/a^3(2ax \cos ax - 2 \sin ax + a^2x^2 \sin ax)$$

$$\int \sin ax \sin bx \, dx = \frac{\sin(a-b)x}{2(a-b)} - \frac{\sin(a+b)x}{2(a+b)} \qquad a^2 \neq b^2$$

$$\int \sin ax \cos bx \, dx = -\left[\frac{\cos(a-b)x}{2(a-b)} + \frac{\cos(a+b)x}{2(a+b)}\right] \qquad a^2 \neq b^2$$

$$\int \cos ax \cos bx \, dx = \frac{\sin(a-b)x}{2(a-b)} + \frac{\sin(a+b)x}{2(a+b)} \qquad a^2 \neq b^2$$

$$\int e^{ax} \, dx = 1/ae^{ax}$$

$$\int xe^{ax} \, dx = \frac{e^{ax}}{a^2}(ax - 1)$$

$$\int x^2 e^{ax} \, dx = \frac{e^{ax}}{a^3}(a^2x^2 - 2ax + 2)$$

$$\int e^{ax} \sin bx \, dx = \frac{e^{ax}}{a^2 + b^2}(a \sin bx - b \cos bx)$$

$$\int e^{ax} \cos bx \, dx = \frac{e^{ax}}{a^2 + b^2}(a \cos bx + b \sin bx)$$

Table A-7 Definite integrals

$$\int_0^\infty x^n e^{-ax}\, dx = \frac{n!}{a^{n+1}} = \frac{\Gamma(n+1)}{a^{n+1}}$$

$$\int_0^\infty e^{-r^2 x^2}\, dx = \frac{\sqrt{\pi}}{2r}$$

$$\int_0^\infty x e^{-r^2 x^2}\, dx = \frac{1}{2r^2}$$

$$\int_0^\infty x^2 e^{-r^2 x^2}\, dx = \frac{\sqrt{\pi}}{4r^3}$$

$$\int_0^\infty x^n e^{-r^2 x^2}\, dx = \frac{\Gamma[(n+1)/2]}{2r^{n+1}}$$

$$\int_0^\infty \frac{\sin ax}{x}\, dx = \frac{\pi}{2},\ 0,\ -\frac{\pi}{2} \qquad \text{for } a>0,\ a=0,\ a<0$$

$$\int_0^\infty \frac{\sin^2 x}{x}\, dx = \frac{\pi}{2}$$

$$\int_0^\infty \frac{\sin^2 ax}{x^2}\, dx = |a|\,\frac{\pi}{2}$$

$$\int_0^\pi \sin^2 mx\, dx = \int_0^\pi \sin^2 x\, dx = \int_0^\pi \cos^2 mx\, dx = \int_0^\pi \cos^2 x\, dx = \frac{\pi}{2} \qquad m \text{ an integer}$$

$$\int_0^\pi \sin mx \sin nx\, dx = \int_0^\pi \cos mx \cos nx\, dx = 0 \qquad m \neq n \quad m,\, n \text{ integers}$$

$$\int_0^\pi \sin mx \cos nx\, dx = \begin{cases} \dfrac{2m}{m^2 - n^2} & \text{if } m+n \text{ odd} \\[2mm] 0 & \text{if } m+n \text{ even} \end{cases}$$

Table A-8 Fourier transform of operations

Operation	$x(t)$	$X(\omega)$	$X(f)$
Complex conjugate	$x^*(t)$	$X^*(-\omega)$	$X^*(-f)$
Reversal	$x(-t)$	$X(-\omega)$	$X(-f)$
Symmetry	$X(f)$	$2\pi x(-\omega)$	$x(-f)$
Scaling	$x(at)$	$\dfrac{1}{|a|}X(\omega/a)$	$\dfrac{1}{|a|}X(f/a)$
Time delay	$x(t-t_0)$	$X(\omega)e^{-jt_0\omega}$	$X(f)e^{-j2\pi t_0 f}$
Time differentiation	$\dfrac{d^n}{dt^n}x(t)$	$(j\omega)^n X(\omega)$	$(j2\pi f)^n X(f)$
Time integration	$\displaystyle\int_{-\infty}^{t}x(\lambda)\,d\lambda$	$\dfrac{1}{j\omega}X(\omega)+\pi X(0)\delta(\omega)$	$\dfrac{1}{j2\pi f}X(f)+\tfrac{1}{2}X(0)\delta(f)$
Time correlation	$R(\tau)=\displaystyle\int x(t)y^*(t+\tau)\,dt$	$X(\omega)Y^*(\omega)$	$X(f)Y^*(f)$
Frequency translation	$x(t)e^{j\omega_0 t}$	$X(\omega-\omega_0)$	$X(f-f_0)$
Frequency differentiation	$(-j)^n t^n x(t)$	$\dfrac{d^n}{d\omega^n}X(\omega)$	$\left(\dfrac{1}{2\pi}\right)^n\dfrac{d^n}{df^n}X(f)$
Frequency convolution	$x(t)y(t)$	$\dfrac{1}{2\pi}X(\omega)*Y(\omega)$	$X(f)*Y(f)$

Other Fourier transform relationships:
   Parseval's theorem

$$\int_{-\infty}^{\infty}x(t)y^*(t)\,dt=\frac{1}{2\pi}\int_{-\infty}^{\infty}X(\omega)Y^*(\omega)\,d\omega=\int_{-\infty}^{\infty}X(f)Y^*(f)\,df$$

$$x(0)=\int_{-\infty}^{\infty}X(f)\,df$$

$$X(0)=\int_{-\infty}^{\infty}x(t)\,dt$$

Table A-9 Fourier transforms

Name	$x(t)$	$X(\omega)$	$X(f)$
Rect t	$\begin{cases}1 & \|t\|\leq\frac{1}{2}\\ 0 & \|t\|>\frac{1}{2}\end{cases}$	$\text{sinc}(\omega/2\pi)$	$\text{sinc}\,f$
Sinc t	$\dfrac{\sin\pi t}{\pi t}$	$\text{rect}\left(\dfrac{\omega}{2\pi}\right)$	$\text{rect}\,f$
Exponential	$e^{-\alpha t}u(t)$	$\dfrac{1}{\alpha+j\omega}$	$\dfrac{1}{\alpha+j2\pi f}$
Two-sided exponential	$e^{-\alpha\|t\|}$	$\dfrac{2\alpha}{\alpha^2+\omega^2}$	$\dfrac{2\alpha}{\alpha^2+4\pi^2 f^2}$
Trian t	$\begin{cases}1-\|t\| & \|t\|\leq 1\\ 0 & \|t\|>1\end{cases}$	$\text{sinc}^2\dfrac{\omega}{2\pi}$	$\text{sinc}^2 f$
Gaussian	$e^{-\pi t^2}$	$e^{-\omega^2/4\pi}$	$e^{-\pi f^2}$
Impulse	$\delta(t)$	1	1
Step	$u(t)$	$\pi\delta(\omega)+\dfrac{1}{j\omega}$	$\frac{1}{2}\delta(f)+\dfrac{1}{j2\pi f}$
Sgn t	$t/\|t\|$	$\dfrac{2}{j\omega}$	$\dfrac{1}{j\pi f}$
Constant	K	$2\pi K\delta(\omega)$	$K\delta(f)$
Cosine	$\cos\omega_0 t$	$\pi\delta(\omega+\omega_0)+\pi\delta(\omega-\omega_0)$	$\frac{1}{2}\delta(f+f_0)+\frac{1}{2}\delta(f-f_0)$
Sine	$\sin\omega_0 t$	$j\pi\delta(\omega+\omega_0)-j\pi\delta(\omega-\omega_0)$	$\dfrac{j}{2}\delta(f+f_0)-\dfrac{j}{2}\delta(f-f_0)$
Complex exponential	$e^{j\omega_0 t}$	$2\pi\delta(\omega-\omega_0)$	$\delta(f-f_0)$
Impulse train	$\sum_\infty \delta(t-nT)$	$\dfrac{2\pi}{T}\sum_\infty\delta\left(\omega-\dfrac{2\pi n}{T}\right)$	$\dfrac{1}{T}\sum_\infty\delta\left(f-\dfrac{n}{T}\right)$
Periodic wave	$x(t)=\sum_\infty x_T(t-nT)$	$\dfrac{2\pi}{T}\sum_\infty X_T\left(\dfrac{2\pi n}{T}\right)\delta\left(\omega-\dfrac{2\pi n}{T}\right)$	$\dfrac{1}{T}\sum_\infty X_T\left(\dfrac{n}{T}\right)\delta\left(f-\dfrac{n}{T}\right)$
	$=\sum_\infty a_n e^{j2\pi nt/T}$	$2\pi\sum_\infty a_n\delta\left(\omega-\dfrac{2\pi n}{T}\right)$	$\sum_\infty a_n\delta\left(f-\dfrac{n}{T}\right)$
Ramp	$tu(t)$	$j\pi\delta'(\omega)-\dfrac{1}{\omega^2}$	$\dfrac{j}{4\pi}\delta'(f)-\dfrac{1}{4\pi^2 f^2}$
Power	t^n	$2\pi(j)^n\delta^{(n)}(\omega)$	$\left(\dfrac{j}{2\pi}\right)^n\delta^{(n)}(f)$

Table A-10 Correlation function–spectral density pairs

$R_x(\tau)$	$S_x(\omega)$	$S_x(f)$
$e^{-\alpha\lvert\tau\rvert}$	$\dfrac{2\alpha}{\alpha^2 + \omega^2}$	$\dfrac{2\alpha}{\alpha^2 + (2\pi f)^2}$
$2W \operatorname{sinc} W\tau$	$\operatorname{rect}\left(\dfrac{\omega}{2\pi W} - \dfrac{1}{2}\right)$	$\operatorname{rect}\left(\dfrac{f}{W} - \dfrac{1}{2}\right)$ (ideal low-pass)
$2B \operatorname{sinc}\left(\dfrac{B\tau}{2}\right) \cos \omega_c t$	$+\operatorname{rect}\left(\dfrac{\omega + \omega_c}{2\pi B} - \dfrac{1}{2}\right)$ $+\operatorname{rect}\left(\dfrac{\omega + \omega_c}{2\pi B} - \dfrac{1}{2}\right)$	$+\operatorname{rect}\left(\dfrac{f + f_c}{B} - \dfrac{1}{2}\right)$ $+\operatorname{rect}\left(\dfrac{f - f_c}{B} - \dfrac{1}{2}\right)$ (ideal bandpass)
$\alpha e^{-\alpha\lvert\tau\rvert} \cos \omega_c t$	$\dfrac{\alpha^2}{\alpha^2 + (\omega + \omega_c)^2}$ $+\dfrac{\alpha^2}{\alpha^2 + (\omega - \omega_c)^2}$	$\dfrac{\alpha^2}{\alpha^2 + 4\pi^2(f + f_c)^2}$ $+\dfrac{\alpha^2}{\alpha^2 + 4\pi^2(f - f_c)^2}$ (Butterworth bandpass)
1	$2\pi\delta(\omega)$	$\delta(f)$
$\dfrac{N_0}{2}\delta(\tau)$	$N_0/2$	$N_0/2$ (white noise)
$\cos \omega_c t$	$\pi\delta(\omega + \omega_c)$ $+\pi\delta(\omega - \omega_c)$	$\tfrac{1}{2}\delta(f + f_c)$ $+\tfrac{1}{2}\delta(f - f_c)$

Table A-11 Probability functions

Discrete distributions

Binomial

$$P_r(x) = \binom{n}{x} p^x q^{n-x} \qquad x = 0, 1, 2, \ldots, n$$

$$= 0 \qquad\qquad \text{otherwise}$$

$$0 < p < 1, \quad q = 1 - p$$

$$p(x) = \sum_{k=0}^{n} \binom{n}{k} p^k q^{n-k} \delta(x - k)$$

$$\bar{x} = np$$

$$\sigma_x^2 = npq$$

Poisson

$$P_r(x) = \frac{a^x e^{-a}}{x!} \qquad x = 0, 1, 2, \ldots$$

$$\bar{x} = a$$

$$\sigma_x^2 = a$$

Continuous distributions

Exponential

$$p(x) = ae^{-ax} \qquad x > 0$$

$$= 0 \qquad\qquad \text{otherwise}$$

$$\bar{x} = a^{-1}$$

$$\sigma_x^2 = a^{-2}$$

Gaussian (normal)

   PDF of random noise with mean $\bar{x}$ and variance σ_x^2.

$$p(x) = \frac{1}{\sigma_x \sqrt{2\pi}} \exp\left[-\frac{(x - \bar{x})^2}{2\sigma_x^2} \right] \qquad -\infty \le x \le \infty$$

$$E\{x\} = \bar{x}$$

$$E\{(x - \bar{x})\} = \sigma_x^2$$

Bivariate Gaussian (normal)

$$p(x, y) = \frac{1}{2\pi\sigma_x\sigma_y\sqrt{1 - \rho^2}} \exp\left\{ -\frac{1}{2(1 - \rho^2)} \left[\left(\frac{x - \bar{x}}{\sigma_x}\right)^2 + \left(\frac{y - \bar{y}}{\sigma_y}\right)^2 - \frac{2\rho}{\sigma_x\sigma_y}(x - \bar{x})(y - \bar{y}) \right] \right\}$$

$$E\{x\} = \bar{x}$$

$$E\{y\} = \bar{y}$$

$$E\{(x - \bar{x})\} = \sigma_x^2$$

$$E\{(y - \bar{y})\} = \sigma_y^2$$

$$E\{(x - \bar{x})(y - \bar{y})\} = \sigma_x\sigma_y\rho$$

(continued)

Table A-11   Probability functions (continued)

Rayleigh

PDF of the envelope of Gaussian random noise having zero mean and variance σ_n^2.

$$p(r) = \frac{r}{\sigma_n^2} \exp[-r^2/\sigma_n^2] \qquad r \geq 0$$

$$E\{r\} = \bar{r} = \sigma_n \sqrt{\pi/2}$$

$$E\{(r - \bar{r})^2\} = \sigma_r^2 = \left(2 - \frac{\pi}{2}\right)\sigma_n^2$$

Rician

PDF of the envelope of a sinusoid with amplitude A plus zero mean Gaussian noise with variance σ^2

$$p(r) = \frac{r}{\sigma^2} \exp\left[-\frac{r^2 + A^2}{2\sigma^2}\right] I_0\left(\frac{Ar}{\sigma^2}\right) \qquad r \geq 0$$

For $A/\sigma \gg 1$, this is closely approximated by the following Gaussian PDF.

$$p(r) \simeq \frac{1}{\sigma\sqrt{2\pi}} \exp\left[-\frac{(r - A)^2}{2\sigma^2}\right]$$

Uniform

$$p(x) = \frac{1}{b - a} \qquad a < x < b$$

$$= 0 \qquad \text{elsewhere}$$

$$\bar{x} = \frac{a + b}{2}$$

$$\sigma_x^2 = \frac{(b - a)^2}{12}$$

THE Q FUNCTION

In computing error probabilities, it is often necessary to find the area under the tail of the normal or Gaussian PDF as shown in Fig. B-1.

Mathematically this can be expressed as

$$P = \int_{x_0}^{\infty} \frac{1}{\sigma\sqrt{2\pi}}\, e^{-(x-m)^2/2\sigma^2}\, dx \tag{B-1}$$

Making the substitution

$$y = \frac{x-m}{\sigma}$$

this becomes

$$P = \int_{(x_0-m)}^{\infty} \frac{1}{\sqrt{2\pi}}\, e^{-y^2/2}\, dy \tag{B-2}$$

This integral cannot be evaluated in closed form but is related to the error function that is frequently encountered in statistics. Because of its frequent occurrence in analysis of communication engineering problems, this integral is often designated the Q function; that is,

$$Q(\alpha) = \int_{\alpha}^{\infty} \frac{1}{\sqrt{2\pi}}\, e^{-y^2/2}\, dy \tag{B-3}$$

Clearly Eq. (B-1) or (B-2) can be evaluated as

$$P = Q\left(\frac{x_0-m}{\sigma}\right) \tag{B-4}$$

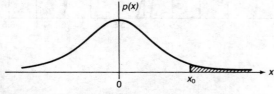

Figure B-1 Normal probability density function.

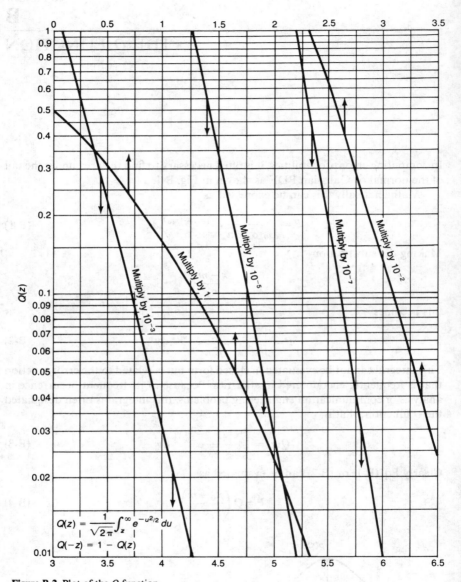

Figure B-2 Plot of the Q function.

$$Q(z) = \frac{1}{\sqrt{2\pi}} \int_z^\infty e^{-u^2/2} \, du$$

$$Q(-z) = 1 - Q(z)$$

The Q function is related to the widely tabulated error function in the following way.

$$\text{erf}(\alpha) = \frac{2}{\sqrt{\pi}} \int_0^\alpha e^{-x^2} \, dx \tag{B-5}$$

$$Q(\alpha) = \frac{1}{2}\left[1 - \text{erf}\left(\frac{\alpha}{\sqrt{2}}\right)\right] = \frac{1}{2}\text{erfc}\left(\frac{x}{\sqrt{2}}\right) \tag{B-6}$$

$$\text{erfc}(\alpha) = 2Q(\sqrt{2}\alpha) \tag{B-7}$$

$$\text{erf}(\alpha) = 1 - 2Q(\sqrt{2}\alpha) \tag{B-8}$$

The Q function is bounded by two analytical expressions as follows:

$$\left(1 - \frac{1}{\alpha^2}\right)\frac{1}{\alpha\sqrt{2\pi}} e^{-\alpha^2/2} \leq Q(\alpha) \leq \frac{1}{\alpha\sqrt{2\pi}} e^{-\alpha^2/2} \tag{B-9}$$

For values of α greater than about three, both of these bounds closely approximate $Q(\alpha)$.

Another upper bound that is sometimes used to approximate the Q function is

$$Q(\alpha) \leq \frac{1}{2} e^{-\alpha^2/2} \tag{B-10}$$

To aid in evaluating the Q function, a detailed graph extending over many decades is given in Fig. B-2. To use the graph to find $Q(\alpha)$, the point along the abscissa corresponding to α is found on either the upper or lower scale. The graph is then entered along a vertical line going to the intersection with the curve whose arrow is pointing toward the abscissa where α was found. This curve will have a scale factor indicated on it, and this scale factor is used to multiply the magnitude of $Q(\alpha)$ read from the ordinate corresponding to the intersection. As an example, let $\alpha = 4.75$. This value of α is near the center of the lower abscissa. Moving upward into the graph, the intersection with the "multiply by 10^{-5}" gives an ordinate value of 0.103. Therefore, $Q(4.75) = 0.103 \times 10^{-5} = 1.03 \times 10^{-6}$. The process is reversed to find α given $Q(\alpha)$.

Some additional properties of $Q(\alpha)$ are as follows:

$$Q(-\alpha) = 1 - Q(\alpha) \tag{B-11}$$

$$Q(0) = \frac{1}{2} \tag{B-12}$$

NUMERICAL ANSWERS TO SELECTED PROBLEMS

CHAPTER 1

1.2-1	Approx. 18 dB
1.2-2	20%
1.2-4	0.91 frames/sec; 95.4 lines/sec
1.3-1	20.372 μV
1.3-2	4.619 dB
1.3-4	10.1 dB; 1.43×10^{-3} v (rms)
1.4-1	62.59 dB; 72.71 dB
1.4-2	8.218 m; 0.475°
1.4-3	27.61 dB
1.4-4	300 MHz; 600 MHz
1.4-5	52.935 watts
1.4-6	10 watts; 100 m²
1.5-1	39.06%
1.5-2	95–1195 kHz; 0–740 kHz
1.5-7	180 kHz; 0.26/66.7%; 0.05/ 12.8%
1.6-1	49.93 dB; 35.74 dB
1.6-2	35 kHz; 56 MHz
1.6-3	$\dfrac{1}{2\pi}$ V; $\dfrac{500}{\pi}$ Hz

CHAPTER 2

2.2-1	0; 1×10^{-8} V^2
2.2-2	$\sqrt{5}$ V (rms)
2.2-3	4430 Hz
2.2-4	1750 Hz; 1750 Hz
2.2-7	0.5601 A^2 watts; 0.5196 A^2 watts
2.3-5	0.0023; 0.1773; 0.00858; 0.01832

2.4-4	108.5%
2.5-3	$\sqrt{2}$ V (rms)
2.6-3	10 dB
2.6-5	9.11 dB
2.7-1	0 dB; 3.78 dB
2.7-2	40 kHz

CHAPTER 3

3.1-1	-87.0 dB; -88.24 dB
3.1-3	$\leq 5.71°$
3.1-4	1.8×10^{-13} watts
3.1-5	1.8×10^{-13} watts
3.1-6	35.03 dB
3.1-8	10^6 Hz
3.1-9	10^6 Hz
3.1-10	0.5 V^2; 12,000 Hz
3.2-1	425.421 kHz; 59.02 dB
3.2-2	3577.71 Hz; 14.287 dB
3.2-3	14 dB
3.2-6	55.18 dB
3.2-7	17.39 dB
3.2-8	28.565 dB; 38.98 dB; 75.62 dB
3.2-9	178.9 kHz
3.2-10	66.71 dB
3.2-11	23.82 dB
3.2-12	6.98 dB
3.2-14	16.22
3.2-15	> 0.75
3.2-16	16.22
3.2-17	9.55×10^{-7}
3.2-18	1.25
3.2-21	3.28792×10^{-2} rad²

CHAPTER 4

4.1-1	$-1.951; 0.0016$
4.1-2	$0; 0.1587$
4.2-2	$2.5; 0.03125$
4.2-3	0.0228
4.2-4	0.34
4.2-5	0.23975
4.2-6	2.21
4.2-7	$0.239 \ V; 4.17 \times 10^{-5}$
4.2-8	2.87×10^{-7}
4.2-9	4×10^{-6}
4.3-1	2.768×10^{-3}
4.3-2	3.09×10^{-3}
4.4-1	$2.87 \times 10^{-7}; 8.18 \times 10^{-6}$
4.4-2	47.3%
4.4-3	1.82×10^{-3}
4.5-1	0
4.5-3	6.285×10^{-9}
4.5-4	0
4.6-1	2.027×10^{-6}
4.6-2	5.21×10^{-6}

CHAPTER 5

5.1-1	$109.95 \ kb/s$
5.1-2	$12.53 \ watts$
5.1-3	$500 \ kb/s$
5.1-4	$14.12 \ mW$
5.1-5	$1.58 \times 10^{-3}; 0.154$
5.1-6	$37.06 \ kb/s$
5.1-7	$715.15 \ kHz; 0.135$
5.1-9	$71.5 \ kHz; 2.4 \times 10^{-4};$ $50 \ kHz; 7.83 \times 10^{-4}$
5.1-10	7.15×10^{-3}
5.2-1	3.37×10^{-3}
5.2-2	$50 \ kHz; 2.2 \times 10^{-2}$
5.2-4	2.2×10^{-2}
5.2-5	$3.37 \times 10^{-3}; 2.27 \times 10^{-5}$
5.2-6	$1 \ MHz; 0.303$
5.2-7	$500 \ kHz; 0.159$
5.3-1	$92.4 \ kb/s$

CHAPTER 6

6.2-1	$1.1 \ Mhz; \leq 11$
6.3-2	$0.67; 0.44$
6.3-3	$13.67; 4.67$
6.4-1	7.58×10^{-4}
6.4-4	$\leq 5.18 \times 10^{-6}$
6.5-1	$\leq 5.74 \times 10^{-7}; 3.73 \times 10^{-6};$ 2.62×10^{-6}
6.5-2	$\leq 1.15 \times 10^{-6}; \geq 20.79 \ mW$
6.5-3	$25.75 \ kbs$
6.6-1	$\leq 8.61 \times 10^{-7}$
6.6-2	$\leq 1.81 \times 10^{-7}$
6.6-3	6.81×10^{-3}
6.6-4	$90 \ kHz; 20 \ kHz$
6.6-5	$22.6; 1.12 \times 10^{-3}; 7.33°$

CHAPTER 7

7.1-1	$10; 56.8 \ dB; 2.87 \times 10^{-7};$ $80 \ kb/s$
7.1-2	$44.2 \ dB$
7.2-1	$133.9 \ kb/s; 1.44 \ Mb/s$
7.3-1	$9.007 \ dB$
7.3-2	$15.8 \ dB$
7.3-3	$3.3; 1.685$

CHAPTER 8

8.1-1	$21.2 \ kHz; 1.98 \ MHz$
8.2-1	$2.1 \ Mb/s; 36$
8.2-3	$48.2 \ dB$
8.2-5	$44.08 \ dB; -15.84 \ dB$
8.2-6	$10\%; 8.86 \ MHz$
8.3-1	$256; 78.64 \ MHz$
8.3-2	$256 \ MHz; 50.1 \ dB; 225,000;$ $25 \ kHz$
8.3-3	$2.15 \times 10^9 \ b/s$
8.3-4	256
8.3-5	$49.1 \ dB$
8.4-1	$10; 33.1 \ dB$

8.4-2	97.66 watts; 3.3 Mhz; 30.1 dB		**11.2-2**	120 MHz; 40,000; 80,000
8.6-1	2048		**11.2-3**	29.86 MHz
8.6-2	48.16 dB; 629.15 MHz		**11.2-4**	200,000; $217-344$; 56.1 MHz; 57,240
8.6-3	10 μsec; 1024; 98.3 MHz		**11.2-5**	2000; $11,477-13,524$; 4
8.8-1	12.5 kHz		**11.2-6**	2074; 22.9 MHz; 4.148 MHz
8.8-2	64			

CHAPTER 9

9.3-1	18
9.3-2	6
9.3-3	630
9.6-1	648 Mb/s
9.6-2	-26.02 dB
9.8-2	0.748 dB; 4 dB
9.8-3	51.5%
9.9-1	196.86 ft.
9.9-2	1.5×10^{-11} watts
9.10-1	60.28 dB
9.10-2	-17 dB; 23 dB
9.10-3	14.31 dB
9.11-1	21.23 dB; 46.2 dB; 3.98×10^{-10} watts
9.11-2	26 dB; 2 watts

CHAPTER 10

10.2-1	6×10^{-9} watts
10.2-2	0.00313
10.2-3	4096; 264 kb/s
10.3-1	2048; 0.00667 sec; 3.255 μsec
10.3-2	9.83 MHz; 40.96 kW

CHAPTER 11

11.1-2	15
11.1-4	1.67×10^7; 0.56 min.
11.1-5	100 MHz; 36; 200 MHz; 37 dB

CHAPTER 12

12.3-1	75
12.3-2	10
12.8-1	260 sec
12.8-2	0.13 sec; 3080 Hz; 1.008 sec; 6320 Hz; 414.6 sec; 6405.5 sec
12.8-3	0.018 sec; 2010.8 Hz; 0.0721 sec; 2043.2 Hz; 57.22 sec; 228.9 sec
12.8-4	5; 0.99498
12.8-5	2.882 sec
12.8-6	0.48 sec
12.9-1	13; 20 sec
12.9-2	160 sec
12.13-1	5

CHAPTER 13

13.2-2	10^{-4}
13.2-4	29.51 dB; 33.52 dB
13.2-5	22; 23
13.2-6	163; 150
13.2-7	17%; 3.94 dB
13.2-8	6.67×10^{-6}
13.2-9	17; 4; 5.78 dB
13.3-2	6.4; 95.9; 1618; 1; 863
13.3-3	5.214; 1726
13.4-1	72 MHz
13.5-3	0.7369
13.5-4	2.92×10^{-11} watts

INDEX